Studies in Logic
Mathematical Logic and Foundations
Volume 81

Factual and Plausible Reasoning

Volume 71
Fathoming Formal Logic: Volume I. Theory and Decision Procedures for Propositional Logic
Odysseus Makridis

Volume 72
Fathoming Formal Logic: Volume II. Semantics and Proof Theory for Predicate Logic
Odysseus Makridis

Volume 73
Measuring Inconsistency in Information
John Grant and Maria Vanina Mrtinez, eds.

Volume 74
Dictionary of Argumentation. An Introduction to Argumentation Studies
Christian Plantin. With a Foreword by J. Anthony Blair

Volume 75
Theory of Effective Propositional Paraconsistent Logics
Arnon Avron, Ofer Arieli and Anna Zamansky

Volume 76
Argumentation and Inference. Proceedings of the 2nd European Conference on Argumentation. Volume I
Steve Oswald and Didier Maillat, eds.

Volume 77
Argumentation and Inference. Proceedings of the 2nd European Conference on Argumentation. Volume II
Steve Oswald and Didier Maillat, eds.

Volume 78
Logic and Philosophy of Logic. Recent Trends in Latin America and Spain
Max A. Freund, Max Fernández de Castro and Marco Ruffino, eds.

Volume 79
Games Iteration Numbers. A Philosophical Introduction to Computability Theory
Luca M. Possati

Volume 80
Logics of Proofs and Justifications
Roman Kuznets and Thomas Studer

Volume 81
Factual and Plausible Reasoning
David Billington

Studies in Logic Series Editor
Dov Gabbay   dov.gabbay@kcl.ac.uk

Factual and Plausible Reasoning

David Billington

© Individual author and College Publications, 2019
All rights reserved.

ISBN 978-1-84890-303-6

College Publications
Scientific Director: Dov Gabbay
Managing Director: Jane Spurr

http://www.collegepublications.co.uk

All rights reserved. No part of this publication may be reproduced, stored in a retrieval system or transmitted in any form, or by any means, electronic, mechanical, photocopying, recording or otherwise without prior permission, in writing, from the publisher.

*This book is dedicated to
the Muse of Mathematics
for her gift of seemingly external
insights, ideas, and connections,
that become conscious only
as sleep departs or
as daydreams approach.*

Preface

The genesis of this book.

In July 1988 I visited Donald Nute at the University of Georgia, USA. He introduced me to his Defeasible Logic which was first published in 1987 [69]. Having studied only classical logic I was intrigued by such a different logic and continued to visit Professor Nute until August 2010. After almost 10 years working on Defeasible Logic, in December 1997, I decided to invent by own logic for plausible reasoning from scratch. Not surprisingly it turned out to be almost the same as Defeasible Logic, but with the intention of reasoning with propositions rather than just literals as is the case with Defeasible Logic. This logic, which I call Propositional Plausible Logic (PPL), was finished in April 2015 and is the subject of Chapters 5, 6, and 7.

Once the detailed mathematics of PPL was completed I began to consider whether PPL did the right things or not. Considering various tricky examples had been done while creating PPL. What was required was a theorem which said that PPL did plausible reasoning. But that meant defining what plausible reasoning was. It could be argued that the literature (particularly on non-monotonic reasoning) contains some attempts at such a definition. However, examples show that whatever reasoning these attempts characterise, it is not plausible reasoning. Chapter 4 contains my attempt to define what it means for a logic to do plausible reasoning.

Propositional Plausible Logic allows the indisputable facts of a reasoning situation to be specified. The specification and manipulation of these facts requires classical propositional logic, which is the subject of Chapter 2. However, it can happen that the 'facts' given to PPL are inconsistent. What should PPL do in such a case? Ideally PPL should ask the user to correct the situation. But such a correction may not be possible, perhaps due to ignorance or lack of time or money. In this faulty situation PPL should use a logic that behaves rationally in the presence of inconsistent 'facts'. Such logics are the subject of Chapter 3. The mathematics for this book was completed in January 2019.

Intended readership.

Factual reasoning is reasoning with statements that are certain, called facts. Classical propositional logic is often used for such reasoning. Our version of a classical propositional logic is presented in Chapter 2. However classical propositional logic has two faults. Its most serious fault is the irrational way it behaves when the statements are inconsistent. The second, less serious, fault is that there is an intuitive understanding of the meaning of 'follows from' that classical propositional logic does not capture. Various new consistent subsets of a set of inconsistent statements are investigated in Chapter 3. This yields new more rational propositional logics for factual reasoning that capture the missing intuitive meaning of 'follows from'.

Logics that do factual reasoning have properties that can be expressed by using consequence functions. A consequence function is meant to be a function whose input is a set F of formulas and whose output is the set of consequences of F. However, there is no adequate definition of what a consequence function is. In Chapter 3 a new definition of what a consequence function should be is proposed and shown to have many desirable properties.

Defeasible statements are statements that are likely, or probable, or usually true, but may occasionally be false. Moreover we shall not distinguish between different degrees of likelihood. Hence there are no numbers, like probabilities, involved.

Plausible reasoning is reasoning with statements that are either facts or defeasible statements. Many principles of plausible reasoning are suggested and several important plausible reasoning examples are considered. A propositional logic is defined that satisfies all the principles and reasons correctly with all the examples. As far as we are aware, this is the only such logic.

The intended readership of this book is anyone who is interested in factual reasoning or plausible reasoning as outlined above. However, this is a research-level monograph in mathematical logic and so a corresponding level of mathematical knowledge and maturity is required.

Particularly in a monograph on pure mathematics, there is a contest between left and right, or full, justification, and left only, or ragged-right, justification. Full justification looks neater provided the spacing is not too abnormal and the breaking of words and mathematical expressions is not too frequent or ugly. This can be facilitated by adding needless words, or by using shorter or longer synonyms, for example 'so', 'hence', and 'therefore'. But the disadvantage of this is that the same mathematical task is done, or set out, slightly differently in different places. Left only justification allows proper spacing between words and minimises the number of broken mathematical expressions. It also allows the mathematics to be set out to facilitate understanding, and so that patterns are more easily recognised. The right balance will depend on the reader, but an attempt to get a reasonable compromise has been made.

Almost certainly there are errors in this book. But I hope they are able to be easily corrected by the reader.

Preface

How to read this book.

Although this book strives to define things before using them, it is not necessary to start at the beginning and read sequentially. As usual important results are called theorems, whereas less important or purely technical results are called lemmas. Each result is followed immediately by its proof, which can be skipped. To facilitate this the end of each proof is marked in bold by EndProofLemma# or EndProofTheorem# where # is the number of the result. The more technical lemmas may be skimmed. Furthermore a sensible reading strategy is to start at any point of interest, probably Chapter 5, or 4, or 3, and only refer to earlier parts as required. The index should facilitate this reading strategy.

Acknowledgments.

I am pleased to thank the following colleagues for their assistance: Donald Nute for introducing me to his ideas about defeasible reasoning, for providing excellent working conditions when I visited him, and for his hospitality; Andrew Rock for his implementations of various versions of PPL; Grigoris Antoniou and Andrew Rock for their creative discussions; Michael Maher for his helpful comments on drafts of some of my work; Vladimir Estivill-Castro for realising where PPL could be used, and Andrew Rock and René Hexel for realising how to use it; and Patrick Marchisella for gently telling me that I was wrong and for his meticulous reading of some of my work. All the aforementioned also participated in worthwhile discussions with me, thank you all. There are many others that I am happy to thank for being that valuable resource: the intelligent person without formal training in logic.

Griffith University *David Billington*
Brisbane, Australia
March 2019

Contents

1	**Preliminaries**		1
	1.1	Introduction	1
	1.2	Prerequisites	2
		1.2.1 Sets	2
		1.2.2 Relations and Functions	4
		1.2.3 Ordinals and Cardinals	6
		1.2.4 Rooted Trees	7
2	**Propositional Resolution Logic (PRL)**		9
	2.1	Introduction	9
	2.2	Syntax and Semantics	10
	2.3	Negation	21
	2.4	Resolution	28
	2.5	Singletons	34
	2.6	Core	46
	2.7	Normal Forms	54
		2.7.1 Rewrite Functions	54
		2.7.2 Equivalence Is Preserved	57
		2.7.3 Termination	58
		2.7.4 Correctness	64
	2.8	Stable Normal Forms	66
	2.9	Proof Using Resolution	74
3	**Rational Reasoning**		77
	3.1	Introduction	77
	3.2	Four Satisfiable Subsets of a Set of Clauses	78
		3.2.1 Maximal Satisfiable, Minimal Unsatisfiable	78
		3.2.2 No Errors, Some Non-errors	82
		3.2.3 Pre-cumulativity	85
		3.2.4 Disjunction	91

3.3 Rational Propositional Logics 100
    3.3.1 Rational Clauses Functions 100
    3.3.2 Follows From and Tautologies 102
    3.3.3 Two Hierarchies of Rational Logics 102
 3.4 Consequence Functions 108
    3.4.1 A Review 109
    3.4.2 Rational Consequence Functions 113
    3.4.3 Properties of Rational Consequence Functions 116

4 Principles of Plausible Reasoning 127
 4.1 Introduction 127
 4.2 Representation 130
 4.3 Evidence and Non-Monotonicity 131
 4.4 Conjunction 132
 4.5 Disjunction 133
 4.6 Supraclassicality 134
 4.7 Right Weakening 135
 4.8 Consistency 135
 4.9 Multiple Intuitions: Ambiguity 137
 4.10 Decisiveness 138
 4.11 Truth Values 139
 4.12 Correctness 140
 4.13 Some Non-monotonic Logics 140

5 Propositional Plausible Logic (PPL) 145
 5.1 Plausible Descriptions 145
 5.2 The Proof Relation and the Proof Algorithms 151
 5.3 The Proof Theory 158
 5.4 A Truth Theory 170

6 Properties of Propositional Plausible Logic 173
 6.1 Conjunction and Right Weakening 173
 6.2 Consistency 174
 6.3 Truth Values 184
 6.4 Proof Algorithm Hierarchy 186

7 Examples 199
 7.1 Introduction 199
 7.2 The Non-Monotonicity Example 200
 7.3 The Ambiguity Puzzle 202
 7.4 The 3-lottery Example 203
 7.5 The Left Factual Disjunction Example 205
 7.6 The 4-lottery Example 215
 7.7 PPL, the Signpost Examples, and the Principles 219
 7.8 Priority 221

	7.9	Warning Rules 222
		7.9.1 The Red Light Example 222
		7.9.2 Unwanted Chaining 223
	7.10	Self Attack 224

8 Conclusion 229
8.1 Review 229
8.1.1 Classical Propositional Logic 229
8.1.2 Rational Reasoning 230
8.1.3 Plausible Reasoning 230
8.1.4 Propositional Plausible Logic (PPL) 231
8.2 Future Work 232
8.2.1 Propositional Resolution Logic (PRL) 233
8.2.2 Rational Reasoning 233
8.2.3 Plausible Reasoning 233
8.2.4 Propositional Plausible Logic (PPL) 234

References 237

Index 243

Chapter 1
Preliminaries

Abstract The sections of Chapter 1 Preliminaries are: 1.1 Introduction, and 1.2 Prerequisites.

The Introduction gives an overview and a context of each of the chapters in this book. It also gives a high level view of what is new in each chapter.

The second section contains the mathematical prerequisites for, and some of the notation used in, this book. It has four subsections: 1.2.1 Sets, 1.2.2 Relations and Functions, 1.2.3 Ordinals and Cardinals, and 1.2.4 Rooted Trees.

1.1 Introduction

This book grew out of research on two very different propositional logics, one for reasoning about statements that are plausible and the other for reasoning about statements that are certain, that is, facts. Hence the title. Some chapters, (2, 5, 6, 7), restrict their concern to just propositional logics, whereas other chapters, (3, 4), are concerned with more general logics. An overview of each of the chapters follows.

After an introduction and some prerequisites in Chapter 1 Preliminaries, we introduce our version of classical propositional logic in Chapter 2 Propositional Resolution Logic. The inference mechanism that we find most useful is resolution, and so the structure of propositions, which we call formulas, is based on sets of symbols rather than the usual sequences of symbols. Most texts on classical propositional logic define conjunctive normal form but do not give an explicit algorithm for transforming a formula into an equivalent formula in conjunctive normal form. We give such an algorithm and prove that it works. As far as we know this algorithm is new. A formula may have many different conjunctive normal forms, so we define a new special conjunctive normal form, called stable conjunctive normal form, that is unique.

It is well-known that classical propositional logics can prove every formula from an inconsistent set of formulas. We regard this as unintuitive and irrational, so in Chapter 3 Rational Reasoning we define new propositional logics that behave intuitively

1

and rationally when reasoning with an inconsistent set of formulas. We also consider whether or not tautologies follow from a set of formulas.

The consequence functions of the logics defined in Chapter 3 do not satisfy any of the definitions of a consequence function that are in the literature. Also there are functions that do satisfy all of the definitions of a consequence function that are in the literature, but are intuitively not consequence functions. Hence the definitions of a consequence function that are in the literature do not characterise our idea of a consequence function. Chapter 3 concludes with a new definition of a consequence function, called a rational consequence function, and a list of some of its properties.

The rest of this book is concerned with plausible reasoning. Chapter 3 shows that any sensible logic should be non-monotonic, whether it does plausible reasoning or not. The literature on non-monotonic reasoning seems to conflate or confuse non-monotonic reasoning with plausible reasoning. So Chapter 4 Principles of Plausible Reasoning gives a list of new principles that we hope characterises what it means for a logic to do plausible reasoning.

In Chapter 5 Propositional Plausible Logic (PPL) we define a logic that satisfies all the principles of plausible reasoning given in Chapter 4. As far as we know, this is the only propositional logic that satisfies all the principles in Chapter 4. Some important properties of PPL are proved in Chapter 6 Properties of Propositional Plausible Logic. We apply PPL to some important examples in Chapter 7 Examples. Earlier versions of Chapters 4, 5, 6, and 7 appear in [13].

Chapter 8 Conclusion, reviews what has been done in Chapters 2 to 7. It also contains suggestions for further research.

1.2 Prerequisites

The reader who wants to understand the proofs in this book should be familiar with at least the following concepts and topics: counter-examples, proof by mathematical induction, sets, classical propositional logic, the ordinals less than $\omega + \omega$, the countable cardinals, and rooted trees. Considerably less mathematical knowledge and ability is needed if the reader ignores the proofs. However, in order to reduce confusion, we shall explicitly mention the following topics and notation.

1.2.1 Sets

The phrase 'if and only if' is used so often and is so long that, as usual, we shall abbreviate it by 'iff'. The notation $x \in X$ means that x is an element of, or in, the set X. We shall sometimes abbreviate 'for all x in X' by '$\forall x \in X$', and 'there exists x in X' by '$\exists x \in X$'. X is a subset of Y is denoted by $X \subseteq Y$ and is defined by $X \subseteq Y$ iff if

1.2 Prerequisites

$x \in X$ then $x \in Y$. The notation $X \subset Y$ means $X \subseteq Y$ and $X \neq Y$, and denotes that X is a *strict subset* of Y.

Let X and Y be any sets. The *intersection* of X and Y is denoted by $X \cap Y$ and is defined by $X \cap Y = \{z : z \in X \text{ and } z \in Y\}$. The *union* of X and Y is denoted by $X \cup Y$ and is defined by $X \cup Y = \{z : z \in X \text{ or } z \in Y\}$. The *subtraction*, X minus Y, is denoted by $X - Y$ and is defined by $X - Y = \{z \in X : z \notin Y\}$. A common alternative notation for $X - Y$ is $X \setminus Y$.

The binary operations union and intersection of two sets can be generalised to unary operations on one set, as follows. Let $\mathcal{S}$ be a set such that every element of $\mathcal{S}$ is also a set. The *unary union* of $\mathcal{S}$, $\bigcup \mathcal{S}$, is defined by $\bigcup \mathcal{S} = \{s : \text{there exists } S \text{ in } \mathcal{S} \text{ such that } s \in S\}$. The *unary intersection* of $\mathcal{S}$, $\bigcap \mathcal{S}$, is defined by $\bigcap \mathcal{S} = \{s : \text{for all } S \text{ in } \mathcal{S}, s \in S\}$. So if each of A and B is a set then $\bigcap \{A, B\} = A \cap B$, and $\bigcup \{A, B\} = A \cup B$. In general, if $A_1, A_2, A_3, \ldots$ are sets then $\bigcap \{A_1, A_2, A_3, \ldots\} = A_1 \cap A_2 \cap A_3 \cap \ldots$, and $\bigcup \{A_1, A_2, A_3, \ldots\} = A_1 \cup A_2 \cup A_3 \cup \ldots$.

In arithmetic we are familiar with "multiplying brackets". The intersection and union operations satisfy similar properties. For example, if each of $A_1, A_2, A_3, B_1, B_2,$ and B_3 is a set then $(A_1 \cup B_1) \cap (A_2 \cup B_2) \cap (A_3 \cup B_3)$
$= (A_1 \cap A_2 \cap A_3) \cup (A_1 \cap A_2 \cap B_3)$
$\cup (A_1 \cap B_2 \cap A_3) \cup (A_1 \cap B_2 \cap B_3)$
$\cup (B_1 \cap A_2 \cap A_3) \cup (B_1 \cap A_2 \cap B_3)$
$\cup (B_1 \cap B_2 \cap A_3) \cup (B_1 \cap B_2 \cap B_3)$.
We shall need the following generalisation of this.

Lemma 1.2.1. *Let $\mathcal{A}$ be a set such that every element of $\mathcal{A}$ is also a set. For each set A in $\mathcal{A}$ let B_A be a set. If $S \subseteq \mathcal{A}$ let $I_S = \bigcap(S \cup \{B_A : A \in \mathcal{A} - S\})$. Let $M = \{A \cup B_A : A \in \mathcal{A}\}$. Then $\bigcap M = \bigcup \{I_S : S \subseteq \mathcal{A}\}$.*

Proof

Suppose the conditions of the lemma hold.

Take any x in $\bigcap M$. Let $S_x = \{A \in \mathcal{A} : x \in A\}$. Then $x \in \bigcap \{B_A : A \in \mathcal{A} - S_x\}$. So $x \in I_{S_x}$. Hence $x \in \bigcup \{I_S : S \subseteq \mathcal{A}\}$. Therefore $\bigcap M \subseteq \bigcup \{I_S : S \subseteq \mathcal{A}\}$.

Conversely, take any e in $\bigcup \{I_S : S \subseteq \mathcal{A}\}$. Then there exists S_e such that $S_e \subseteq \mathcal{A}$ and $e \in I_{S_e}$. So $e \in \bigcap (S_e \cup \{B_A : A \in \mathcal{A} - S_e\})$. Take any A in $\mathcal{A}$. Then either $A \in S_e$ or $A \in \mathcal{A} - S_e$. If $A \in S_e$ then $e \in A$; and if $A \in \mathcal{A} - S_e$ then $e \in B_A$. So in both cases $e \in A \cup B_A$. Hence for each A in $\mathcal{A}$, $e \in A \cup B_A$. So $e \in \bigcap M$. Therefore $\bigcup \{I_S : S \subseteq \mathcal{A}\} \subseteq \bigcap M$.

Thus $\bigcap M = \bigcup \{I_S : S \subseteq \mathcal{A}\}$.

EndProofLemma1.2.1

If X is a set then the set of all subsets of X is called the *power set* of X and is denoted by $\mathscr{P}(X)$. So $\mathscr{P}(X) = \{S : S \subseteq X\}$. Instead of $\{S \in \mathscr{P}(X) : \ldots\}$ we may write the more comprehensible $\{S \subseteq X : \ldots\}$. The *empty set*, which is denoted by $\{\}$, has no elements. A *singleton* is a set that has exactly one element. The number of elements in, or cardinality of, a set S is denoted by $|S|$. We shall say more about cardinality later in this section.

The set of all integers is denoted by $\mathbb{Z}$, and the set of all positive integers is denoted by $\mathbb{Z}^+$. So $\mathbb{Z}^+ = \{i \in \mathbb{Z} : i > 0\}$. There are two common definitions of the set $\mathbb{N}$ of natural numbers. One equates $\mathbb{N}$ with $\mathbb{Z}^+$, and the other adds zero to $\mathbb{Z}^+$. So care is needed to avoid confusion. The areas of discrete mathematics, logic, set theory, and computer science almost always include zero as a natural number, hence the following definition. The set $\mathbb{N}$ of *natural numbers* is defined by $\mathbb{N} = \mathbb{Z}^+ \cup \{0\}$. If m and n are integers then we define four different *integer intervals* as follows.

$[m..n] = \{i \in \mathbb{Z} : m \leq i \leq n\}$. $\qquad [m..n) = \{i \in \mathbb{Z} : m \leq i < n\}$.
$(m..n] = \{i \in \mathbb{Z} : m < i \leq n\}$. $\qquad (m..n) = \{i \in \mathbb{Z} : m < i < n\}$.

The *double dot* notation $m..n$ means include all the integers between m and n; a square bracket means include the end point next to that bracket; and a parenthesis means exclude the end point next to that parenthesis. The double dot notation can also be used even if some integers in an integer interval are missing. For example, $[m..n, k] = [m..n] \cup \{k\}$ and $[m..n, k..l] = [m..n] \cup [k..l]$. It is sometimes convenient to be able to use the double dot notation for infinite 'integer intervals'. For example $[0..\infty) = \mathbb{N}$ and $(-\infty..\infty) = \mathbb{Z}$, where ∞ denotes an informal idea of infinity. The appropriate formal definition of ∞ is in Subsection 1.2.3 Ordinals and Cardinals.

If S is a finite set of integers then $\min S$ denotes the minimum integer in S, and $\max S$ denotes the maximum integer in S. If S is empty then it may be convenient to define $\min\{\}$ and $\max\{\}$; but the most appropriate values depend on the context, so care is needed.

1.2.2 Relations and Functions

The *Cartesian product* of two sets, X and Y, is denoted by $X \times Y$ and is defined by $X \times Y = \{(x, y) : x \in X \text{ and } y \in Y\}$. A *binary relation* from a set X to a set Y is a subset of $X \times Y$. If $R \subseteq X \times Y$ then the infix notation xRy, and the prefix notation $R(x, y)$, both mean $(x, y) \in R$. A binary relation *on* a set X is a subset of $X \times X$. To indicate that a relation does not hold we usually put a slash / through the symbol denoting the relation. For example $\neq, \notin, \nsubseteq, \not>, \ldots$

A *function* f from a set X to a set Y is a binary relation from X to Y such that for all x in X, $|\{y \in Y : (x, y) \in f\}| = 1$. Let f be a function from X to Y. Suppose $n \in \mathbb{Z}^+$. Then f is said to be *n*to1 iff for all y in Y, $|\{x \in X : f(x) = y\}| \leq n$. Later we shall use the idea of a 2to1 function and also a 1to1 function. We say f is an *onto* function iff for all y in Y there exists x in X such that $f(x) = y$. Also f is a *bijection* iff f is 1to1 and onto.

Suppose f is a function from X to X. Define f^0 to be the function from X to X such that for all x in X, $f^0(x) = x$. Take any n in $\mathbb{N}$ and suppose the function f^n is defined. Define f^{n+1} to be the function from X to X such that for all x in X, $f^{n+1}(x) = f(f^n(x))$. So $f^1 = f$.

A *binary operation on* the set X is a function from $X \times X$ to X. A binary operation is also called a *binary operator*. If $*$ is a binary operation on X and $x \in X$ and $y \in X$ then $*(x, y)$ [prefix notation] is usually written as $x * y$ [infix notation].

1.2 Prerequisites

A function f from X to X is *idempotent* iff for all x in X, $f(f(x)) = f(x)$. That is, applying f once is enough, more applications will not change the answer. A binary operation $*$ on X is *idempotent* iff for all x in X, $x*x = x$. A function f from $\mathscr{P}(X)$ to $\mathscr{P}(X)$ is *inclusive* iff for all $S \subseteq X$, $S \subseteq f(S)$.

An *infinite sequence* of elements of the set E is a function from $\mathbb{Z}^+$ to E, and is often denoted by $(e_1, e_2, ...)$ or $(e_i : i \in \mathbb{Z}^+)$. A *finite sequence* of elements of the set E is a function from $[1..n]$ to E, where $n \in \mathbb{N}$. A sequence S from $[1..n]$ to E is often denoted by $(e_1, e_2, ..., e_n)$ or by $(e_i : i \in [1..n])$; and the *length* of S, denoted by $|S|$, is n. The empty sequence is denoted by $()$ and has length 0. So a sequence is 1to1 iff it has no repeated elements. For the rest of this paragraph S, S_1, and S_2 denote sequences. If S_1 is finite then the concatenation of S_1 onto the left of S_2 is denoted by $S_1{+}{+}S_2$. We define $e{+}S = (e){+}{+}S$, and if S is finite then we define $S{+}e = S{+}{+}(e)$. Define $Set(S)$ to be the set of all elements of S. Then e is an element of S, denoted by $e \in S$, means $e \in Set(S)$; and e is not an element of S, $e \notin S$, means $e \notin Set(S)$. Suppose $S = S_1{+}{+}S_2$. Then S_1 is a *prefix* of S and S_2 is a *suffix* of S; moreover S_1 is a *proper prefix* of S iff $S \neq S_1$, and S_2 is a *proper suffix* of S iff $S \neq S_2$. Also we define $S - S_1 = S_2$ and $S - S_2 = S_1$. The sequence $(a_1, ..., a_m)$ is a *subsequence* of the sequence $(b_1, ..., b_n)$ iff there is a 1to1 function s from $[1..m]$ to $[1..n]$ such that (i) if $\{i, j\} \subseteq [1..m]$ and $i < j$ then $s(i) < s(j)$; and (ii) for each i in $[1..m]$, $a_i = b_{s(i)}$.

Let $(e_i : i \in [1..n])$ be a sequence of n numbers. If n is a positive integer then the sum of $(e_i : i \in [1..n])$ is denoted by $\sum(e_i : i \in [1..n])$ and is defined by $\sum(e_i : i \in [1..n]) = e_1 + ... + e_n$. The sum of the empty sequence is define by $\sum() = 0$.

Suppose R is a binary relation on a set X. R is *reflexive* iff for all x in X, xRx. R is *symmetric* iff if $\{x, y\} \subseteq X$ and xRy then yRx. R is *antisymmetric* iff if $\{x, y\} \subseteq X$ and xRy and yRx then $x = y$. R is *transitive* iff if $\{x, y, z\} \subseteq X$ and xRy and yRz then xRz. R is an *equivalence relation* iff R is reflexive and symmetric and transitive. R is a *partial order* iff R is antisymmetric and transitive. R is *cyclic* iff there exists a finite sequence, $(x_1, x_2, ..., x_n)$ where $n \geq 1$, of elements of X such that $x_n R x_1$ and for all i in $[1..n-1]$, $x_i R x_{i+1}$. A relation is *acyclic* iff it is not cyclic.

Suppose R is a binary relation on X and f is a function from X to X. Then f is *monotonic* with respect to R iff whenever $\{x_1, x_2\} \subseteq X$ and $x_1 R x_2$ then $f(x_1) R f(x_2)$. A monotonic function preserves R and so is sometimes called a morphism with respect to R. If f is a function from $X \times X$ to X then f is *monotonic* with respect to R iff whenever $\{x_1, x_2, x_3, x_4\} \subseteq X$ and $x_1 R x_2$ and $x_3 R x_4$ then $f(x_1, x_3) R f(x_2, x_4)$. Usually R will be the subset relation $\subseteq$, but sometimes it will be just reflexive and transitive.

In arithmetic the equation $x \times (y + z) = (x \times y) + (x \times z)$ is an example of distributivity and we say $\times$ distributes over $+$. It enables $x\times$ to be pushed into the brackets (when transforming the left side), or taken out of the brackets as a common factor (when transforming the right side). This property can be generalised to relate two operations or functions. Let u be a unary function and f and g be binary functions. Then f *distributes over* g iff $f(x, g(y, z)) = g(f(x, y), f(x, z))$. Also u *distributes over* f iff $u(f(x, y)) = f(u(x), u(y))$.

Distributivity involves two functions. But it can be generalised to involve three functions, in which case it is called De Morgan's law. Let u be a unary function, and f, g, and h be binary functions. The binary *De Morgan's law* is $f(x,g(y,z)) = h(f(x,y),f(x,z))$. The much more common unary *De Morgan's law* is $u(f(x,y)) = g(u(x),u(y))$.

1.2.3 Ordinals and Cardinals

We usually regard natural numbers as independent mathematical objects. But occasionally we shall need some infinite numbers, and then this more general idea of number needs to be based on sets. These generalised numbers are called ordinal numbers or simply ordinals; their definition follows.

Definition 1.2.2.
1) A set T is *transitive* iff every element of T is a strict subset of T.
2) A set α is an *ordinal* iff α is transitive and every element of α is transitive.
3) If α and β are ordinals then $\alpha < \beta$ iff $\alpha \in \beta$ iff $\beta > \alpha$.

A set T is called transitive because if $x \in y$ and $y \in T$ then $x \in T$. It can be shown that if α, β, and γ are any ordinals then
  i) If $\alpha < \beta$ and $\beta < \gamma$ then $\alpha < \gamma$.
  ii) Exactly one of the following three statements holds: $\alpha < \beta$, $\alpha = \beta$, $\alpha > \beta$.
  iii) Every set of ordinals has a least element. That is, if X is a set of ordinals then there is an ordinal l in X such that if $x \in X$ then either $l < x$ or $l = x$.
This means that there is a smallest ordinal, and a second smallest ordinal, and so on.

The first few ordinals, their usual notation, and their definition are as follows.

$0 = \{\}$
$1 = \{0\}$
$2 = \{0,1\}$
$3 = \{0,1,2\}$
$\vdots$
$n = \{0,1,2,...,n-1\}$
$\vdots$
$\omega = \{0,1,2,...\} = \mathbb{N}$
$\omega+1 = \omega \cup \{\omega\} = \{0,1,2,...,\omega\}$
$\omega+2 = \omega+1 \cup \{\omega+1\} = \{0,1,2,...,\omega,\omega+1\}$
$\vdots$

In general if α is an ordinal then the next ordinal after α is $\alpha \cup \{\alpha\}$, called the *successor* of α. Not every ordinal is a successor ordinal, for example 0 and ω are not successor ordinals. The elements of ω are *finite ordinals*, and ordinals that are not finite are called *infinite ordinals*. Ordinals are just what we need for proofs by mathematical induction.

However, ordinals are not what we need for measuring the size, or number of elements, of a set. Two sets are the same size, or have the same cardinality, iff there is a bijection between them. We say a cardinal number, or simply a *cardinal*, is the smallest ordinal in the set of all ordinals of the same cardinality. The cardinality of a set S is the unique cardinal $|S|$ such that there is a bijection between S and $|S|$. The smallest infinite (that is, not finite) cardinal is ω and it is usually denoted by $\aleph_0$ (aleph nought). The elements of $\aleph_0$ are *finite cardinals*, and cardinals that are not finite are called *infinite cardinals*.

So we have three different symbols for the set $\{0,1,2,...\}$. When we are thinking of the elements of this set as independent mathematical objects then we refer to this set as the set of natural numbers and denote it by $\mathbb{N}$. When we are thinking of this set as an ordinal we denote it by ω, and when we are thinking of this set as a cardinal we denote it by $\aleph_0$.

A set is finite iff there is a bijection between it and a finite cardinal. A set is *denumerable* or *countably infinite* iff there is a bijection between it and $\aleph_0$. A set is *countable* iff it is finite or denumerable. A set that is not countable is *uncountable*.

1.2.4 Rooted Trees

Trees, and in particular rooted trees, are a very important mathematical structure. So it is worthwhile defining them and the notation that we shall use.

Definition 1.2.3. A *rooted tree* is a triple (N,A,r) where N is a countable set of *nodes*, $A \subseteq N \times N$ is a set of *arcs*, $r \in N$ is the *root* node, and all of the following three conditions hold.
1) For all n in N, $(n,r) \notin A$.
2) For all c in $N - \{r\}$ there is exactly one p in N such that $(p,c) \in A$.
3) For all n in $N - \{r\}$ there exists l in $\mathbb{Z}^+$ such that for all i in $[1..l-1]$, $(r,n_1) \in A$, $(n_i, n_{i+1}) \in A$, and $n_l = n$.

Let $T = (N,A,r)$ be a rooted tree. Then it can be shown that A is acyclic. Define $|T|$ to be the number of nodes in T, so $|T| = |N|$. We say T is *finite* iff $|T|$ is finite. If $(p,c) \in A$ then p is called the *parent* of c and c is called a *child* of p. Define $|p|$ to be the number of children of p, so $|p| = |\{c \in N : (p,c) \in A\}|$. A node is called a *leaf* iff it has no children. A *path* in T from $p_1 \in N$ to $p_k \in N$ is a sequence of k different nodes $(p_1, p_2, ..., p_k)$ such that $k \geq 2$ and for each i in $[1..k-1]$, (p_i, p_{i+1}) is an arc of T. The *length* of the path $(p_1, p_2, ..., p_k)$ is $k-1$. The set of *ancestors* of $c \in N$ in T is $\{n \in N :$ there is a path in T from n to $c\}$. The set of *descendants* of $p \in N$ in T is $\{n \in N :$ there is a path in T from p to $n\}$.

Let T be a rooted tree. Then Q is a *subtree* of T iff Q is a rooted tree, and every node of Q is a node of T, and every arc of Q is an arc of T. Let p be a node (or point) of T. Then T_p is the subtree of T *generated by* p iff T_p is a subtree of T, p is the root of T_p, and if Q is a subtree of T whose root is p then every node of Q is a node of T_p.

We now define the level of a node and the height of a rooted tree.

Definition 1.2.4. Let $T = (N, A, r)$ be a rooted tree. Define a function *level* from N to $\mathbb{N}$ recursively as follows.
1) $level(r) = 0$.
2) If $(p, c) \in A$ then $level(c) = level(p) + 1$.

If $\{level(n) : n \in N\}$ is finite then the *height* of T is $\max\{level(n) : n \in N\}$; else the height of T is ω.

A relationship between the number of nodes of a rooted tree and its height is given in the following lemma.

Lemma 1.2.5. Let T be a finite rooted tree in which each node has at most c children. Let the height of T be h.
1) If $c = 1$ then $|T| = h+1$.
2) If $c \geq 2$ then $|T| \leq (c^{h+1} - 1)/(c-1)$.
Proof
Suppose the conditions of the lemma hold. Then $|T| \leq 1 + c + c^2 + c^3 + ... + c^h$. If $c = 1$ then $|T| = h+1$. So suppose $c \geq 2$. Then $|T|(c-1) = c|T| - |T| = (c + c^2 + c^3 + ... + c^{h+1}) - (1 + c + c^2 + c^3 + ... + c^h) = c^{h+1} - 1$. Hence $|T| = (c^{h+1} - 1)/(c-1)$.
EndProofLemma1.2.5

The following lemma gives a characterisation of when a rooted tree is finite. It follows easily from König's infinity lemma.

Lemma 1.2.6. Let T be a rooted tree. T is finite iff
1) each node in T has only finitely many children, and
2) each path in T has only finitely many nodes.
Proof
Clearly if T is finite then both (1) and (2) must hold.

So suppose T is not finite and (1) holds. Let r be the root of T. Then r has infinitely many descendants. By (1) there is a child of r, say c_1, that has infinitely many descendants. By (1) there is a child of c_1, say c_2, that has infinitely many descendants. In general, if c_i has infinitely many descendants then by (1) there is a child of c_i, say c_{i+1}, that has infinitely many descendants. This process produces a path $(r, c_1, c_2, ...)$ in T that does not have finitely many nodes. Hence (2) fails.

Thus the lemma is proved.
EndProofLemma1.2.6

Chapter 2
Propositional Resolution Logic (PRL)

Abstract The sections of Chapter 2 Propositional Resolution Logic (PRL) are: 2.1 Introduction, 2.2 Syntax and Semantics, 2.3 Negation, 2.4 Resolution, 2.5 Singletons, 2.6 Core, 2.7 Normal Forms, 2.8 Stable Normal Forms, and 2.9 Proof Using Resolution.

The syntax of a formula is based on sets rather than sequences. This makes formulas simpler and easier to use with resolution. But it allows formulas which have singleton sets. This complication is dealt with in Section 2.5 Singletons. Section 2.6 Core removes unnecessary clauses from a set of clauses.

A new algorithm for re-writing a formula until it is in conjunctive normal form or disjunctive normal form is proved to be correct in Section 2.7 Normal Forms. Stable conjunctive normal form and stable disjunctive normal form are new normal forms. In Section 2.8 Stable Normal Forms it is shown that each formula has exactly one stable conjunctive normal form and exactly one stable disjunctive normal form.

2.1 Introduction

Classical propositional logic deals with propositions, which are statements that are either true or false but not both. For instance 'Whales breathe air.' is true, '$3+2=6$' is false, and the truth of 'Extra-terrestrial life exists.' is not known. Not every sentence is a statement. For instance 'Shut the door.' is a command, and 'Is 1 a prime?' is a question. Neither are statements.

Classical propositional logic has (at least) three tasks. The first is to define what a statement, or formula, is. This is done by giving the syntax, or rules of formation, of all formulas. Typically complex formulas are formed from simpler formulas. The second is to give a truth value (either true or false) to every formula by giving a semantics, or rules of evaluation, for all formulas. Typically the truth value of each formula can be calculated from the truth values of its simplest subformulas. The third task is reasoning; that is, defining the conditions under which a formula follows from, or is implied by, a set of formulas. If the reasoning conditions are purely syntactic (that is, only use the structure of the formulas) then these conditions are referred to as the proof theory of the logic. Alternatively the reasoning conditions could be purely semantic (that is, only using the truth values of the formulas).

There are many different classical propositional logics, including axiomatic systems, natural deduction systems, and resolution systems. In this chapter we shall define a new classical propositional logic called Propositional Resolution Logic, which is often abbreviated to PRL.

2.2 Syntax and Semantics

The formulas of PRL are made from a suitable set of symbols called an alphabet, which we now define.

Definition 2.2.1. An *alphabet* for PRL consists of the following three parts.
1) A countable set, *Atm*, of (propositional) *atoms*.
2) The set $\{\neg, \wedge, \vee\}$ of *connectives* for PRL denoting negation, conjunction, and disjunction, respectively.
3) The set of punctuation symbols consisting of the comma and both braces (curly brackets).

We also need to make sure that *Atm* does not contain any misleading, or potentially ambiguous, symbols. In particular, *Atm* must not contain any sequence of connectives or punctuation symbols.

Now that our alphabet is sufficiently determined we can define a formula.

Definition 2.2.2.
1) If a is an atom, $a \in Atm$, then a is a *formula*.
2) If f is a formula then $\neg f$ is a *formula*.
3) If F is a finite set of formulas then $\wedge F$ is a *formula*.
4) If F is a finite set of formulas then $\vee F$ is a *formula*.
5) Every formula can be built by a finite number of applications of (1), (2), (3), and (4).

The set of all formulas is denoted by *Fml*.

So $Atm \subseteq Fml$. Parts (3) and (4) of the above definition is the main difference between sequence-based formulas and set-based formulas. In particular conjunction $\wedge$ and disjunction $\vee$ are not necessarily binary operations. Also the set F in Definition 2.2.2(3,4) may be empty; making $\wedge\{\}$ and $\vee\{\}$ formulas. By Definition 2.2.2(2), the following are formulas:
$\neg\wedge\{\}, \neg\neg\wedge\{\}, \neg\neg\neg\wedge\{\}, \neg\neg\neg\neg\wedge\{\}, ...$
$\neg\vee\{\}, \neg\neg\vee\{\}, \neg\neg\neg\vee\{\}, \neg\neg\neg\neg\vee\{\}, ...$
So *Fml* is infinite, even if there are no atoms; that is, $Atm = \{\}$. But it can be shown that *Fml* has the same cardinality as $\mathbb{N}$, so *Fml* is countable.

2.2 Syntax and Semantics

Although formulas are our main concern, it is convenient to write $\wedge F$ and $\vee F$ even though the set of formulas F may be infinite. We shall call such expressions quasi-formulas ('quasi' means 'as if') because they behave as if they are ordinary formulas.

Definition 2.2.3. If $n \in \mathbb{N}$ then define $\neg^n$ to be the juxtaposition of n negation signs. Define the set, $QFml$, of all *quasi-formulas* by
$QFml = Fml \cup \{\neg^n \wedge F : F \subseteq Fml \text{ and } n \in \mathbb{N}\} \cup \{\neg^n \vee F : F \subseteq Fml \text{ and } n \in \mathbb{N}\}$.

So every formula is also a quasi-formula, $\neg^0 \wedge F = \wedge F$, and $\neg^0 \vee F = \vee F$. Also $\neg^n \wedge F$ and $\neg^n \vee F$ are ordinary formulas iff F is finite. A quasi-formula is built from smaller components by applying a connective.

Definition 2.2.4. The set, $Cp(f)$, of *components* of the quasi-formula f is defined as follows.
1) If a is an atom then $Cp(a) = \{\}$.
2) $Cp(\neg f) = \{f\}$.
3) If F is a set of formulas then $Cp(\wedge F) = Cp(\vee F) = F$.
We say g is a *component* of the quasi-formula f iff $g \in Cp(f)$. If F is a set of formulas then $Cp(F) = \bigcup \{Cp(f) : f \in F\}$.

Note that f is never a component of itself. Also if F is a set of formulas then $Cp(F)$ is the set of all component formulas of formulas in F.

A subformula of a formula f is any formula that is 'inside' or 'internal to' f. We shall use the term 'subformula of f' even though f may be a quasi-formula. This is to avoid clumsy terms like 'sub-quasi-formula'.

Definition 2.2.5. Suppose f is a quasi-formula and $n \in \mathbb{N}$. The set $Subformula(f)$ of subformulas of f is defined recursively as follows.
$Subformula(f, 0) = \{f\}$.
$Subformula(f, n+1) = \bigcup \{Cp(g) : g \in Subformula(f, n)\}$.
$Subformula(f) = \bigcup \{Subformula(f, n) : n \in \mathbb{N}\}$.
Define g to be a *subformula of f* iff $g \in Subformula(f)$. A *strict subformula* of f is a subformula of f which is not f.

Sometimes it is helpful to represent of a formula f by a tree, $Tree(f)$, of subformulas of f. The idea is that the root of $Tree(f)$ is f. And if g is a node of $Tree(f)$ then g has exactly $|Cp(g)|$ children and each component of g is a child of g. This concept of $Tree(f)$ is probably clear. Unfortunately a precise definition of $Tree(f)$ requires some notational complications. The problem is that a subformula of f can appear in many places in f, so nodes cannot be subformulas of f, but something that is labelled by a subformula of f.

Definition 2.2.6. If f is a quasi-formula then the *tree of f* is denoted by $Tree(f)$ and is defined as follows.
1) The *name* of the root of $Tree(f)$ is $n(0)$ and the *label* of $n(0)$ is f.

2) If $n(0,...,i)$ is the *name* of a node of $Tree(f)$ labelled by g then $n(0,...,i,j)$ is the *name* of a child of $n(0,...,i)$ iff $j \in \{k \in \mathbb{Z} : 0 \leq k < |Cp(g)|\}$. Each component of g is the *label* of exactly one child of $n(0,...,i)$.

There are some particularly useful kinds of formulas which are defined below.

Definition 2.2.7.
1) The set, *Lit*, of all *literals* is defined by $Lit = Atm \cup \{\neg a : a \in Atm\}$.
2) A *clause* is either a literal or the disjunction, $\vee L$, of a finite set, L, of literals.
3) The set of all clauses is denoted by *Cls*; so $Cls = Lit \cup \{\vee L : L \subseteq Lit \text{ and } L \text{ is finite}\}$.
4) If F is a set of formulas then $Cls(F) = \{f \in F : f \text{ is a clause}\}$.
5) $\vee\{\}$ is the *empty clause* or *falsum*.
6) A clause c is a *unit clause* iff either c is a literal or $c = \vee L$ and $|L| = 1$.
7) A *meet* is either a literal or the conjunction, $\wedge L$, of a finite set, L, of literals.
8) The set of all meets is denoted by *Mts*; so $Mts = Lit \cup \{\wedge L : L \subseteq Lit \text{ and } L \text{ is finite}\}$.
9) If F is a set of formulas then $Mts(F) = \{f \in F : f \text{ is a meet}\}$.
10) $\wedge\{\}$ is the *empty meet* or *verum*.
11) A meet m is a *unit meet* iff either m is a literal or $m = \wedge L$ and $|L| = 1$.

So $Cls \cap Mts = Lit$, $Cls(F)$ is the set of clauses in F, and $Mts(F)$ is the set of meets in F. The word 'unit' means that there is only one literal involved.

A note on notation
There is no commonly used name for a conjunction of a finite set of literals. Names that have been used include: 'dual-clause', 'conjunctive clause', 'clause', and 'term'. None of these are ideal: 'dual-clause' and 'conjunctive clause' are poor because they are long and depend on the idea of a clause, which is unnecessary; 'clause' and 'term' already have different well established meanings. Given its meaning in partially ordered sets, 'meet' seems a good choice.
End note

The idea of one clause being a part, or subclause, of another clause is important. To define this concept we introduce the idea of the set of literals in a clause or meet, and then give some of its properties.

Definition 2.2.8. Let l be a literal, L be a set of literals, and G be either a set of clauses or a set of meets.
1) $Lit(l) = \{l\}$.
2) $Lit(\vee L) = L = Lit(\wedge L)$.
3) $Lit(G) = \{Lit(g) : g \in G\}$.

Lemma 2.2.9. Let C be any set of clauses, M be any set of meets, L be any finite set of literals, and l be any literal.
1) If $|L| \neq 1$ then $L \in Lit(C)$ iff $\vee L \in C$.
2) If $|L| \neq 1$ then $L \in Lit(M)$ iff $\wedge L \in M$.
3) $\{l\} \in Lit(C)$ iff either $l \in C$ or $\vee\{l\} \in C$.
4) $\{l\} \in Lit(M)$ iff either $l \in M$ or $\wedge\{l\} \in M$.

2.2 Syntax and Semantics

5) Suppose g is either a clause or a meet. Then $|Lit(g)| = 1$ iff there is a literal l such that $g \in \{l, \vee\{l\}, \wedge\{l\}\}$.

Proof

These results follow directly from Definition 2.2.8.
EndProofLemma2.2.9

We can now define the 'being a part of' idea as it applies to clauses and meets, as well as introduce a notation for it.

Definition 2.2.10. Let c_1 and c_2 be any clauses, and m_1 and m_2 be any meets. Let C be any set of clauses and M be any set of meets.
 1) c_1 is a *subclause* of c_2, denoted $c_1 \leq c_2$, iff $Lit(c_1) \subseteq Lit(c_2)$.
 2) c_2 is a *superclause* of c_1, denoted $c_2 \geq c_1$, iff $Lit(c_2) \supseteq Lit(c_1)$.
 3) c_1 is a *strict subclause* of c_2, denoted $c_1 < c_2$, iff $Lit(c_1) \subset Lit(c_2)$.
 4) c_2 is a *strict superclause* of c_1, denoted $c_2 > c_1$, iff $Lit(c_2) \supset Lit(c_1)$.
 5) m_1 is a *submeet* of m_2, denoted $m_1 \leq m_2$, iff $Lit(m_1) \subseteq Lit(m_2)$.
 6) m_2 is a *supermeet* of m_1, denoted $m_2 \geq m_1$, iff $Lit(m_2) \supseteq Lit(m_1)$.
 7) m_1 is a *strict submeet* of m_2, denoted $m_1 < m_2$, iff $Lit(m_1) \subset Lit(m_2)$.
 8) m_2 is a *strict supermeet* of m_1, denoted $m_2 > m_1$, iff $Lit(m_2) \supset Lit(m_1)$.
 9) $Sup(C) = \{c \in C : c \text{ is a strict superclause of a clause in } C\}$.
 10) $Sup(M) = \{m \in M : m \text{ is a strict supermeet of a meet in } M\}$.

We note that a subclause of a clause is not the same as a subformula of a clause. Every subformula of a clause c is a subclause of c. However, $\vee\{a,b\}$ is a subclause of $\vee\{a,b,c\}$, but $\vee\{a,b\}$ is not a subformula of $\vee\{a,b,c\}$.

Similarly, every subformula of a meet m is a submeet of m, but not every submeet of m is a subformula of m.

This completes the development of the syntax of PRL.

As indicated at the start of this chapter, statements, called formulas in PRL, are either true or false but not both. Consider the following three formulas, where a and b are atoms: $\neg a$, $\wedge\{a,b\}$, $\vee\{a,b\}$. The truth value of each formula depends on the truth values of a and b.

In general the truth value of a formula can be calculated from the truth values of its components. Indeed our understanding of what the connectives mean depends on how this calculation is defined. Let f be a formula and F be a finite set of formulas. If we want $\neg$ to mean 'not' then there is only one way of calculating the truth value of $\neg f$ from the truth value of f. Similarly, if we want $\wedge$ to mean 'and' then there is only one way of calculating the truth value of $\wedge F$ from the truth values of the elements of F. However there are at least three different meanings for 'or'. *Inclusive disjunction* makes $\vee F$ true iff at least one element of F is true. Another reasonable meaning is for $\vee F$ to be true iff exactly one element of F is true. *Exclusive disjunction* makes $\vee F$ true iff an odd number of elements of F are true. As is traditional, we shall only be concerned with inclusive disjunction.

To avoid confusion we shall use NOT, AND, and OR for the negation, conjunction, and disjunction of truth values, respectively. Their definition follows.

Definition 2.2.11. There are only two truth values: the true truth value denoted by **T**, and the false truth value denoted by **F**. Suppose $S \subseteq \{\mathbf{T}, \mathbf{F}\}$.
1) $\text{NOT}\,\mathbf{T} = \mathbf{F}$.
2) $\text{NOT}\,\mathbf{F} = \mathbf{T}$.
3) $\text{NOT}\,S = \{\text{NOT}\,s : s \in S\}$.
4) $\text{AND}\,S = \mathbf{F}$ iff $\mathbf{F} \in S$.
5) $\text{AND}\,S = \mathbf{T}$ iff $\mathbf{F} \notin S$.
6) $\text{OR}\,S = \mathbf{T}$ iff $\mathbf{T} \in S$.
7) $\text{OR}\,S = \mathbf{F}$ iff $\mathbf{T} \notin S$.

In particular we have the following lemma.

Lemma 2.2.12.
1) $\text{AND}\{\} = \mathbf{T}$.
2) $\text{OR}\{\} = \mathbf{F}$.
3) $\text{AND}\{\mathbf{T}\} = \mathbf{T} = \text{OR}\{\mathbf{T}\}$.
4) $\text{AND}\{\mathbf{F}\} = \mathbf{F} = \text{OR}\{\mathbf{F}\}$.
5) $\text{AND}\{\mathbf{T}, \mathbf{F}\} = \mathbf{F}$.
6) $\text{OR}\{\mathbf{T}, \mathbf{F}\} = \mathbf{T}$.
Proof
  These results follow directly from Definition 2.2.11(4 – 7).
EndProofLemma2.2.12

The truth value of a formula is calculated by functions called valuations, which we now define not just for formulas but for quasi-formulas.

Definition 2.2.13. Any function, say v, from *QFml* to $\{\mathbf{T}, \mathbf{F}\}$, is called a *valuation* iff 1, 2, 3, and 4 all hold.
1) If $F \subseteq QFml$ then $v(F) = \{v(f) : f \in F\}$.
2) If $\wedge F$ is a quasi-formula then $v(\wedge F) = \text{AND}\,v(F) = \text{AND}\{v(f) : f \in F\}$.
3) If $\vee F$ is a quasi-formula then $v(\vee F) = \text{OR}\,v(F) = \text{OR}\{v(f) : f \in F\}$.
4) If $\neg f$ is a quasi-formula then $v(\neg f) = \text{NOT}\,v(f)$.
The set of all valuations is denoted by *Val*.

By Definition 2.2.13, once a valuation has assigned a truth value to all the atoms in a quasi-formula then the truth value of the quasi-formula is determined.

If a valuation, v, makes a formula, f, true then it is often said that v satisfies, or is a model for, f. This relation is denoted by the double turnstile symbol $\models$. As well as these ideas the following definition also introduces the idea of falsification.

Definition 2.2.14. Let v be a valuation, f be a quasi-formula, F be a set of quasi-formulas, and suppose $X \in \{f, F\}$. We read $v \models X$ as v satisfies X. As usual, the negation of $v \models X$ is written $v \not\models X$.
1) $v \models f$ means $v(f) = \mathbf{T}$.
2) $v \models F$ means for all f in F, $v \models f$.
3) X is *satisfiable* means there is a valuation which satisfies X.

2.2 Syntax and Semantics

4) X is *unsatisfiable* means X is not satisfiable.
5) v *falsifies* f means $v(f) = \mathbf{F}$.
6) v *falsifies* F iff for all f in F, v falsifies f.
7) X is *falsifiable* means there is a valuation that falsifies X.
8) X is *unfalsifiable* means X is not falsifiable.

Although there is no standard notation for 'falsifies', we do have the following equivalences, $v(f) = \mathbf{F}$ iff $v \not\models f$ iff $v \models \neg f$. But be warned that 'falsifies' does not mean 'does not satisfy'. For example suppose $v(f) = \mathbf{T}$ and $v(g) = \mathbf{F}$. Then v does not satisfy $\{f,g\}$ and v does not falsify $\{f,g\}$. So although a symbol for 'falsifies' would be useful, we cannot use $\not\models$.

We noted after Definition 2.2.13 that the truth value of a quasi-formula is determined by the truth values of its atoms. But what if the quasi-formula has no atoms? Then the truth value of such a quasi-formula would not depend on the valuation used. This and related ideas are the subject of the next definition and lemma.

Definition 2.2.15. Let f be any quasi-formula, F be any set of quasi-formulas, and suppose $X \in \{f, F\}$.
1) f is a *tautology* iff for all valuations v, $v(f) = \mathbf{T}$.
2) The set of all tautologies is denoted by *Taut*.
3) F is *tautology-free* iff $F \cap Taut = \{\}$.
4) f is a *contradiction* iff for all valuations v, $v(f) = \mathbf{F}$.
5) The set of all contradictions is denoted by *Contrad*.
6) X is *contingent* iff there are two valuations v_1 and v_2 such that $v_1 \models X$ and $v_2 \not\models X$.

So f is contingent iff f is not a tautology and f is not a contradiction.

Lemma 2.2.16. Let v be any valuation, L be a set of literals, f be a formula, and F be a set of formulas.
1) $v(\wedge\{\}) = \mathbf{T}$ and so $\wedge\{\}$ is a tautology.
   $v(\vee\{\}) = \mathbf{F}$ and so $\vee\{\}$ is a contradiction.
2) $v(\vee\{f\}) = v(f) = v(\wedge\{f\})$.
3) If there exists an atom a such that $\{a, \neg a\} \subseteq L$ then $v(\vee L) = \mathbf{T}$ and $v(\wedge L) = \mathbf{F}$.
4) L is not contingent iff either $L = \{\}$ or there is an atom a such that $\{a, \neg a\} \subseteq L$.
5) $\wedge L$ is contingent iff L is contingent iff $\vee L$ is contingent.
6) Suppose $F' \subseteq F$.
   .1) If $v(\wedge F') = \mathbf{F}$ then $v(\wedge F) = \mathbf{F}$.
   .2) If $v(\wedge F) = \mathbf{T}$ then $v(\wedge F') = \mathbf{T}$.
   .3) If $v(\vee F') = \mathbf{T}$ then $v(\vee F) = \mathbf{T}$.
   .4) If $v(\vee F) = \mathbf{F}$ then $v(\vee F') = \mathbf{F}$.
7) $v \models F$ iff $v \models \wedge F$.

Proof
   (1) $v(\wedge\{\}) = \text{AND}\{\} = \mathbf{T}$. Also $v(\vee\{\}) = \text{OR}\{\} = \mathbf{F}$.
   (2) $v(\vee\{f\}) = \text{OR}\{v(f)\} = v(f)$ and $v(\wedge\{f\}) = \text{AND}\{v(f)\} = v(f)$.

(3) Suppose there exists an atom a such that $\{a, \neg a\} \subseteq L$. Either $v(a) = \mathbf{T}$ or $v(\neg a) = \mathbf{T}$. So $v(\vee L) = \text{OR}\{v(l) : l \in L\} = \mathbf{T}$. Also either $v(a) = \mathbf{F}$ or $v(\neg a) = \mathbf{F}$. So $v(\wedge L) = \text{AND}\{v(l) : l \in L\} = \mathbf{F}$.

(4) Suppose L is not contingent. That is, $\vee L$ is not contingent. If every valuation falsifies $\vee L$ then $L = \{\}$. If every valuation satisfies $\vee L$ then, by part 1, $L \neq \{\}$ and there must be an atom a such that $\{a, \neg a\} \subseteq L$.

Conversely, suppose either $L = \{\}$ or there exists an atom a such that $\{a, \neg a\} \subseteq L$. If $L = \{\}$ then by part (1), every valuation makes $\vee L$ false, and so L is not contingent. So suppose there exists an atom a such that $\{a, \neg a\} \subseteq L$. By part (3) every valuation makes $\vee L$ true, and so L is not contingent.

(5) The first equivalence follows directly from the definitions.

Suppose L is not contingent. By part (4), either $L = \{\}$ or there is an atom a such that $\{a, \neg a\} \subseteq L$. Let v be any valuation. In the first case $v(\vee L) = v(\vee \{\}) = \mathbf{F}$. In the second case $v(\vee L) = \mathbf{T}$. So in both cases $\vee L$ is not contingent.

Conversely suppose L is contingent. By part (4), $L \neq \{\}$ and there is no atom a such that $\{a, \neg a\} \subseteq L$. Therefore there is a valuation that satisfies L, and there is a valuation that falsifies L. Hence $\vee L$ is contingent.

(6) Suppose $F' \subseteq F$.

(.1) If $v(\wedge F') = \mathbf{F}$ then $\text{AND}\{v(f) : f \in F'\} = \mathbf{F}$.
So $\mathbf{F} \in \{v(f) : f \in F'\} \subseteq \{v(f) : f \in F\}$. Hence $\mathbf{F} = \text{AND}\{v(f) : f \in F\} = v(\wedge F)$.

(.2) This is the contrapositive of (.1).

(.3) If $v(\vee F') = \mathbf{T}$ then $\text{OR}\{v(f) : f \in F'\} = \mathbf{T}$.
So $\mathbf{T} \in \{v(f) : d \in F'\} \subseteq \{v(f) : f \in F\}$. Hence $\mathbf{T} = \text{OR}\{v(f) : f \in F\} = v(\vee F)$.

(.4) This is the contrapositive of (.3).

(7) This follows straightforwardly from the definitions.

EndProofLemma2.2.16

From Lemma 2.2.16(1) we see that our set-based notation for conjunction and disjunction provides a natural notation for the smallest tautology and the smallest contradiction. Such a notation is lacking in the usual sequence-based notation. Mostly our set-based notation is simpler than the usual sequence-based notation. The sequence-based formula $(a \vee (b \vee a))$ becomes $\vee\{a, b\}$ in our set-based notation. There is no proliferation of parentheses and hence no need for a precedence order of operations to reduce the plethora of parentheses. The use of sets removes the unnecessary repetition and ordering found in the usual sequence-based notation. However, Lemma 2.2.16(2) shows the only case where our set-based notation introduces an unnecessary complication. In general, if f is a formula then $\vee\{f\}$, f, and $\wedge\{f\}$ are three different formulas. But there is no need for this distinction because all three always have the same truth value. We shall say more about this at the beginning of Section 2.5.

Necessary and sufficient conditions for a clause and a meet to be contingent are given in Lemma 2.2.16(4 and 5). Suppose $F' \subseteq F$. Lemma 2.2.16(6) shows that the relationship between $\wedge F'$ and $\wedge F$, and the relationship between $\vee F'$ and $\vee F$, are as expected.

As indicated at the start of this chapter, reasoning can be defined syntactically or semantically. The usual syntactic characterisation requires more concepts and so is

2.2 Syntax and Semantics

postponed until later. However the usual semantic characterisation is given in the next definition. It is very useful mathematically, but it does have some strange properties that are considered in Chapter 3.

Definition 2.2.17. Let f and g be two quasi-formulas, and F and G be two sets of quasi-formulas. Suppose $X \in \{f, F\}$ and $Y \in \{g, G\}$. We write $X \models Y$, and say X implies Y, to mean every valuation that satisfies X also satisfies Y. In symbols, $X \models Y$ iff for all v in Val, if $v \models X$ then $v \models Y$. We write $X \equiv Y$, and say X is equivalent to Y, to mean $X \models Y$ and $Y \models X$.

Some properties of $\models$ and $\equiv$ are given in the next lemma. Some of these properties have names which we mention in parentheses.

Lemma 2.2.18. Let f, g, and h be quasi-formulas, and F, G, and H be sets of quasi-formulas. Suppose $X \in \{f, F\}$, $Y \in \{g, G\}$, and $Z \in \{h, H\}$.
1) $X \models X$ and $X \equiv X$. (Reflexivity)
2) If $X \equiv Y$ then $Y \equiv X$. (Symmetry)
3) If $X \models Y$ and $Y \models Z$ then $X \models Z$. (Transitivity)
   If $X \equiv Y$ and $Y \equiv Z$ then $X \equiv Z$. (Transitivity)
4) $f \equiv g$ iff for all valuations v, $v(f) = v(g)$.
5) $F \models G$ iff for all h, (if $G \models h$ then $F \models h$).
   $F \equiv G$ iff for all h, ($F \models h$ iff $G \models h$).
6) $F \models G$ iff $F \models \wedge G$ iff $\wedge F \models G$ iff $\wedge F \models \wedge G$.
7) $F \equiv G$ iff $\wedge F \equiv \wedge G$.
8) $F \equiv \wedge F$.
9) Let g be a subformula of f. Let g' be a formula such that $g \equiv g'$. Let f' be the result of replacing exactly one occurrence of g in f by g'. Then $f \equiv f'$.

Proof
   Let f, g, and h be quasi-formulas, and F, G, and H be sets of quasi-formulas. Suppose $X \in \{f, F\}$, $Y \in \{g, G\}$, and $Z \in \{h, H\}$.
   (1, 2) (1) and (2) follow immediately from Definition 2.2.17.
   (3) The transitivity of $\models$ follows immediately from Definition 2.2.17.
   Suppose $X \equiv Y$ and $Y \equiv Z$. Then $X \models Y$, $Y \models X$, $Y \models Z$, and $Z \models Y$. So $X \models Z$ and $Z \models X$. Thus $X \equiv Z$.
   (4) $f \equiv g$ iff $f \models g$ and $g \models f$
iff for all valuations v, ($v(f) = \mathbf{T}$ iff $v(g) = \mathbf{T}$)
iff for all valuations v, $v(f) = v(g)$.
   (5) Suppose $F \models G$. Then for all h, (if $G \models h$ then $F \models h$).
   Conversely, suppose that for all h, (if $G \models h$ then $F \models h$). Take any valuation v such that $v \models F$. We must show that $v \models G$. Take any g in G. Then $G \models g$ and so $F \models g$. Hence $v \models g$. Therefore $v \models G$. Thus $F \models G$.
   So we have proved that $F \models G$ iff for all h, (if $G \models h$ then $F \models h$).
   Therefore $G \models F$ iff for all h, (if $F \models h$ then $G \models h$).
   These two results show that $F \equiv G$ iff for all h, ($F \models h$ iff $G \models h$).

(6) Let v be any valuation. By Lemma 2.2.16(7), $v \models F$ iff $v \models \wedge F$. Therefore $F \models G$ iff $F \models \wedge G$ iff $\wedge F \models G$ iff $\wedge F \models \wedge G$.

(7) By Lemma 2.2.18(6), $F \equiv G$ iff $F \models G$ and $G \models F$ iff $\wedge F \models \wedge G$ and $\wedge G \models \wedge F$ iff $\wedge F \equiv \wedge G$.

(8) This follows immediately from Lemma 2.2.16(7).

(9) Let g be a subformula of f. Let g' be a formula such that $g \equiv g'$. Let f' be the result of replacing exactly one occurrence of g in f by g'. Let g_0 denote the occurrence of g in f that is replaced. Then $g_0 = g \equiv g'$.

Special case: $g_0 = f$.
Then $f' = g'$. So $f = g_0 \equiv g' = f'$.

The proof is by induction on the structure of f. Suppose the result holds for all strict subformulas of f. We show it holds for f by considering three cases.

Case 1: f is an atom.
Then $g_0 = f$. So by the special case, $f \equiv f'$.

Case 2: $f = \neg h$.
If $g_0 = \neg h$ then by the special case, $f \equiv f'$. So suppose g_0 is a strict subformula of f. Then g_0 is a subformula of h. Let h' be the result of replacing g_0 in h by g'. By the induction hypothesis, $h \equiv h'$. So $\neg h \equiv \neg h'$. Hence $f = \neg h \equiv \neg h' = f'$.

Case 3: $f = \Diamond H$, where $\Diamond \in \{\wedge, \vee\}$ and H is a set of formulas.
If $g_0 = \Diamond H$ then by the special case, $f \equiv f'$. So suppose g_0 is a strict subformula of f. Then for some h in H, g_0 is a subformula of h. Let h' be the result of replacing g_0 in h by g'. By the induction hypothesis, $h \equiv h'$. Let $H' = (H - \{h\}) \cup \{h'\}$. Then $f = \Diamond H \equiv \Diamond H' = f'$.

Thus by structural induction, for all quasi-formulas f, $f \equiv f'$.
EndProofLem2.2.18

Lemma 2.2.18(1, 2, 3) shows that $\equiv$ is an equivalence relation, and $\models$ is a preorder, that is reflexive and transitive. The next lemma considers the $\equiv$ relation when restricted to sets of clauses and sets of meets. It also characterises subsets of a set of clauses that are equivalent ($\equiv$) to the whole set.

Lemma 2.2.19. Let C be any set of clauses and M be any set of meets.
1) $C - (Sup(C) \cup Taut) \equiv C$.
2) $\vee(M - (Sup(M) \cup Contrad)) \equiv \vee M$.
3) Suppose $C' \subseteq C$. Then $C - C' \equiv C$ iff $C - C' \models C'$.
4) Suppose $S \subseteq C$. Then $S \equiv C$ iff $S \models C - S$.
Proof

Let C be any set of clauses and M be any set of meets.

(1) Since $C - (Sup(C) \cup Taut) \subseteq C$ we have $C \models C - (Sup(C) \cup Taut)$. Conversely, suppose v is a valuation such that $v \models C - (Sup(C) \cup Taut)$. Then $v \models (Sup(C) \cup Taut)$ and so $v \models C$. Therefore $C - (Sup(C) \cup Taut) \models C$. Thus $C - (Sup(C) \cup Taut) \equiv C$.

(2) Since $M - (Sup(M) \cup Contrad) \subseteq M$ we have $\vee(M - (Sup(M) \cup Contrad)) \models \vee M$.

2.2 Syntax and Semantics

Conversely, suppose v is a valuation such that $v \models \vee M$. Then there exists m in M such that $v(m) = \mathbf{T}$. Hence $m \notin Contrad$. Either $m \in M - Sup(M)$ or $m \in Sup(M)$. If $m \in M - Sup(M)$ then $v \models \vee(M - (Sup(M) \cup Contrad))$. So suppose $m \in Sup(M)$. Then there exists m' in $M - Sup(M)$ such that m' is a strict submeet of m. Hence $v(m') = \mathbf{T}$ and so $v \models \vee(M - (Sup(M) \cup Contrad))$. Therefore $\vee M \models \vee(M - (Sup(M) \cup Contrad))$.

Thus $\vee(M - (Sup(M) \cup Contrad)) \equiv \vee M$.

(3) Suppose $C' \subseteq C$. Then $C \models C'$.

Suppose $C - C' \equiv C$. Then $C - C' \models C$. But $C \models C'$ so $C - C' \models C'$.

Conversely suppose $C - C' \models C'$. We have $C - C' \models C - C'$ and so $C - C' \models C$. But $C \models C - C'$ and so $C - C' \equiv C$.

Thus $C - C' \equiv C$ iff $C - C' \models C'$.

(4) This follows from part 3 by replacing $C - C'$ with S.

EndProofLem2.2.19

Let f be a formula and F be a set of formulas. If $F \models f$ then we often say f is a semantic consequence of F. Also if $f \models \vee F$ then we often say f is a semantic implicant of $\vee F$. The following definition introduces some useful notation.

Definition 2.2.20. Let F be any set of formulas.
1) The set of $\models$-consequences of F, $C_\models(F)$, is defined by $C_\models(F) = \{f \in Fml : F \models f\}$.
   If f is a formula then $C_\models(f) = C_\models(\{f\})$.
2) $ClsC_\models(F) = Cls(C_\models(F)) = \{f \in Cls : F \models f\}$.
   If f is a formula then $ClsC_\models(f) = ClsC_\models(\{f\})$.
3) The set of *semantic implicants* of $\vee F$, $Imp(\vee F)$, is defined by
   $Imp(\vee F) = \{f \in Fml : f \models \vee F\}$.
4) $MtsImp(\vee F) = Mts(Imp(\vee F)) = \{f \in Mts : f \models \vee F\}$.

Some properties of $C_\models$ and Imp are given in the following lemma.

Lemma 2.2.21. Let each of F and G be any set of formulas. Let each of f and g be any formula. Let v be any valuation.
1) $C_\models(\{\}) = Taut \subseteq C_\models(F)$.
2) $F \subseteq C_\models(F)$. (Inclusion)
3) $F \equiv C_\models(F)$.
4) $F \models G$ iff $C_\models(F) \supseteq C_\models(G)$.
5) $F \equiv G$ iff $C_\models(F) = C_\models(G)$.
   So if $F \equiv G$ then $C_\models(F) = C_\models(G)$. (Left Equivalence)
6) If $F \subseteq G$ then $C_\models(F) \subseteq C_\models(G)$. (Monotonicity)
7) If $F \subseteq C_\models(G)$ then $C_\models(F \cup G) = C_\models(G)$. (Cumulativity)
8) $C_\models(C_\models(F)) = C_\models(F)$. (Idempotence)
9) $C_\models(\vee\{f,g\}) = C_\models(f) \cap C_\models(g)$.
10) If $f \equiv F$ then $F \subseteq C_\models(f)$.
11) $Imp(\vee\{\}) = Contrad \subseteq Imp(\vee F)$.
12) $F \subseteq Imp(\vee F)$.
13) If $v \models Imp(\vee F)$ then $v \models F$ and $v \models \vee F$.

14) If $F \subseteq G$ then $Imp(\vee F) \subseteq Imp(\vee G)$.
15) If $F \subseteq Imp(\vee G)$ then $Imp(\vee(F \cup G)) = Imp(\vee G)$.
16) $Imp(\vee Imp(\vee F)) = Imp(\vee F)$.

Proof

Let each of F and G be any set of formulas. Let each of f and g be any formula. Let v be any valuation.

(1) If $\mathfrak{t}$ is a tautology then $F \models \mathfrak{t}$. Hence $Taut \subseteq C_\models(F)$ and so $Taut \subseteq C_\models(\{\})$. If $f \in C_\models(\{\})$ then $\{\} \models f$. By Definition 2.2.14, every valuation satisfies $\{\}$, and so every valuation satisfies f. Hence f is a tautology and so $C_\models(\{\}) \subseteq Taut$.

(2) Take any f in F. Then each valuation which satisfies every formula in F must satisfy f. So $F \models f$ and hence $f \in C_\models(F)$. Thus $F \subseteq C_\models(F)$.

(3) If $v \models C_\models(F)$ then by part (2), $v \models F$. Conversely suppose $v \models F$. Then v satisfies every formula in $C_\models(F)$. Thus part (3) holds.

(4) $F \models G$ iff for all $h \in Fml$, (if $G \models h$ then $F \models h$) by Lemma 2.2.18(5),
iff $\{h \in Fml : F \models h\} \supseteq \{h \in Fml : G \models h\}$
iff $C_\models(F) \supseteq C_\models(G)$.

(5) $F \equiv G$ iff for all $h \in Fml$, $(F \models h$ iff $G \models h)$ by Lemma 2.2.18(5),
iff $\{h \in Fml : F \models h\} = \{h \in Fml : G \models h\}$
iff $C_\models(F) = C_\models(G)$.

(6) Suppose $F \subseteq G$. Take any f in $C_\models(F)$. Then $F \models f$ and so each valuation which satisfies every formula in F also satisfies f. Let v be any valuation which satisfies every formula in G. Then v satisfies every formula in F, and so v satisfies f. Therefore $G \models f$ and so $f \in C_\models(G)$. Thus $C_\models(F) \subseteq C_\models(G)$.

(7) By part (6), $C_\models(G) \subseteq C_\models(F \cup G)$.

Suppose $F \subseteq C_\models(G)$. Then $G \models F$. We show $C_\models(F \cup G) \subseteq C_\models(G)$. Take any f in $C_\models(F \cup G)$. Then $F \cup G \models f$. Let v be any valuation which satisfies G. Since $G \models F$, v satisfies $F \cup G$; and so v satisfies f. Hence $G \models f$, and so $f \in C_\models(G)$. Therefore $C_\models(F \cup G) \subseteq C_\models(G)$.

Thus $C_\models(F \cup G) = C_\models(G)$.

(8) By letting $F' = C_\models(F)$ in part (7), we get $C_\models(C_\models(F) \cup F) = C_\models(F)$. By part (2), $C_\models(F) \cup F = C_\models(F)$. Therefore $C_\models(C_\models(F)) = C_\models(F)$.

(9) $C_\models(\vee\{f,g\})$
$= \{h \in Fml : \vee\{f,g\} \models h\}$
$= \{h \in Fml : f \models h \text{ and } g \models h\}$
$= \{h \in Fml : f \models h\} \cap \{h \in Fml : g \models h\}$
$= C_\models(f) \cap C_\models(g)$.

(10) Suppose $f \equiv F$. Then by part 2 and part 5, $F \subseteq C_\models(F) = C_\models(f)$.

(11) If $\mathfrak{c}$ is a contradiction then $\mathfrak{c} \models \vee F$. Hence $Contrad \subseteq Imp(\vee F)$ and so $Contrad \subseteq Imp(\vee\{\})$. If $f \in Imp(\vee\{\})$ then $f \models \vee\{\}$. By Lemma 2.2.16(1), no valuation satisfies $\vee\{\}$, and so no valuation satisfies f. Hence f is a contradiction and so $Imp(\vee\{\}) \subseteq Contrad$. Thus $Imp(\vee\{\}) = Contrad$.

(12) Take any f in F. Then each valuation which satisfies f must satisfy $\vee F$. So $f \models \vee F$ and hence $f \in Imp(\vee F)$. Thus $F \subseteq Imp(\vee F)$.

2.3 Negation

(13) If $v \models Imp(\vee F)$ then by part (12), $v \models F$; and by part (11), $F \neq \{\}$. Hence $v \models \vee F$.

(14) Suppose $F \subseteq G$. Take any f in $Imp(\vee F)$. Then $f \models \vee F$. Hence $f \models \vee G$ and so $f \in Imp(\vee G)$. Thus $Imp(\vee F) \subseteq Imp(\vee G)$.

(15) By part (14), $Imp(\vee G) \subseteq Imp(\vee(F \cup G))$.

Suppose $F \subseteq Imp(\vee G)$. Then for all g in F, $g \models \vee G$. We show $Imp(\vee(F \cup G)) \subseteq Imp(\vee G)$. Take any f in $Imp(\vee(F \cup G))$. Then $f \models \vee(F \cup G)$. Let v be any valuation which satisfies f. Then there exists g in $F \cup G$ such that v satisfies g. If $g \in G$ then v satisfies $\vee G$. So suppose $g \in F$. Then, from above, $g \models \vee G$. So in both cases v satisfies $\vee G$, and so $f \models \vee G$. Hence $f \in Imp(\vee G)$. Therefore $Imp(\vee(F \cup G)) \subseteq Imp(\vee G)$.

Thus $Imp(\vee(F \cup G)) = Imp(\vee G)$.

(16) In part (15), replace F by $Imp(\vee F)$ and then replace G by F to get the following. If $Imp(\vee F) \subseteq Imp(\vee F)$ then $Imp(\vee(Imp(\vee F) \cup F)) = Imp(\vee F)$. By part (12), $Imp(\vee F) \cup F = Imp(\vee F)$. Therefore $Imp(\vee Imp(\vee F)) = Imp(\vee F)$.

EndProofLemma2.2.21

As Definition 2.2.20 shows, we are only interested in Imp applied to a disjunction of a set of formulas, as this gives a dual of $C_\models$. However we could have defined Imp', say, by $Imp'(F) = Imp(\vee F)$. Then Lemma 2.2.21(12) shows that Imp' is inclusive; Lemma 2.2.21(14) shows that Imp' is monotonic; Lemma 2.2.21(15) shows that Imp' is cumulative; and Lemma 2.2.21(16), shows that Imp' is idempotent.

2.3 Negation

We start this section on negation by extending negation to sets of formulas.

Definition 2.3.1. If F is a set of formulas then $\neg F = \{\neg f : f \in F\}$.

The first result in this section is De Morgan's law for $\neg$, $\wedge$, and $\vee$. It says that the negation sign can be 'pushed inwards' provided conjunctions and disjunction are interchanged.

Lemma 2.3.2. Let F be any set of formulas and v any valuation.
1) $v(\neg F) = \text{NOT}\, v(F)$.
2) $v(\neg \wedge F) = v(\vee \neg F)$.   Hence $\neg \wedge F \equiv \vee \neg F$.   (De Morgan's law)
3) $v(\neg \vee F) = v(\wedge \neg F)$.   Hence $\neg \vee F \equiv \wedge \neg F$.   (De Morgan's law)

Proof
  (1) $v(\neg F) = v(\{\neg f : f \in F\}) = \{v(\neg f) : f \in F\} = \{\text{NOT}\, v(f) : f \in F\} = \text{NOT}\,\{v(f) : f \in F\} = \text{NOT}\, v(F)$.

  (2) $v(\neg \wedge F) = \mathbf{T}$ iff $\text{NOT}\, v(\wedge F) = \mathbf{T}$ iff $v(\wedge F) = \mathbf{F}$ iff $\text{AND}\{v(f) : f \in F\} = \mathbf{F}$ iff $\mathbf{F} \in \{v(f) : f \in F\}$ iff $\mathbf{T} \in \{\text{NOT}\, v(f) : f \in F\}$ iff $\mathbf{T} \in \{v(\neg f) : f \in F\}$ iff $\text{OR}\{v(\neg f) : f \in F\} = \mathbf{T}$ iff $v(\vee\{\neg f : f \in F\}) = \mathbf{T}$ iff $v(\vee \neg F) = \mathbf{T}$.

$v(\neg\wedge F) = \mathbf{F}$ iff $\text{NOT}\, v(\wedge F) = \mathbf{F}$ iff $v(\wedge F) = \mathbf{T}$ iff $\text{AND}\{v(f) : f \in F\} = \mathbf{T}$ iff $\mathbf{F} \notin \{v(f) : f \in F\}$ iff $\mathbf{T} \notin \{\text{NOT}\, v(f) : f \in F\}$ iff $\mathbf{T} \notin \{v(\neg f) : f \in F\}$ iff $\text{OR}\{v(\neg f) : f \in F\} = \mathbf{F}$ iff $v(\vee\{\neg f : f \in F\}) = \mathbf{F}$ iff $v(\vee \neg F) = \mathbf{F}$.

(3) $v(\neg\vee F) = \mathbf{T}$ iff $\text{NOT}\, v(\vee F) = \mathbf{T}$ iff $v(\vee F) = \mathbf{F}$ iff $\text{OR}\{v(f) : f \in F\} = \mathbf{F}$ iff $\mathbf{T} \notin \{v(f) : f \in F\}$ iff $\mathbf{F} \notin \{\text{NOT}\, v(f) : f \in F\}$ iff $\mathbf{F} \notin \{v(\neg f) : f \in F\}$ iff $\text{AND}\{v(\neg f) : f \in F\} = \mathbf{T}$ iff $v(\wedge\{\neg f : f \in F\}) = \mathbf{T}$ iff $v(\wedge \neg F) = \mathbf{T}$.

$v(\neg\vee F) = \mathbf{F}$ iff $\text{NOT}\, v(\vee F) = \mathbf{F}$ iff $v(\vee F) = \mathbf{T}$ iff $\text{OR}\{v(f) : f \in F\} = \mathbf{T}$ iff $\mathbf{T} \in \{v(f) : f \in F\}$ iff $\mathbf{F} \in \{\text{NOT}\, v(f) : f \in F\}$ iff $\mathbf{F} \in \{v(\neg f) : f \in F\}$ iff $\text{AND}\{v(\neg f) : f \in F\} = \mathbf{F}$ iff $v(\wedge\{\neg f : f \in F\}) = \mathbf{F}$ iff $v(\wedge \neg F) = \mathbf{F}$.
EndProofLemma2.3.2

By repeatedly applying De Morgan's law to a formula we can push all negation signs inwards until they are next to atoms or other negation signs. Removing double negation signs then gives a formula in negation normal form (nnf). The following definitions make this precise.

Definition 2.3.3.
1) If l is a literal then l is an *nnf-formula*.
2) If F is a finite set of nnf-formulas then $\wedge F$ is an *nnf-formula*, and $\vee F$ is an *nnf-formula*.
3) Every nnf-formula can be built by a finite number of applications of (1) and (2).
The set of all nnf-formulas is denoted by *NnfFml*.

So every clause is an nnf-formula and every meet is an nnf-formula.

Definition 2.3.4. The set, *NnfQFml*, of all *nnf-quasi-formulas* is defined by
$NnfQFml = NnfFml \cup \{\wedge F : F \subseteq NnfFml\} \cup \{\vee F : F \subseteq NnfFml\}$.
A quasi-formula f is in *negation normal form* iff $f \in NnfQFml$.

So a quasi-formula is in negation normal form iff only its atoms are negated. We now define a function that pushes negations inwards as far as possible and removes double negations. The result is an nnf-quasi-formula.

Definition 2.3.5.
1) If l is a literal then $nnf(l) = l$.
2) If $\neg\neg f$ is a quasi-formula then $nnf(\neg\neg f) = nnf(f)$.
3) If F is a set of formulas then $nnf(F) = \{nnf(f) : f \in F\}$.
4) If F is a set of formulas then
$nnf(\wedge F) = \wedge nnf(F)$,
$nnf(\vee F) = \vee nnf(F)$,
$nnf(\neg\wedge F) = \vee nnf(\neg F)$, and
$nnf(\neg\vee F) = \wedge nnf(\neg F)$.

We want to prove that the truth value of $nnf(f)$ and f are the same. But before we can do this, we must first assign to each quasi-formula f an ordinal $|f|$, called the complexity of f, and then use induction on these ordinals.

2.3 Negation

Definition 2.3.6. The *complexity* of a quasi-formula, f, is an ordinal, $|f|$, defined inductively as follows.
1) If a is an atom then $|a| = 0$.
2) If $\neg f$ is a quasi-formula then $|\neg f| = |f| + 1$.
3) $|\wedge\{\}| = 1 = |\vee\{\}|$.
4) If F is a non-empty finite set of formulas then $|\wedge F| = |\vee F| = \max\{|f| : f \in F\} + 1$.
5) If F is an infinite set of formulas then $|\wedge F| = |\vee F| = \omega + 1$.

The following technical lemma is needed before we can prove that $nnf(f) \equiv f$.

Lemma 2.3.7. Let f be any formula and F be any set of formulas.
1) $|f| = 0$ iff f is an atom.
2) f is a formula iff $|f| \in \mathbb{N}$.
3) If $f \in F$ then $|f| < |\wedge F|$, $|f| < |\vee F|$, $|\neg f| < |\neg \wedge F|$, and $|\neg f| < |\neg \vee F|$.

Proof
(1) This follows from Definition 2.3.6.

(2) By Definition 2.2.3, the only quasi-formulas that are not formulas have one of the following forms $\wedge F$, $\vee F$, $\neg \wedge F$, and $\neg \vee F$, where F is an infinite set of formulas. By Definition 2.3.6(5,2), if F is an infinite set of formulas then $|\wedge F| = |\vee F| = \omega + 1$, and $|\neg \wedge F| = |\neg \vee F| = \omega + 2$. So if f is a quasi-formula that is not a formula then $|f| \notin \mathbb{N}$.

Conversely, we prove by induction that if f is a formula then $|f| \in \mathbb{N}$. Take any formula f_0 and suppose that if g is a quasi-formula and $|g| < |f_0|$ then $|g| \in \mathbb{N}$. We show that $|f_0| \in \mathbb{N}$.

Case 1: f_0 is an atom.
Then $|f_0| = 0 \in \mathbb{N}$.

Case 2: $f_0 = \neg g$.
Then $|f_0| = |\neg g| = |g| + 1$. So $|g| < |f_0|$ and hence $|g| \in \mathbb{N}$. Therefore $|g| + 1 \in \mathbb{N}$ and so $|f_0| \in \mathbb{N}$.

Case 3: either $f_0 = \wedge\{\}$ or $f_0 = \vee\{\}$.
Then $|f_0| = 1 \in \mathbb{N}$.

Case 4: either $f_0 = \wedge G$ or $f_0 = \vee G$, where G is a non-empty finite set of formulas.
Then $|f_0| = \max\{|g| : g \in G\} + 1$. So for each g in G, $|g| < |f_0|$ and hence $|g| \in \mathbb{N}$. Therefore $(\max\{|g| : g \in G\} + 1) \in \mathbb{N}$ and so $|f_0| \in \mathbb{N}$.

Thus $|f_0| \in \mathbb{N}$. So by induction, if f is a formula then $|f| \in \mathbb{N}$.

(3) Suppose $f \in F$. By Lemma 2.3.7(2) and Definition 2.3.6(4,5), $|f| < |\wedge F|$ and $|f| < |\vee F|$. So $|\neg f| = |f| + 1 < |\wedge F| + 1 = |\neg \wedge F|$. Similarly, $|\neg f| = |f| + 1 < |\vee F| + 1 = |\neg \vee F|$.

EndProofLemma2.3.7

Lemma 2.3.8. If f is a quasi-formula then $nnf(f) \equiv f$.
Proof
Let v be any valuation. We prove by induction that if f is a quasi-formula then $v(nnf(f)) = v(f)$. Take any quasi-formula f_0 and suppose that if g is a quasi-formula such that $|g| < |f_0|$ then $v(nnf(g)) = v(g)$. We show that $v(nnf(f_0)) = v(f_0)$.

Case 1: f_0 is an atom.
Then $nnf(f_0) = f_0$ and so $v(nnf(f_0)) = v(f_0)$.
Case 2: $f_0 = \neg g$.
Then $|g| < |f_0|$ and hence $v(nnf(g)) = v(g)$.
Case 2.1: g is an atom.
Then f_0 is a literal, hence $nnf(f_0) = f_0$, and so $v(nnf(f_0)) = v(f_0)$.
Case 2.2: $g = \neg h$.
Then $|h| < |g| < |f_0|$. So $v(nnf(h)) = v(h)$. Also $f_0 = \neg g = \neg\neg h$. So $nnf(f_0) = nnf(\neg\neg h) = nnf(h)$. Furthermore $v(f_0) = v(\neg\neg h) = \text{NOT NOT}\, v(h) = v(h)$. Thus $v(nnf(f_0)) = v(nnf(h)) = v(h) = v(f_0)$.
Case 2.3: $g = \wedge F$, where F is a set of formulas.
Then $nnf(f_0) = nnf(\neg g) = nnf(\neg \wedge F) = \vee nnf(\neg F) = \vee nnf(\{\neg f : f \in F\}) = \vee\{nnf(\neg f) : f \in F\}$. By Lemma 2.3.7(3), for all f in F, $|\neg f| < |\neg \wedge F| = |\neg g| = |f_0|$. Hence for all f in F, $v(nnf(\neg f)) = v(\neg f)$. So by De Morgan's law, Lemma 2.3.2(2), $v(nnf(f_0)) = v(\vee\{nnf(\neg f) : f \in F\}) = \text{OR}\{v(nnf(\neg f)) : f \in F\} = \text{OR}\{v(\neg f) : f \in F\} = v(\vee\{\neg f : f \in F\}) = v(\vee \neg F) = v(\neg \wedge F) = v(\neg g) = v(f_0)$.
Case 2.4: $g = \vee F$, where F is a set of formulas.
Then $nnf(f_0) = nnf(\neg g) = nnf(\neg \vee F) = \wedge nnf(\neg F) = \wedge nnf(\{\neg f : f \in F\}) = \wedge\{nnf(\neg f) : f \in F\}$. By Lemma 2.3.7(3), for all f in F, $|\neg f| < |\neg \vee F| = |\neg g| = |f_0|$. Hence for all f in F, $v(nnf(\neg f)) = v(\neg f)$. So by De Morgan's law, Lemma 2.3.2(3), we have $v(nnf(f_0)) = v(\wedge\{nnf(\neg f) : f \in F\}) = \text{AND}\{v(nnf(\neg f)) : f \in F\} = \text{AND}\{v(\neg f) : f \in F\} = v(\wedge\{\neg f : f \in F\}) = v(\wedge \neg F) = v(\neg \vee F) = v(\neg g) = v(f_0)$.
Case 3: $f_0 = \wedge F$, where F is a set of formulas.
Then $nnf(f_0) = nnf(\wedge F) = \wedge nnf(F) = \wedge\{nnf(f) : f \in F\}$. By Lemma 2.3.7(3), for all f in F, $|f| < |\wedge F| = |f_0|$. So for all f in F, $v(nnf(f)) = v(f)$. Hence $v(nnf(f_0)) = v(\wedge\{nnf(f) : f \in F\}) = \text{AND}\{v(nnf(f)) : f \in F\} = \text{AND}\{v(f) : f \in F\} = v(\wedge\{f : f \in F\}) = v(\wedge F) = v(f_0)$.
Case 4: $f_0 = \vee F$, where F is a set of formulas.
Then $nnf(f_0) = nnf(\vee F) = \vee nnf(F) = \vee\{nnf(f) : f \in F\}$. By Lemma 2.3.7(3), for all f in F, $|f| < |\vee F| = |f_0|$. So for all f in F, $v(nnf(f)) = v(f)$. Hence $v(nnf(f_0)) = v(\vee\{nnf(f) : f \in F\}) = \text{OR}\{v(nnf(f)) : f \in F\} = \text{OR}\{v(f) : f \in F\} = v(\vee\{f : f \in F\}) = v(\vee F) = v(f_0)$.
Therefore $v(nnf(f_0)) = v(f_0)$ is proved.
Thus by induction for all quasi-formulas f, $v(nnf(f)) = v(f)$.
But v was arbitrary, so $nnf(f) \equiv f$.
EndProofLemma2.3.8

We can now define the complement notation, $\sim$.

Definition 2.3.9.
1) If f is a quasi-formula then the *complement* of f, $\sim f$, is defined by $\sim f = nnf(\neg f)$.
2) Let S be a set such that for each element s in S, $\sim s$ is defined. Then define $\sim S$ by $\sim S = \{\sim s : s \in S\}$.

Notice that if f is a quasi-formula then $\sim f$ is also a quasi-formula. We now list and prove many properties of the complement function.

2.3 Negation

Lemma 2.3.10. Let F be a set of formulas, C be a set of clauses, M be a set of meets, G be either a set of clauses or a set of meets, f and g be two a quasi-formulas, and v be a valuation.
1) If l is a literal then $\sim l$ is a literal and $l = \sim\sim l$.
2) $\sim F = nnf(\neg F)$.
3) $\sim \wedge F = \vee \sim F$, and $\sim \vee F = \wedge \sim F$. So $\sim \wedge \{\} = \vee \{\}$, and $\sim \vee \{\} = \wedge \{\}$.
4) If $g \in G$ then $g = \sim\sim g$. So $G = \sim\sim G$.
5) $v(\sim f) = \text{NOT}\, v(f)$. So $v(f) = \text{NOT}\, v(\sim f)$.
6) $f \equiv \sim\sim f$.
7) $f \equiv g$ iff $\sim f \equiv \sim g$.
8) F is satisfiable iff $\sim F$ is falsifiable.
9) $F \models \sim f$ iff $f \models \vee \sim F$. So $F \models f$ iff $\sim f \models \vee \sim F$.
10) $\sim C_\models(F) = Imp(\vee \sim F)$. So $C_\models(F) = \sim Imp(\vee \sim F)$.
11) $\sim ClsC_\models(F) = MtsImp(\vee \sim F)$. So $ClsC_\models(F) = \sim MtsImp(\vee \sim F)$.
12) If c is a clause then $\sim Lit(c) = Lit(\sim c)$ and so $\sim Lit(C) = Lit(\sim C)$.
13) If m is a meet then $\sim Lit(m) = Lit(\sim m)$ and so $\sim Lit(M) = Lit(\sim M)$.
14) Suppose $\{g, g'\} \subseteq G$. Then
$g = g'$ iff $\sim g = \sim g'$,
$g \in G$ iff $\sim g \in \sim G$, and
$|G| = |\sim G|$.

Proof

Let F, C, M, G, f, g, and v be as in the statement of the lemma.

(1) Let a be an atom. Then $\sim a = nnf(\neg a) = \neg a$. Also $\sim \neg a = nnf(\neg \neg a) = nnf(a) = a$. So if l is a literal then $\sim l$ is a literal. We have $\sim\sim a = \sim nnf(\neg a) = \sim \neg a = a$, from before. Also $\sim\sim \neg a = \sim nnf(\neg \neg a) = \sim nnf(a) = \sim a = \neg a$, from before. So if l is a literal then $l = \sim\sim l$.

(2) $nnf(\neg F) = nnf(\{\neg f : f \in F\}) = \{nnf(\neg f) : f \in F\} = \{\sim f : f \in F\} = \sim F$.

(3) By part (2) we have $\sim \wedge F = nnf(\neg \wedge F) = \vee nnf(\neg F) = \vee \sim F$. And also $\sim \vee F = nnf(\neg \vee F) = \wedge nnf(\neg F) = \wedge \sim F$.

(4) Let L be a finite set of literals. By parts (3) and (1), $\sim\sim \vee L = \sim \wedge \sim L = \sim \wedge \{\sim l : l \in L\} = \vee \sim \{\sim l : l \in L\} = \vee \{\sim\sim l : l \in L\} = \vee \{l : l \in L\} = \vee L$. Similarly, $\sim\sim \wedge L = \sim \vee \sim L = \sim \vee \{\sim l : l \in L\} = \wedge \sim \{\sim l : l \in L\} = \wedge \{\sim\sim l : l \in L\} = \wedge \{l : l \in L\} = \wedge L$. So, with part (1), we get $g = \sim\sim g$.
Hence $\sim\sim G = \sim\sim \{g : g \in G\} = \{\sim\sim g : g \in G\} = \{g : g \in G\} = G$.

(5) By Lemma 2.3.8, $v(\sim f) = v(nnf(\neg f)) = v(\neg f) = \text{NOT}\, v(f)$. So $\text{NOT}\, v(\sim f) = \text{NOT NOT}\, v(f) = v(f)$.

(6) By Lemma 2.3.10(5), $v(\sim\sim f) = v(\sim(\sim f)) = \text{NOT}\, v(\sim f) = \text{NOT NOT}\, v(f) = v(f)$. Hence $f \equiv \sim\sim f$.

(7) $f \equiv g$ iff for all v in Val, $v(f) = v(g)$
iff for all v in Val, $\text{NOT}\, v(f) = \text{NOT}\, v(g)$
iff for all v in Val, $v(\sim f) = v(\sim g)$
iff $\sim f \equiv \sim g$.

(8) $\sim F$ is falsifiable iff there exists v in Val such that $v(\vee \sim F) = \mathbf{F}$
iff there exists v in Val such that $v(\vee \{\sim f : f \in F\}) = \mathbf{F}$

iff there exists v in *Val* such that $\text{OR}\{v(\sim f) : f \in F\} = \mathbf{F}$
iff there exists v in *Val* such that $\text{OR}\{\text{NOT}\,v(f) : f \in F\} = \mathbf{F}$
iff there exists v in *Val* such that $\mathbf{T} \notin \{\text{NOT}\,v(f) : f \in F\}$
iff there exists v in *Val* such that $\mathbf{F} \notin \{v(f) : f \in F\}$
iff there exists v in *Val* such that $\text{AND}\{v(f) : f \in F\} = \mathbf{T}$
iff there exists v in *Val* such that $v(\wedge F) = \mathbf{T}$
iff F is satisfiable.

(9) Suppose $F \models \sim f$. If v satisfies f then v does not satisfy $\wedge F$. So there is a g in F such that $v(g) = \mathbf{F}$ and hence $v(\sim g) = \mathbf{T}$. That is, $f \models \vee \sim F$.

Conversely suppose $f \models \vee \sim F$. If v satisfies every formula in F then v does not satisfy f, and so v satisfies $\sim f$. That is, $F \models \sim f$.

Now let $\sim f = g$. Then $\sim f \models \vee \sim F$ iff $g \models \vee \sim F$ iff $F \models \sim g$ iff $F \models \sim \sim f$ iff $F \models f$.

(10) By part (9), $\sim C_\models(F) = \{\sim f : f \in Fml \text{ and } F \models f\} = \{f \in Fml : F \models \sim f\}$
$= \{f \in Fml : f \models \vee \sim F\} = Imp(\vee \sim F)$.

(11) We use part (10) and $ClsC_\models(F) = C_\models(F) \cap Cls$. $\sim ClsC_\models(F) = \sim C_\models(F) \cap Mts$
$= Imp(\vee \sim F) \cap Mts = MtsImp(\vee \sim F)$.

(12) Let c be a clause. There are two cases. If c is the literal l then $\sim Lit(c) = \sim Lit(l) = \sim \{l\} = \{\sim l\} = Lit(\sim l) = Lit(\sim c)$. If $c = \vee L$, where L is a finite set of literals, then $Lit(\sim c) = Lit(\sim \vee L) = Lit(\wedge \sim L) = \sim L = \sim Lit(c)$.

So in both cases $\sim Lit(c) = Lit(\sim c)$.

Let $C = \{c : c \in C\}$. Then $\sim Lit(C) = \sim Lit(\{c : c \in C\}) = \sim \{Lit(c) : c \in C\} = \{\sim Lit(c) : c \in C\} = \{Lit(\sim c) : c \in C\} = Lit(\{\sim c : c \in C\}) = Lit(\sim C)$.

(13) Let m be a meet. There are two cases. If m is the literal l then $\sim Lit(m) = \sim Lit(l) = \sim \{l\} = \{\sim l\} = Lit(\sim l) = Lit(\sim m)$. If $m = \wedge L$, where L is a finite set of literals, then $Lit(\sim m) = Lit(\sim \wedge L) = Lit(\vee \sim L) = \sim L = \sim Lit(m)$.

So in both cases $\sim Lit(m) = Lit(\sim m)$.

Let $M = \{m : m \in M\}$. Then $\sim Lit(M) = \sim Lit(\{m : m \in M\}) = \sim \{Lit(m) : m \in M\}$
$= \{\sim Lit(m) : m \in M\} = \{Lit(\sim m) : m \in M\} = Lit(\{\sim m : m \in M\}) = Lit(\sim M)$.

(14) Suppose $\{g, g'\} \subseteq G$. If $g = g'$ then $\sim g = \sim g'$. Conversely, if $\sim g = \sim g'$ then $\sim \sim g = \sim \sim g'$. So by part (4), $g = g'$. Thus $g = g'$ iff $\sim g = \sim g'$.

Therefore $|G| = |\sim G|$.

If $g \in G$ then $\sim g \in \{\sim g : g \in G\} = \sim G$. Conversely, if $\sim g \in \sim G = \{\sim g : g \in G\}$ then there exists $g' \in G$ such that $\sim g = \sim g'$. Hence $g = g'$ and so $g \in G$.

Thus $g \in G$ iff $\sim g \in \sim G$.

EndProofLemma2.3.10

'Compactness' is the notion that if a property holds for all finite parts then it holds for the whole. The last result in this section is the semantic compactness theorem.

Theorem 2.3.11 (semantic compactness). Let F be any set of formulas.

1) (compactness for satisfiability)

   F is satisfiable iff every finite subset of F is satisfiable.

2) (compactness for falsifiability)

   F is falsifiable iff every finite subset of F is falsifiable.

2.3 Negation

Proof

Let F be any set of formulas.

(1) Let us say that a set of formulas is *finitely satisfiable* iff every finite subset of it is satisfiable.

If F is satisfiable then F is finitely satisfiable.

Conversely suppose F is finitely satisfiable. If F is finite then F is satisfiable. So suppose F is infinite. We start by extending F to a maximal finitely satisfiable set G, then we define a valuation which satisfies G and hence F.

Since Fml is countably infinite, there is a bijection between $\mathbb{Z}^+$ and Fml. Hence $Fml = \{f_i : i \in \mathbb{Z}^+\}$. Define $G_0 = F$ and if $G_n \cup \{f_{n+1}\}$ is finitely satisfiable then $G_{n+1} = G_n \cup \{f_{n+1}\}$; else $G_{n+1} = G_n \cup \{\neg f_{n+1}\}$.

We show by induction on n that each G_n is finitely satisfiable. By supposition G_0 is finitely satisfiable. Suppose G_n is finitely satisfiable and prove G_{n+1} is finitely satisfiable. If $G_n \cup \{f_{n+1}\}$ is finitely satisfiable then G_{n+1} is finitely satisfiable. So suppose $G_n \cup \{f_{n+1}\}$ is not finitely satisfiable. Then there is a finite subset H of G_n such that H is satisfiable and $H \cup \{f_{n+1}\}$ is not satisfiable. Take any finite subset S of $G_n \cup \{\neg f_{n+1}\}$. If $S \subseteq G_n$ then S is satisfiable; otherwise $S = H_n \cup \{\neg f_{n+1}\}$ where $H_n \subseteq G_n$. So $H \cup H_n$ is a finite subset of G_n and so is satisfiable. But $H \cup H_n \cup \{f_{n+1}\}$ is not satisfiable, and so $H \cup H_n \cup \{\neg f_{n+1}\}$ is satisfiable. Hence $H_n \cup \{\neg f_{n+1}\}$ is satisfiable. Thus by induction each G_n is finitely satisfiable.

Let $G = \bigcup \{G_n : n \in \mathbb{N}\}$. Then G is finitely satisfiable because each finite subset of G is also a finite subset of some G_n and so satisfiable. Also for each formula f either $f \in G$ or $\neg f \in G$ but not both.

Define a valuation v as follows. If a is an atom then $v(a) = \mathbf{T}$ iff $a \in G$. We now show by induction on the structure of f that if f is any formula then $v(f) = \mathbf{T}$ iff $f \in G$. The statement is true if f is an atom. Take any formula f and suppose that if e is a component of f then $v(e) = \mathbf{T}$ iff $e \in G$.

If $f = \neg e$ then $v(f) = \mathbf{T}$ iff $v(\neg e) = \mathbf{T}$ iff $v(e) = \mathbf{F}$ iff $e \notin G$ iff $\neg e \in G$ iff $f \in G$.

Now let E be a finite set of formulas. If $f = \wedge E$ then $v(f) = \mathbf{T}$ iff $v(\wedge E) = \mathbf{T}$ iff for all e in E, $v(e) = \mathbf{T}$ iff for all e in E, $e \in G$ iff $\wedge E \in G$ (because G is finitely satisfiable) iff $f \in G$.

If $f = \vee E$ then $v(f) = \mathbf{T}$ iff $v(\vee E) = \mathbf{T}$ iff for some e in E, $v(e) = \mathbf{T}$ iff for some e in E, $e \in G$ iff $\vee E \in G$ (because G is finitely satisfiable) iff $f \in G$.

Thus v satisfies G and hence F.

(2) Let us say that a set of formulas is *finitely falsifiable* iff every finite subset of it is falsifiable. By Lemma 2.3.10(8), F is falsifiable iff $\sim F$ is satisfiable iff $\sim F$ is finitely satisfiable iff F is finitely falsifiable.

EndProofTheorem2.3.11

2.4 Resolution

Resolution is usually defined just for clauses, but it can be defined for meets as well, (in which case it is sometimes called dual resolution). Although we will be mainly concerned with resolving clauses, we shall also consider resolving meets because that can be useful for finding causes or diagnoses. So it makes sense to define resolution for sets of literals which could be formed into either clauses or meets; we shall called such sets 'stems'.

Definition 2.4.1. A *stem* is any finite set of literals. If S is a set of stems then define
$\sim S = \{\sim s : s \in S\}$,
$\wedge.S = \{\wedge s : s \in S\}$, and
$\vee.S = \{\vee s : s \in S\}$.

We note that $\wedge.\{\} = \{\} = \vee.\{\}$ whereas $\wedge\{\}$ and $\vee\{\}$ are formulas.

The result of a resolution of two stems is a stem called a resolvent. A sequence of resolutions is a derivation. These and other concepts are defined in the next two definitions.

Definition 2.4.2. Let L and L' be two stems, and l be a literal such that $l \in L$ and $\sim l \in L'$. Then the *resolvent of L and L' on l* is $(L - \{l\}) \cup (L' - \{\sim l\})$, and is denoted by $res(l; L, L')$. We say K is a *resolvent of L and L'* iff there exists a literal l such that $res(l; L, L') = K$; in which case we say that K is a *resolution-child* of L, and also of L', and that L and L' are *resolvable*.

Definition 2.4.3. Let S be a set of stems. A stem s_n is *resolution-derivable from S* iff there is a finite sequence of stems $(s_1, ..., s_n)$ such that for each i in $[1..n]$, either $s_i \in S$ or s_i is a resolvent of two preceding stems. The sequence $(s_1, ..., s_n)$ is called a *resolution-derivation* of s_n from S. The set of all stems which are resolution-derivable from S is denoted by $Res(S)$.

It is sometimes more convenient to resolve two clauses rather than their corresponding stems. The following definitions formalise this.

Definition 2.4.4. Let G be either a set of clauses or a set of meets. The following notation is convenient: $ResLit(G) = Res(Lit(G))$. If $\{g_1, g_2\} \subseteq G$ then g_1 and g_2 are *resolvable* iff $Lit(g_1)$ and $Lit(g_2)$ are resolvable.

Definition 2.4.5. Let c be any clause and C be any set of clauses.
1) If l is a literal then define $\check{}(l) = \check{l} = \vee\{l\}$.
2) If c is not a literal then define $\check{}(c) = \check{c} = c$.
3) Define $\check{}(C) = \check{C} = \{\check{c} : c \in C\}$.

The check symbol, $\check{}$, reminds us that $\check{c}$ always starts with the disjunction symbol $\vee$.

2.4 Resolution

Definition 2.4.6. Let C be a set of clauses. Define $Res(C) = \vee.ResLit(C)$. If $\{c_1, c_2\} \subseteq \check{C}$ then c_3 is a *resolvent* of c_1 and c_2 iff there exists a literal l in $Lit(c_1) \cup Lit(c_2)$ such that $c_3 = \vee res(l; Lit(c_1), Lit(c_2))$. A clause c_n is *resolution-derivable* from $\check{C}$ iff there is a finite sequence of clauses $(c_1, ..., c_n)$ such that for each i in $[1..n]$, either $c_i \in \check{C}$ or c_i is a resolvent of two preceding clauses. The sequence $(c_1, ..., c_n)$ is called a *resolution-derivation* of c_n from $\check{C}$.

If S is a set of stems and $(s_1, ..., s_n)$ is a resolution-derivation of s_n from S then $(\vee s_1, ..., \vee s_n)$ is a resolution-derivation of $\vee s_n$ from $\vee.S$. If C is a set of clauses and $(c_1, ..., c_n)$ is a resolution-derivation of c_n from $\check{C}$ then $(Lit(c_1), ..., Lit(c_n))$ is a resolution-derivation of $Lit(c_n)$ from $Lit(C)$. So $Res(C)$ is the set of all clauses which are resolution-derivable from $\check{C}$. Hence $\check{C} \subseteq Res(C)$.

Similarly the resolution of two meets, rather than their corresponding stems, is formalised by the following definitions.

Definition 2.4.7. Let m be any meet and M be any set of meets.
1) If l is a literal then define $\hat{}(l) = \hat{l} = \wedge\{l\}$.
2) If m is not a literal then define $\hat{}(m) = \hat{m} = m$.
3) Define $\hat{}(M) = \hat{M} = \{\hat{m} : m \in M\}$.

The hat symbol, $\hat{}$, reminds us that $\hat{m}$ always starts with the conjunction symbol $\wedge$.

Definition 2.4.8. Let M be a set of meets. Define $Res(M) = \wedge.ResLit(M)$. If $\{m_1, m_2\} \subseteq \hat{M}$ then m_3 is a *resolvent* of m_1 and m_2 iff there exists a literal l in $Lit(m_1) \cup Lit(m_2)$ such that $m_3 = \wedge res(l; Lit(m_1), Lit(m_2))$. A meet m_n is *resolution-derivable* from $\hat{M}$ iff there is a finite sequence of meets $(m_1, ..., m_n)$ such that for each i in $[1..n]$, either $m_i \in \hat{M}$ or m_i is a resolvent of two preceding meets. The sequence $(m_1, ..., m_n)$ is called a *resolution-derivation* of m_n from $\hat{M}$.

If S is a set of stems and $(s_1, ..., s_n)$ is a resolution-derivation of s_n from S then $(\wedge s_1, ..., \wedge s_n)$ is a resolution-derivation of $\wedge s_n$ from $\wedge.S$. If M is a set of meets and $(m_1, ..., m_n)$ is a resolution-derivation of m_n from $\hat{M}$ then $(Lit(m_1), ..., Lit(m_n))$ is a resolution-derivation of $Lit(m_n)$ from $Lit(M)$. So $Res(M)$ is the set of all meets which are resolution-derivable from $\hat{M}$. Hence $\hat{M} \subseteq Res(M)$.

When no ambiguity results we often abbreviate 'resolution-child' to 'child', 'resolution-derivable' to 'derivable', and 'resolution-derivation' to 'derivation'.

We shall conclude this section with three technical lemmas that will require the following ideas.

Definition 2.4.9. Let S be a set of stems and suppose $D = (L_1, ..., L_n)$ is a derivation of L_n from S.

A *path in D starting at K_1 and ending at K_k* is a subsequence $(K_1, ..., K_k)$ of D such that $k \geq 2$ and for all i in $[1..k-1]$, K_{i+1} is a child of K_i.

L_j is a *resolution-descendant* of L_i iff there is a path in D starting at L_i and ending at L_j.

If $L_i = \{\sim l_i\}$ then a *maximal l_i-less path in D starting at L_i* is a path $(L_i, K_1, ..., K_k)$ in D starting at L_i such that for all j in $[1..k]$, $l_i \notin K_j$ and if K_{k+1} is in D and K_{k+1} is a child of K_k then $l_i \in K_{k+1}$.

If $L_i = \{\sim l_i\}$ then the *reach* of L_i, denoted $Reach(L_i)$, is defined by $Reach(L_i) = \{L_j : L_j \neq L_i$ and L_j is in a maximal l_i-less path in D starting at $L_i\}$.

As before, when no ambiguity results we often abbreviate 'resolution-descendant' to 'descendant'.

Lemma 2.4.10. Let S be a set of stems. Suppose $D = (L_1, ..., L_n)$ is a shortest derivation of L_n from S.
1) D has no repeated elements.
2) For all i in $[1..n-1]$, there exists j in $[i+1..n]$ such that L_j is a child of L_i.
3) For all i in $[1..n-1]$, L_n is a descendant of L_i.
4) If $i \in [1..n-1]$ and $L_i = \{\sim l_i\}$ then $L_n \in Reach(L_i)$.
Proof

Suppose S and D are as in the statement of the lemma.

(1) If $i \in [1..n-1]$ and $L_i = L_n$ then $(L_1, ..., L_i)$ is a shorter derivation of L_n from S. If $i \in [1..n-1]$ and $j \in [1..n-1]$ and $i < j$ and $L_i = L_j$ then deleting L_j from D gives a shorter derivation of L_n from S.

(2) If such a j did not exist then deleting L_j from D gives a shorter derivation of L_n from S.

(3) follows from part (2).

(4) Suppose there is an i in $[1..n-1]$ such that $L_i = \{\sim l_i\}$. Assume $L_n \notin Reach(L_i)$. Replace each L_j in $Reach(L_i)$ by $L_j \cup \{l_i\}$. The resulting sequence $D^* = (L_1, ..., L_i, L_{i+1}^*, ..., L_{n-1}^*, L_n)$ where $L_j^* \in \{L_j, L_j \cup \{l_i\}\}$ is still a derivation of L_n from S of length n. But L_i now has no child in D^* and so can be deleted from D^* to give a derivation of L_n from S of length $n-1$. (Indeed if L_j was a child of L_i then $L_j^* = L_j \cup \{l_i\}$ and so L_j^* can also be deleted from D^*, because it is already an element of $(L_1, ..., L_{i-1})$, to make an even shorter derivation of L_n from S.) This contradicts the minimality of D and so part (4) holds.
EndProofLemma2.4.10

Lemma 2.4.11. Let S be a set of stems and L a set of literals. Let G be either a set of clauses or a set of meets. If G is a set of clauses then let $\Diamond = \vee$. If G is a set of meets then let $\Diamond = \wedge$.
1) $S \subseteq Res(S)$, and so $\Diamond.Lit(G) \subseteq Res(G)$.
2) $G \cap Lit = \{\}$ iff $G = \Diamond.Lit(G)$.
3) If $G \cap Lit = \{\}$ then $G \subseteq Res(G)$.
4) $Res(G) \cap Lit = \{\}$.
5) If $S' \subseteq S$ then $Res(S') \subseteq Res(S)$, and so if $G' \subseteq G$ then $Res(G') \subseteq Res(G)$.
6) $Res(Res(S)) = Res(S)$, and so $Res(Res(G)) = Res(G)$.
7) $Lit(Res(G)) = ResLit(G)$.
8) $\bigcup Res(S) = \bigcup S$, and so $\bigcup Lit(Res(G)) = \bigcup Lit(G)$.
9) If S is finite then $Res(S)$ is finite, and so if G is finite then $Res(G)$ is finite.

2.4 Resolution   31

10) If $s_0 \in Res(S)$ and L is finite then $s_0 \cup L \in Res(\{s \cup L : s \in S\})$.
11) If $\Diamond s_0 \in Res(G)$ and L is finite then $\Diamond(s_0 \cup L) \in Res(\{\Diamond(Lit(g) \cup L) : g \in G\})$.
12) If L is not empty and $K_0 \in Res(S \cup \{\{\sim l\} : l \in L\})$ and $K_0 \notin Res(S)$ and $K_0 \notin \{\{\sim l\} : l \in L\}$ then there exists k in L such that $K_0 \cup \{k\} \in Res(S \cup \{\{\sim l\} : l \in L - \{k\}\})$.
13) If $\{\} \in Res(S \cup \{\{\sim l\} : l \in L\})$ and L is contingent or empty then there is a finite subset K of L such that $K \in Res(S)$.
14) If $\Diamond\{\} \in Res(G \cup \{\Diamond\{\sim l\} : l \in L\})$ and L is contingent or empty then there is a finite subset K of L such that $\Diamond K \in Res(G)$.
15) $\sim Res(S) = Res(\sim S)$, and so $Res(S) = \sim Res(\sim S)$.
16) $\sim Res(G) = Res(\sim G)$, and so $Res(G) = \sim Res(\sim G)$.

Proof

Let S, L, G, and $\Diamond$ be as in the statement of the lemma.

(1) $S \subseteq Res(S)$ follows immediately from Definition 2.4.3. Since $Lit(G)$ is a set of stems, $\Diamond.Lit(G) \subseteq \Diamond.ResLit(G) = Res(G)$.

(2) Since $Lit(G)$ is a set of stems, $\Diamond.Lit(G)$ does not contain any literals. So if $G = \Diamond.Lit(G)$ then G contains no literals. Conversely suppose G does not contain any literals. Then $\Diamond.Lit(G) = \Diamond.\{Lit(g) : g \in G\} = \{\Diamond Lit(g) : g \in G\} = \{g : g \in G\} = G$.

(3) This follows directly from parts (1) and (2).

(4) $Res(G) = \Diamond.ResLit(G)$ and so $Res(G)$ contains no literals.

(5) Suppose $S' \subseteq S$. Then $Res(S') \subseteq Res(S)$ follows immediately from Definition 2.4.3. Suppose $G' \subseteq G$. Then $Lit(G') \subseteq Lit(G)$. But $Lit(G')$ and $Lit(G)$ are both sets of stems, so $ResLit(G') \subseteq ResLit(G)$. Hence $Res(G') = \Diamond.ResLit(G') \subseteq \Diamond.ResLit(G) = Res(G)$.

(6) From parts (1) and (5) we have $Res(S) \subseteq Res(Res(S))$. Conversely suppose $L_n \in Res(Res(S))$. Then there is a derivation $(L_1, ..., L_n)$ of L_n from $Res(S)$. So for each i in $[1..n]$, if L_i is not a resolvent of two previous elements then there is a derivation $D(L_i)$ of L_i from S. Replacing each such L_i with $D(L_i)$ gives a derivation of L_n from S. Hence $L_n \in Res(S)$, and so $Res(Res(S)) \subseteq Res(S)$.

Thus $Res(Res(S)) = Res(S)$.

Recall $Res(G) = \Diamond.ResLit(G)$. Since $Lit(G)$ is a set of stems, $Res(ResLit(G)) = ResLit(G)$. Hence $Res(Res(G)) = \Diamond.ResLit(Res(G)) = \Diamond.ResLit(\Diamond.ResLit(G)) = \Diamond.Res(ResLit(G)) = \Diamond.ResLit(G) = Res(G)$.

(7) We have $Lit(Res(G)) = \{Lit(g) : g \in Res(G)\} = \{Lit(g) : g \in \Diamond.ResLit(G)\} = \{L : L \in ResLit(G)\} = ResLit(G)$.

(8) If s_1 and s_2 are resolvable stems then any resolvent of s_1 and s_2 is a subset of $s_1 \cup s_2$. Hence $\bigcup Res(S) = \bigcup S$. Since $Lit(G)$ is a set of stems, by part (7), $\bigcup Lit(Res(G)) = \bigcup Res(Lit(G)) = \bigcup Lit(G)$.

(9) By part 8, $Res(S) \subseteq \mathscr{P}(\bigcup Res(S)) = \mathscr{P}(\bigcup S)$. Therefore $|Res(S)| \leq 2^{|\bigcup S|}$. Since stems are finite, if S is finite then $|\bigcup S|$ is finite and so $Res(S)$ is finite.

Suppose G is finite. Then $Lit(G) = \{Lit(g) : g \in G\}$ is a finite set of stems. So $ResLit(G)$ is finite and hence $\Diamond.ResLit(G)$ is finite. That is $Res(G)$ is finite.

(10) If the stem K is a resolvent of the stems M and N then $K \cup L$ is a resolvent of $M \cup L$ and $N \cup L$. The result follows by induction on the length of a derivation.

(11) Suppose $\Diamond s_0 \in Res(G)$ and L is finite. Let $S = Lit(G) = \{Lit(g) : g \in G\}$. Then $s_0 \in ResLit(G) = Res(S)$. By part (10), $s_0 \cup L \in Res(\{s \cup L : s \in S\})$, so $\Diamond(s_0 \cup L) \in \Diamond.Res(\{s \cup L : s \in S\}) = \Diamond.Res(\{s \cup L : s \in Lit(G)\}) = \Diamond.ResLit(\{\Diamond(s \cup L) : s \in Lit(G)\}) = Res(\{\Diamond(s \cup L) : s \in Lit(G)\}) = Res(\{\Diamond(Lit(g) \cup L) : g \in G\})$.

(12) Suppose L is not empty and $K_0 \in Res(S \cup \{\{\sim l\} : l \in L\})$ and $K_0 \notin Res(S)$ and $K_0 \notin \{\{\sim l\} : l \in L\}$. Let $D = (L_1, ..., L_n = K_0)$ be a shortest derivation of K_0 from $S \cup \{\{\sim l\} : l \in L\}$. If each L_i is either in S or is the child of two previous elements of D then $K_0 \in Res(S)$. So there is an L_i such that $L_i \notin S$ and L_i is not the child of two previous elements of D. Then $L_i \in \{\{\sim l\} : l \in L\}$. Let $L_i = \{\sim l_i\}$ and $l_i = k$. Then $k \in L$. Since $L_n \notin \{\{\sim l\} : l \in L\}$, $L_i \neq L_n$ and so by Lemma 2.4.10(4), $L_n \in Reach(L_i)$. Replace each L_j in $Reach(L_i)$ by $L_j \cup \{l_i\}$. The resulting sequence $D^* = (L_1, ..., L_i, L_{i+1}^*, ..., L_{n-1}^*, L_n \cup \{l_i\})$ where $L_j^* \in \{L_j, L_j \cup \{l_i\}\}$ is a derivation of $L_n \cup \{l_i\}$ from $S \cup \{\{\sim l\} : l \in L\}$. But L_i now has no child in D^* and so can be deleted from D^* to give a derivation of $L_n \cup \{l_i\}$ from $S \cup \{\{\sim l\} : l \in L - \{l_i\}\}$. Thus $K_0 \cup \{k\} \in Res(S \cup \{\{\sim l\} : l \in L - \{k\}\})$.

(13) Suppose $\{\} \in Res(S \cup \{\{\sim l\} : l \in L\})$ and L is contingent or empty. Then there is a finite derivation of $\{\}$ from $S \cup \{\{\sim l\} : l \in L\}$. So only a finite subset of $\{\{\sim l\} : l \in L\}$ was used in the derivation. So there is a finite subset L' of L such that $\{\} \in Res(S \cup \{\{\sim l\} : l \in L'\})$ and L' is empty or contingent.

If either L' is empty or $\{\} \in Res(S)$ then let $K = \{\}$. So suppose L' is not empty and $\{\} \notin Res(S)$. By part (12) with $L = L'$ and $K_0 = \{\}$ there exists k_1 in L' such that $\{k_1\} \in Res(S \cup \{\{\sim l\} : l \in L' - \{k_1\}\})$.

If either $L' - \{k_1\}$ is empty or $\{k_1\} \in Res(S)$ then let $K = \{k_1\}$. So suppose $L' - \{k_1\}$ is not empty and $\{k_1\} \notin Res(S)$. If $\{k_1\} \in \{\{\sim l\} : l \in L' - \{k_1\}\}$ then $\sim k_1 \in L' - \{k_1\}$ and so L' is not contingent. Therefore $\{k_1\} \notin \{\{\sim l\} : l \in L' - \{k_1\}\}$. By part (12) with $L = L' - \{k_1\}$ and $K_0 = \{k_1\}$ there exists k_2 in $L' - \{k_1\}$ such that $\{k_1, k_2\} \in Res(S \cup \{\{\sim l\} : l \in L' - \{k_1, k_2\}\})$.

If either $L' - \{k_1, k_2\}$ is empty or $\{k_1, k_2\} \in Res(S)$ then let $K = \{k_1, k_2\}$. So suppose $L' - \{k_1, k_2\}$ is not empty and $\{k_1, k_2\} \notin Res(S)$. Since $|\{k_1, k_2\}| = 2$ we have $\{k_1, k_2\} \notin \{\{\sim l\} : l \in L' - \{k_1, k_2\}\}$. By part (12) with $L = L' - \{k_1, k_2\}$ and $K_0 = \{k_1, k_2\}$ there exists k_3 in $L' - \{k_1, k_2\}$ such that $\{k_1, k_2, k_3\} \in Res(S \cup \{\{\sim l\} : l \in L' - \{k_1, k_2, k_3\}\})$.

Because L' is finite we can continue this reasoning until eventually either $L' - \{k_1, ..., k_n\}$ is empty or $\{k_1, ..., k_n\} \in Res(S)$ in which case we let $K = \{k_1, ..., k_n\}$.

(14) Suppose $\Diamond\{\} \in Res(G \cup \{\Diamond\{\sim l\} : l \in L\})$ and L is contingent or empty. Then $\Diamond\{\} \in \Diamond.Res(Lit(G \cup \{\Diamond\{\sim l\} : l \in L\})) = \Diamond.Res(Lit(G) \cup \{\{\sim l\} : l \in L\})$. Hence $\{\} \in Res(Lit(G) \cup \{\{\sim l\} : l \in L\})$, and so by part (13), there is a finite subset K of L such that $K \in Res(Lit(G))$. Therefore $\Diamond K \in \Diamond.ResLit(G) = Res(G)$.

(15) The resolvent of $\{k\} \cup L$ and $\{\sim k\} \cup M$ is $L \cup M$. The resolvent of $\{\sim k\} \cup \sim L$ and $\{k\} \cup \sim M$ is $\sim L \cup \sim M$. Since $\sim(L \cup M) = \sim L \cup \sim M$, if s_1 and s_2 are stems then $res(k; s_1, s_2)$ exists iff $res(k; \sim s_1, \sim s_2)$ exists, and also $\sim res(k; s_1, s_2) = res(k; \sim s_1, \sim s_2)$. Therefore $\sim Res(S) = Res(\sim S)$. Hence $Res(S) = \sim Res(\sim S)$.

(16) Let C be a set of clauses and let M be a set of meets. By part (15) with $S = Lit(C)$, and Lemma 2.3.10(12), $\sim Res(C) = \sim \vee.ResLit(C) = \wedge.\sim ResLit(C) =$

2.4 Resolution

$\wedge.Res(\sim Lit(C)) = \wedge.Res(Lit(\sim C)) = Res(\sim C)$. By part (15) with $S = Lit(M)$, and Lemma 2.3.10(13), $\sim Res(M) = \sim\wedge.ResLit(M) = \vee.\sim ResLit(M) = \vee.Res(\sim Lit(M)) = \vee.Res(Lit(\sim M)) = Res(\sim M)$. Thus $\sim Res(G) = Res(\sim G)$ and so $Res(G) = \sim Res(\sim G)$.

EndProofLemma2.4.11

Lemma 2.4.12. Let C be any set of clauses, M be any set of meets, and v be any valuation.
1) $v \models C$ iff $v \models \vee.Lit(C)$.
2) $\wedge Res(C) \equiv Res(C) \equiv C \equiv \wedge C$. So C is satisfiable iff $Res(C)$ is satisfiable.
3) $\vee Res(M) \equiv \vee M$. So M is falsifiable iff $Res(M)$ is falsifiable.
4) If C is finite and $\vee\{\} \notin Res(C)$ then C is satisfiable.
5) C is satisfiable iff $\vee\{\} \notin Res(C)$.
6) If M is finite and $\wedge\{\} \notin Res(M)$ then M is falsifiable.
7) M is falsifiable iff $\wedge\{\} \notin Res(M)$.

Proof

Let C be any set of clauses, M be any set of meets, and v be any valuation. By Definition 2.4.6, $Res(C) = \vee.ResLit(C)$; and by Definition 2.4.8, $Res(M) = \wedge.ResLit(M)$.

(1) This follows from Lemma 2.2.9(1,3).

(2) Suppose $v \models Res(C)$. By Lemma 2.4.11(1), $\vee.Lit(C) \subseteq Res(C)$. So $v \models \vee.Lit(C)$. By part (1), $v \models C$. Hence $Res(C) \models C$.

For the converse we first show that if two resolvable clauses are satisfied by v then their resolvent is also satisfied by v. Let L and L' be two finite sets of literals. Let k be a literal such that $k \notin L$ and $\sim k \notin L'$. Then the resolvent of $\vee(\{k\} \cup L)$ and $\vee(\{\sim k\} \cup L')$ on k is $\vee(L \cup L')$. Suppose v satisfies both $\vee(\{k\} \cup L)$ and $\vee(\{\sim k\} \cup L')$. If $v(k) = \mathbf{T}$ then $v(\sim k) = \mathbf{F}$. Hence there exists l' in L' such that $v(l') = \mathbf{T}$. Therefore v satisfies $\vee(L \cup L')$. If $v(k) = \mathbf{F}$ then there exists l in L such that $v(l) = \mathbf{T}$. Therefore v satisfies $\vee(L \cup L')$. Thus in both cases v satisfies $\vee(L \cup L')$.

A simple induction on the length of a derivation from C shows that if $v \models C$ then $v \models Res(C)$. Hence $C \models Res(C)$.

Therefore $Res(C) \equiv C$. So by Lemma 2.2.18(8), $\wedge Res(C) \equiv Res(C) \equiv C \equiv \wedge C$.

(3) By Lemma 2.3.10(6,3), Lemma 2.4.11(16), and Lemma 2.4.12(2), $\vee Res(M) \equiv \sim\sim\vee Res(M) = \sim[\wedge\sim Res(M)] = \sim[\wedge Res(\sim M)] \equiv \sim[\wedge(\sim M)] = \sim[\vee(M)] \equiv \vee M$. So a valuation makes every meet in M false iff it makes every meet in $Res(M)$ false.

(4) Suppose C is finite and $\vee\{\} \notin Res(C)$. We shall construct a valuation which satisfies every clause in $Res(C)$ and hence every clause in C.

If $Res(C)$ is empty then C is empty and so C is satisfiable. So suppose $Res(C)$ is not empty. Take a smallest clause, $\vee L_1$, in $Res(C)$. It contains at least one literal, say l_1. Define a partial valuation v_1 by $v_1(l_1) = \mathbf{T}$. We show that v_1 does not falsify any clause in $Res(C)$. Assume v_1 falsifies a clause in $Res(C)$. Then the clause must be $\vee\{\sim l_1\}$. But $\vee L_1$ is a smallest clause, so $|L_1| \leq |\{\sim l_1\}| = 1$. Hence $\vee L_1 = \vee\{l_1\}$. Therefore $\vee\{\} \in Res(C)$. This contradiction shows that v_1 does not falsify any clause in $Res(C)$.

Suppose that $v_n(l_i) = \mathbf{T}$ for all i in $[1..n]$, and that v_n does not falsify any clause in $Res(C)$. If v_n satisfies every clause in $Res(C)$ then v_n satisfies every clause in C as $C \subseteq Res(C)$. So suppose there is a clause c in $Res(C)$ such that $v_n(c)$ is undefined.

Take a clause $\vee L_{n+1}$ in $Res(C)$ such that (i) $v_n(\vee L_{n+1})$ is undefined, and (ii) if $\vee L \in Res(C)$ and $v_n(\vee L)$ is undefined then $|\{l \in L_{n+1} : v_n(l) \text{ is undefined}\}| \leq |\{l \in L : v_n(l) \text{ is undefined}\}|$. Take any l_{n+1} in L_{n+1} such that $v_n(l_{n+1})$ is undefined. Define a partial valuation v_{n+1} by for all i in $[1..n+1]$, $v_{n+1}(l_i) = \mathbf{T}$. We show that v_{n+1} does not falsify any clause in $Res(C)$. Assume v_{n+1} falsifies a clause in $Res(C)$. Then the clause must be $\vee(L' \cup \{\sim l_{n+1}\})$ where $\sim l_{n+1} \notin L'$ and $l_{n+1} \notin L'$ and v_n falsifies $\vee L'$. So $\{l \in L' \cup \{\sim l_{n+1}\} : v_n(l) \text{ is undefined}\} = \{\sim l_{n+1}\}$. Hence $\{l \in L_{n+1} : v_n(l) \text{ is undefined}\} = \{l_{n+1}\}$. So $L_{n+1} = K \cup \{l_{n+1}\}$ where v_n falsifies $\vee K$. Since $\vee(L' \cup \{\sim l_{n+1}\}) \in Res(C)$ and $\vee(K \cup \{l_{n+1}\}) \in Res(C)$ we have $\vee(L' \cup K) \in Res(C)$. But v_n falsifies $\vee(L' \cup K)$, contradicting the supposition that v_n does not falsify any clause in $Res(C)$. Therefore v_{n+1} does not falsify any clause in $Res(C)$.

Since C is finite $Cp(C)$ is finite. Let $|Cp(C)| = k$. Then for some i in $[1..k]$, v_i satisfies every clause in $Res(C)$ and hence every clause in C.

(5) If C is satisfiable then by Lemma 2.4.12(2), $Res(C)$ is satisfiable and so $\vee\{\} \notin Res(C)$. Conversely suppose $\vee\{\} \notin Res(C)$. We show that every finite subset of C is satisfiable. Let C' be any finite subset of C. Then $Res(C') \subseteq Res(C)$ and so $\vee\{\} \notin Res(C')$. By Lemma 2.4.12(4), C' is satisfiable. Therefore every finite subset of C is satisfiable. So by compactness (Theorem 2.3.11(1)), C is satisfiable.

(6) Suppose M is finite and $\wedge\{\} \notin Res(M)$. Then $\sim M$ is finite and $\vee\{\} \notin \sim Res(M)$ because $\wedge\{\} \in Res(M)$ iff $\sim \wedge\{\} \in \sim Res(M)$ and $\sim \wedge\{\} = \vee\{\}$. By Lemma 2.4.11(16), $\vee\{\} \notin Res(\sim M)$. So by Lemma 2.4.12(4), $\sim M$ is satisfiable. Thus by Lemma 2.3.10(8), M is falsifiable.

(7) By Lemma 2.4.12(5), Lemma 2.3.10(8), and Lemma 2.4.11(16), M is falsifiable iff $\sim M$ is satisfiable iff $\vee\{\} \notin Res(\sim M)$ iff $\vee\{\} \notin \sim Res(M)$ iff $\wedge\{\} \notin Res(M)$.
EndProofLemma2.4.12

2.5 Singletons

If f is a formula then by Lemma 2.2.16(2), $\vee\{f\}$, f, and $\wedge\{f\}$ are all equivalent. These equivalences can be used to simplify quasi-formulas. First we formalise the notion of a quasi-formula having a singleton. Then we try to remove singletons from quasi-formulas.

Definition 2.5.1. Let f be any quasi-formula.
1) f *has a singleton* iff there is a formula g such that at least one of $\wedge\{g\}$ or $\vee\{g\}$ is a subformula of f.
2) f *has no singletons* iff it is not the case that f has a singleton.

Definition 2.5.2. The function *replace-singletons*, $rs(.)$, is defined as follows.
1) If F is any set of quasi-formulas then $rs(F) = \{rs(f) : f \in F\}$.
2) If a is an atom then $rs(a) = a$.
3) If f is a quasi-formula then $rs(\neg f) = \neg rs(f)$.
4) If f is a formula and $\Diamond \in \{\wedge, \vee\}$, then $rs(\Diamond\{f\}) = rs(f)$.

2.5 Singletons

5) If F is any set of formulas such that $|F| \neq 1$ and $\Diamond \in \{\wedge, \vee\}$, then $rs(\Diamond F) = \Diamond rs(F) = \Diamond \{rs(f) : f \in F\}$.

It might be thought that $rs(f)$ has no singletons in it. But consider the following example. Let a be an atom and $f = \wedge\{a, \vee\{a\}\}$. Then $rs(f) = rs(\wedge\{a, \vee\{a\}\}) = \wedge rs(\{a, \vee\{a\}\}) = \wedge\{rs(a), rs(\vee\{a\})\} = \wedge\{a, rs(a)\} = \wedge\{a\}$. So to make sure that no singletons remain we must apply rs as many times as necessary. The following definition does this.

Definition 2.5.3. The function *no-singletons*, $ns(.)$, is defined as follows. Let f be any formula.
1) If F is any set of quasi-formulas then $ns(F) = \{ns(f) : f \in F\}$.
2) If $rs(f) = f$ then $ns(f) = f$.
3) If $rs(f) \neq f$ then $ns(f) = ns(rs(f))$.
4) If $n \in \mathbb{N}$, $\Diamond \in \{\wedge, \vee\}$, and F is an infinite set of formulas then $ns(\neg^n \Diamond F) = \neg^n \Diamond ns(F) = \neg^n \Diamond \{ns(f) : f \in F\}$.

After sufficiently many applications of rs to a formula the result does not change and it has no singletons. To prove this and other results we need to count the number of connectives and singletons in a formula.

If $\diamond$ is a connective then $N(\diamond, f)$ will denote the number of occurrences of $\diamond$ in the formula f. Similarly, $N(\diamond, F)$ will denote the number of occurrences of $\diamond$ in the finite set of formulas F. The formal definition of this notation follows.

Definition 2.5.4. Suppose f is a formula, and F is a finite set of formulas.
1) If $\diamond \in \{\neg, \wedge, \vee\}$ then $N(\diamond, F) = \Sigma(N(\diamond, f) : f \in F)$.
2) If a is an atom and $\diamond \in \{\neg, \wedge, \vee\}$ then $N(\diamond, a) = 0$.
3) If $\diamond \in \{\wedge, \vee\}$ then $N(\neg, \neg f) = 1 + N(\neg, f)$, and $N(\diamond, \neg f) = N(\diamond, f)$.
4) If $\diamond \in \{\neg, \vee\}$ then $N(\wedge, \wedge F) = 1 + N(\wedge, F)$, and $N(\diamond, \wedge F) = N(\diamond, F)$.
5) If $\diamond \in \{\neg, \wedge\}$ then $N(\vee, \vee F) = 1 + N(\vee, F)$, and $N(\diamond, \vee F) = N(\diamond, F)$.
6) If $C \subseteq \{\neg, \wedge, \vee\}$ then $N(C, f) = \Sigma(N(\diamond, f) : \diamond \in C)$, and $N(C, F) = \Sigma(N(\diamond, F) : \diamond \in C)$.

We observe that $N(C, F) = \Sigma(N(\diamond, f) : \diamond \in C \text{ and } f \in F)$.

If f is a formula and F is a finite set of formulas then the number of singletons in f, or F, will be denoted by $N(s, f)$, or $N(s, F)$, respectively. We now give the formal definition of this notation.

Definition 2.5.5. The function $N(s, .)$ is defined as follows.
1) If F is any finite set of formulas then $N(s, F) = \Sigma(N(s, f) : f \in F)$.
2) If a is an atom then $N(s, a) = 0$.
3) If f is a formula then $N(s, \neg f) = N(s, f)$.
4) If f is a formula then $N(s, \wedge\{f\}) = N(s, \vee\{f\}) = 1 + N(s, f)$.
5) If F is a finite set of formulas such that $|F| \neq 1$ then $N(s, \wedge F) = N(s, \vee F) = N(s, F)$.

6) If $n \in \mathbb{N}$, $\Diamond \in \{\wedge, \vee\}$, and F is an infinite set of formulas then $N(s, \neg^n \Diamond F) = 0$ iff for each f in F, $N(s, f) = 0$.

The following lemma gives some of the properties of the concepts previously defined in this section.

Lemma 2.5.6.
1) If F is any set of quasi-formulas and $F' \subseteq F$ then $ns(F') \subseteq ns(F)$ and $rs(F') \subseteq rs(F)$.
2) Suppose f is a formula, F is a finite set of formulas, $\diamond \in \{\neg, \wedge, \vee\}$, and $C \subseteq \{\neg, \wedge, \vee\}$. Then $N(\diamond, f) \in \mathbb{N}$, $N(\diamond, F) \in \mathbb{N}$, $N(C, f) \in \mathbb{N}$, $N(C, F) \in \mathbb{N}$, $N(s, f) \in \mathbb{N}$, $N(s, F) \in \mathbb{N}$, and $rs(f)$ is a formula.
3) If f is a quasi-formula and $n \in \mathbb{N}$ then $ns(rs^n(f)) = ns(f)$.
4) Let f be a formula.
   4.1) $N(\wedge, f) \geq N(\wedge, rs(f))$ and so $N(\wedge, F) \geq N(\wedge, rs(F))$.
   4.2) $N(\vee, f) \geq N(\vee, rs(f))$ and so $N(\vee, F) \geq N(\vee, rs(F))$.
   4.3) $N(\{\wedge, \vee\}, f) \geq N(\{\wedge, \vee\}, rs(f))$.
   4.4) If $N(s, f) > 0$ then $N(\{\wedge, \vee\}, f) > N(\{\wedge, \vee\}, rs(f))$.
5) If f is any formula then the following are equivalent.
   5.1) $N(s, f) = 0$.
   5.2) $rs(f) = f$.
   5.3) For each i in $\mathbb{Z}^+$, $rs^i(f) = f$.
   5.4) There exists i in $\mathbb{Z}^+$ such that $rs^i(f) = f$.
6) Let f be a formula.
   6.1) If $n \geq N(\{\wedge, \vee\}, f)$ then $ns(f) = rs^n(f)$.
   6.2) If $n \in \mathbb{Z}^+$ then there exists m in $\mathbb{Z}^+$ such that $ns^n(f) = rs^m(f)$.
7) Let f be a formula and F be a finite set of formulas.
   7.1) $N(\wedge, f) \geq N(\wedge, ns(f))$ and so $N(\wedge, F) \geq N(\wedge, ns(F))$.
   7.2) $N(\vee, f) \geq N(\vee, ns(f))$ and so $N(\vee, F) \geq N(\vee, ns(F))$.
   7.3) $N(\{\wedge, \vee\}, f) \geq N(\{\wedge, \vee\}, ns(f))$.
   7.4) If $N(s, f) > 0$ then $N(\{\wedge, \vee\}, f) > N(\{\wedge, \vee\}, ns(f))$.
8) Suppose f is any quasi-formula, F is any set of quasi-formulas, and $i \in \mathbb{Z}^+$. Then $ns^i(f) = ns(f)$ and $ns^i(F) = ns(F)$.
9) Let F be any set of formulas.
   9.1) Let F be finite. Then $rs(F) = F$ iff for all f in F, $rs(f) = f$.
   9.2) $ns(F) = F$ iff for all f in F, $ns(f) = f$.
10) If f is any quasi-formula then the following are equivalent.
    10.1) f has no singletons.
    10.2) $N(s, f) = 0$.
    10.3) $ns(f) = f$.
    10.4) For each i in $\mathbb{Z}^+$, $ns^i(f) = f$.
    10.5) There exists i in $\mathbb{Z}^+$ such that $ns^i(f) = f$.
11) If f is any quasi-formula then $N(s, ns(f)) = 0$.
12) If f is any quasi-formula and $N(s, f) = 0$ then $rs(f) = f$.
13) If f is any quasi-formula then $rs(ns(f)) = ns(f)$.

2.5 Singletons

14) If f is any quasi-formula then $rs(f) \equiv f \equiv ns(f)$.
15) If F is any set of formulas then $\vee rs(F) \equiv \vee F \equiv \vee ns(F)$ and
$rs(F) \equiv \wedge rs(F) \equiv \wedge F \equiv F \equiv ns(F) \equiv \wedge ns(F)$.

Proof

(1)
Let F be any set of quasi-formulas and suppose $F' \subseteq F$.
Then $ns(F') = \{ns(f) : f \in F'\} \subseteq \{ns(f) : f \in F\} = ns(F)$.
Also $rs(F') = \{rs(f) : f \in F'\} \subseteq \{rs(f) : f \in F\} = rs(F)$.

(2)
This part has many subparts. We mark the start of the proof of each subpart by ▷.

▷ Suppose $\diamond \in \{\neg, \wedge, \vee\}$. We prove $N(\diamond, f) \in \mathbb{N}$ by induction on the structure of the formula f. Let $X(f)$ be the statement "$N(\diamond, f) \in \mathbb{N}$".

If a is an atom then $N(\diamond, a) = 0 \in \mathbb{N}$.

Take any formula g. Suppose that for every component f of g we have $N(\diamond, f) \in \mathbb{N}$. We show $N(\diamond, g) \in \mathbb{N}$.

Case 1: $g = \neg f$.
By $X(f)$, $N(\diamond, g) = N(\diamond, \neg f) \in \{N(\diamond, f), 1 + N(\diamond, f)\} \subseteq \mathbb{N}$.

Case 2: $g = \Diamond F$ where $\Diamond \in \{\wedge, \vee\}$ and F is a finite set of formulas.
For each f in F, by $X(f)$, $N(\diamond, f) \in \mathbb{N}$. Hence $N(\diamond, F) = \Sigma(N(\diamond, f) : f \in F) \in \mathbb{N}$. So $N(\diamond, g) = N(\diamond, \Diamond F) \in \{N(\diamond, F), 1 + N(\diamond, F)\} \subseteq \mathbb{N}$.

Thus by induction, for each formula f, $N(\diamond, f) \in \mathbb{N}$.

▷ Since F is finite, $N(\diamond, F) = \Sigma(N(\diamond, f) : f \in F) \in \mathbb{N}$.

▷ Suppose $C \subseteq \{\neg, \wedge, \vee\}$. Then $N(C, f) = \Sigma(N(\diamond, f) : \diamond \in C) \in \mathbb{N}$.

▷ Since F is finite, $N(C, F) = \Sigma(N(\diamond, F) : \diamond \in C) \in \mathbb{N}$.

▷ We prove $N(s, f) \in \mathbb{N}$ by induction on the structure of the formula f. Let $X(f)$ be the statement "$N(s, f) \in \mathbb{N}$".

If a is an atom then $N(s, a) = 0 \in \mathbb{N}$.

Take any formula g. Suppose that for every component f of g we have $N(s, f) \in \mathbb{N}$. We show $N(s, g) \in \mathbb{N}$.

Case 1: $g = \neg f$.
By $X(f)$, $N(s, g) = N(s, \neg f) = N(s, f) \in \mathbb{N}$.

Case 2: $g = \Diamond F$ where $\Diamond \in \{\wedge, \vee\}$ and F is a finite set of formulas.
For each f in F, by $X(f)$, $N(s, f) \in \mathbb{N}$. Hence $N(s, F) = \Sigma(N(s, f) : f \in F) \in \mathbb{N}$. So $N(s, g) = N(s, \Diamond F) \in \{N(s, F), 1 + N(s, F)\} \subseteq \mathbb{N}$.

Thus by induction, for each formula f, $N(s, f) \in \mathbb{N}$.

▷ Since F is finite, $N(s, F) = \Sigma(N(s, f) : f \in F) \in \mathbb{N}$.

▷ We prove $rs(f)$ is a formula by induction on the structure of the formula f. Let $X(f)$ be the following statement. "If f is a formula then $rs(f)$ is a formula."

If a is an atom then $rs(f) = a$. So $X(a)$ is true.

Take any formula g. Then every component of g is a formula. Suppose that for every component f of g, $rs(f)$ is a formula. So for every component f of g, $X(f)$ is true. We show $X(g)$ is true.

Case 1: $g = \neg f$.
Then $rs(g) = rs(\neg f) = \neg rs(f)$. By $X(f)$, $rs(g)$ is a formula; and so $X(g)$ is true.

Case 2: $g = \Diamond F$ where $\Diamond \in \{\wedge, \vee\}$ and F is a finite set of formulas.
Then $rs(g) = rs(\Diamond F) = \Diamond \{rs(f) : f \in F\}$. Since by $X(f)$, each $rs(f)$ is a formula, we have $rs(g)$ is a formula. So $X(g)$ is true.

Thus by induction, for each formula f, $rs(f)$ is a formula.

(3)

Let f be a formula. Let $X(n)$ be the following statement. "If $n \in \mathbb{N}$ then $ns(rs^n(f)) = ns(f)$." We prove $X(n)$ by induction on n.

If $n = 0$ then $rs^0(f) = f$. Hence $X(0)$ holds.

Take any k in $\mathbb{N}$ and suppose $X(k)$ holds. That is, $ns(rs^k(f)) = ns(f)$. We show $X(k+1)$ holds; that is, $ns(rs^{k+1}(f)) = ns(f)$.

Let $rs^k(f) = g$. By part (2), g is a formula. Then $ns(g) = ns(f)$ and $rs(g) = rs^{k+1}(f)$.

Case 1: $rs(g) = g$.
Then $ns(g) = g$. So $ns(rs^{k+1}(f)) = ns(rs(g)) = ns(g) = ns(f)$. Therefore $X(k+1)$ holds.

Case 2: $rs(g) \neq g$ and g is a formula.
Then $ns(g) = ns(rs(g))$. So $ns(rs^{k+1}(f)) = ns(rs(g)) = ns(g) = ns(f)$. Therefore $X(k+1)$ holds.

So by induction, for all n in $\mathbb{N}$, $ns(rs^n(f)) = ns(f)$. Thus if f is any formula and $n \in \mathbb{N}$ then $ns(rs^n(f)) = ns(f)$.

We shall use this result to prove that if f is any quasi-formula and $n \in \mathbb{N}$ then $ns(rs^n(f)) = ns(f)$. Suppose f is a quasi-formula that is not a formula and $n \in \mathbb{N}$. Then $f = \neg^k \Diamond H$, where $k \in \mathbb{N}$, $\Diamond \in \{\wedge, \vee\}$, and H is an infinite set of formulas.

So $ns(rs^n(f)) = ns(rs^n(\neg^k \Diamond H)) = ns(\neg^k \Diamond rs^n(H)) = ns(\neg^k \Diamond \{rs^n(h) : h \in H\})$
$= \neg^k \Diamond ns(\{rs^n(h) : h \in H\}) = \neg^k \Diamond \{ns(rs^n(h)) : h \in H\} = \neg^k \Diamond \{ns(h) : h \in H\} = \neg^k \Diamond ns(H) = ns(\neg^k \Diamond H) = ns(f)$.

Thus if f is a quasi-formula and $n \in \mathbb{N}$ then $ns(rs^n(f)) = ns(f)$.

(4)

(4.1) and (4.2)
Suppose $\Diamond \in \{\wedge, \vee\}$.

Let $X(f)$ be the following statement. "$N(\Diamond, f) \geq N(\Diamond, rs(f))$." We prove $X(f)$ by induction on the structure of the formula f.

If a is an atom then $rs(a) = a$, and $N(\Diamond, a) = 0$. Hence $0 = N(\Diamond, a) = N(\Diamond, rs(a))$. So $X(a)$ holds.

Take any formula g. Suppose that for every component f of g we have $X(f)$. We show $X(g)$ holds.

Case 1: $g = \neg f$.
Then $rs(g) = rs(\neg f) = \neg rs(f)$. By $X(f)$, $N(\Diamond, f) \geq N(\Diamond, rs(f))$. So $N(\Diamond, g) = N(\Diamond, \neg f) = N(\Diamond, f) \geq N(\Diamond, rs(f)) = N(\Diamond, \neg rs(f)) = N(\Diamond, rs(g))$. Therefore $X(g)$ holds.

Case 2: $g = \diamond \{f\}$ where $\diamond \in \{\wedge, \vee\}$.
Then $rs(g) = rs(\diamond \{f\}) = rs(f)$. By $X(f)$, $N(\Diamond, f) \geq N(\Diamond, rs(f))$. So $N(\Diamond, g) = N(\Diamond, \diamond \{f\}) \geq N(\Diamond, f) \geq N(\Diamond, rs(f)) = N(\Diamond, rs(g))$. Therefore $X(g)$ holds.

2.5 Singletons

Case 3: $g = \diamond F$ where $\diamond \in \{\wedge, \vee\}$ and F is a finite set of formulas such that $|F| \neq 1$. Then $rs(g) = rs(\diamond F) = \diamond rs(F)$. By $X(f)$, for each f in F, $N(\Diamond, f) \geq N(\Diamond, rs(f))$.

Subcase 3.1: $\Diamond = \diamond$.
Then $N(\Diamond, g) = N(\Diamond, \diamond F) = 1 + N(\Diamond, F) = 1 + \sum(N(\Diamond, f) : f \in F) \geq 1 + \sum(N(\Diamond, rs(f)) : f \in F) = 1 + N(\Diamond, rs(F)) = N(\Diamond, \diamond rs(F)) = N(\Diamond, rs(g))$. Therefore $X(g)$ holds.

Subcase 3.2: $\Diamond \neq \diamond$.
Then $N(\Diamond, g) = N(\Diamond, \diamond F) = N(\Diamond, F) = \sum(N(\Diamond, f) : f \in F) \geq \sum(N(\Diamond, rs(f)) : f \in F) = N(\Diamond, rs(F)) = N(\Diamond, \diamond rs(F)) = N(\Diamond, rs(g))$.
Therefore $X(g)$ holds.

Thus by induction, for each formula f, $N(\Diamond, f) \geq N(\Diamond, rs(f))$.
So $N(\Diamond, F) = \sum(N(\Diamond, f) : f \in F) \geq \sum(N(\Diamond, rs(f)) : f \in F) = N(\Diamond, \{rs(f) : f \in F\}) = N(\Diamond, rs(F))$.

(4.3)
By (4.1) and (4.2), $N(\{\wedge, \vee\}, f) = N(\wedge, f) + N(\vee, f) \geq N(\wedge, rs(f)) + N(\vee, rs(f)) = N(\{\wedge, \vee\}, rs(f))$.

(4.4)
Let $X(f)$ denote the following statement. "If $N(s, f) > 0$ then $N(\{\wedge, \vee\}, f) > N(\{\wedge, \vee\}, rs(f))$." We prove $X(f)$ by induction on the structure of the formula f.

If a is an atom then $N(s, a) = 0$, $rs(a) = a$, and $N(\{\wedge, \vee\}, a) = 0$. So $X(a)$ holds.

Take any formula g. Suppose that for every component f of g we have $X(f)$. We show $X(g)$ holds.

Case 1: $g = \neg f$.
Then $rs(g) = rs(\neg f) = \neg rs(f)$. Suppose $N(s, g) > 0$ holds. We have $N(s, g) = N(s, \neg f) = N(s, f)$. So $N(s, f) > 0$. By $X(f)$, $N(\{\wedge, \vee\}, f) > N(\{\wedge, \vee\}, rs(f))$. So $N(\{\wedge, \vee\}, g) = N(\{\wedge, \vee\}, \neg f) = N(\{\wedge, \vee\}, f) > N(\{\wedge, \vee\}, rs(f)) = N(\{\wedge, \vee\}, \neg rs(f)) = N(\{\wedge, \vee\}, rs(g))$.
Therefore $X(g)$ holds.

Case 2: $g = \Diamond\{f\}$ where $\Diamond \in \{\wedge, \vee\}$.
Then $rs(g) = rs(\Diamond\{f\}) = rs(f)$. By part (4.3), $N(\{\wedge, \vee\}, g) = N(\{\wedge, \vee\}, \Diamond\{f\}) = 1 + N(\{\wedge, \vee\}, f) > N(\{\wedge, \vee\}, f) \geq N(\{\wedge, \vee\}, rs(f)) = N(\{\wedge, \vee\}, rs(g))$. Therefore $X(g)$ holds.

Case 3: $g = \Diamond F$ where $\Diamond \in \{\wedge, \vee\}$ and F is a finite set of formulas such that $|F| \neq 1$. Then $rs(g) = rs(\Diamond F) = \Diamond rs(F)$. Suppose $N(s, g) > 0$ holds. We have $N(s, g) = N(s, \Diamond F) = N(s, F) = \sum(N(s, f) : f \in F)$. So there exists f_1 in F such that $N(s, f_1) > 0$. By $X(f_1)$, $N(\{\wedge, \vee\}, f_1) > N(\{\wedge, \vee\}, rs(f_1))$. By part (4.3) we get the following: $N(\{\wedge, \vee\}, g) = N(\{\wedge, \vee\}, \Diamond F) = 1 + N(\{\wedge, \vee\}, F) = 1 + \sum(N(\{\wedge, \vee\}, f) : f \in F) > 1 + \sum(N(\{\wedge, \vee\}, rs(f)) : f \in F) = 1 + N(\{\wedge, \vee\}, rs(F)) = N(\{\wedge, \vee\}, \Diamond rs(F)) = N(\{\wedge, \vee\}, rs(g))$. Therefore $X(g)$ holds.

Thus by induction, for each formula f, $X(f)$ holds.

(5)
(5.1) implies (5.2)
Let $X(f)$ denote the following statement. "If f is a formula and $N(s, f) = 0$ then $rs(f) = f$." We prove $X(f)$ by induction on the structure of the formula f.

If a is an atom then $N(s, a) = 0$ and $rs(a) = a$. So $X(a)$ holds.

Take any formula g. Suppose that for every component f of g we have $X(f)$ holds. We show $X(g)$ holds.

Case 1: $g = \neg f$.
Then $N(s,g) = N(s,\neg f) = N(s,f)$ and $rs(g) = rs(\neg f) = \neg rs(f)$. Suppose $N(s,g) = 0$. Then $N(s,f) = 0$. By $X(f)$, $rs(f) = f$. Hence $\neg rs(f) = \neg f$, and so $rs(g) = g$. Therefore $X(g)$ holds.

Case 2: $g = \Diamond\{f\}$ where $\Diamond \in \{\wedge, \vee\}$.
Then $N(s,g) = N(s, \Diamond\{f\}) = 1 + N(s,f)$. So $N(s,g) \neq 0$. Therefore $X(g)$ holds.

Case 3: $g = \Diamond F$ where $\Diamond \in \{\wedge, \vee\}$ and F is a finite set of formulas such that $|F| \neq 1$. Then $N(s,g) = N(s, \Diamond F) = N(s,F) = \Sigma(N(s,f) : f \in F)$ and $rs(g) = rs(\Diamond F) = \Diamond rs(F) = \Diamond\{rs(f) : f \in F\}$. Suppose $N(s,g) = 0$. Then $\Sigma(N(s,f) : f \in F) = 0$. Hence for all f in F, $N(s,f) = 0$. For all f in F, by $X(f)$, $rs(f) = f$. So $rs(g) = \Diamond\{rs(f) : f \in F\} = \Diamond\{f : f \in F\} = \Diamond F = g$. Therefore $X(g)$ holds.

Thus by induction, for each formula f, if $N(s,f) = 0$ then $rs(f) = f$.

(5.2) implies (5.3)

Let f be a formula and suppose $rs(f) = f$. The proof is by induction on i. If $i = 1$ then $rs^i(f) = rs(f) = f$. Take any k in $\mathbb{Z}^+$ and suppose $rs^k(f) = f$. Then $rs^{k+1}(f) = rs(rs^k(f)) = rs(f) = f$. So by induction for each i in $\mathbb{Z}^+$, $rs^i(f) = f$.

(5.3) implies (5.4)

This follows because $\mathbb{Z}^+$ is not empty.

(5.4) implies (5.1)

Suppose $i \in \mathbb{Z}^+$ and $rs^i(f) = f$. By parts (2, 4.3), $N(\{\wedge,\vee\}, f) \geq N(\{\wedge,\vee\}, rs(f)) \geq \ldots \geq N(\{\wedge,\vee\}, rs^i(f)) \geq 0$.

Assume $N(s,f) \neq 0$. By part (4.4) we have, $N(\{\wedge,\vee\}, f) > N(\{\wedge,\vee\}, rs(f)) \geq N(\{\wedge,\vee\}, rs^i(f))$ and so $rs^i(f) \neq f$. This contradiction shows that $N(s,f) = 0$.

(6)

Let f be any formula.

(6.1)

Suppose $n \geq N(\{\wedge,\vee\}, f)$.

Case 1: There exist i in $[1..n]$ such that $N(s, rs^i(f)) = 0$.
Let $rs^i(f) = g$. So $N(s,g) = 0$. Now $rs^n(f) = rs^{n-i}(rs^i(f)) = rs^{n-i}(g)$. If $i = n$ then $rs^n(f) = g$. If $i < n$ then by part (5), $rs^{n-i}(g) = g$. So in both cases, $rs^n(f) = g$. Also by part (5), $rs(g) = g$. Hence $ns(g) = g$. By part (3), $ns(f) = ns(rs^n(f)) = ns(g) = g = rs^n(f)$.

Case 2: For each i in $[1..n]$, $N(s, rs^i(f)) \neq 0$.
By part (2), for each i in $[1..n]$, $N(s, rs^i(f)) > 0$. So by parts (2, 4.4), $n \geq N(\{\wedge,\vee\}, f) > N(\{\wedge,\vee\}, rs(f)) > \ldots > N(\{\wedge,\vee\}, rs^n(f)) > N(\{\wedge,\vee\}, rs^{n+1}(f)) \geq 0$. But this is a contradiction. So this case cannot occur.

(6.2)

Let $X(n)$ be the following statement. "If $n \in \mathbb{Z}^+$ then there exists m in $\mathbb{Z}^+$ such that $ns^n(f) = rs^m(f)$." We prove $X(n)$ by induction on n. If $n = 1$ let $m = N(\{\wedge,\vee\}, f) + 1$. Then $m \in \mathbb{Z}^+$. By part (6.1), $ns^n(f) = ns(f) = rs^m(f)$. So $X(1)$ holds.

Take any k in $\mathbb{Z}^+$ and suppose $X(k)$ holds. We show $X(k+1)$ holds. By $X(k)$, there exists j in $\mathbb{Z}^+$ such that $ns^k(f) = rs^j(f)$. Let $g = rs^j(f)$ and $i = N(\{\wedge,\vee\}, g)$. By

2.5 Singletons

part (6.1), $ns(g) = rs^i(g)$. So $ns^{k+1}(f) = ns(ns^k(f)) = ns(rs^j(f)) = ns(g) = rs^i(g) = rs^i(rs^j(f)) = rs^{i+j}(f)$. Therefore $X(k+1)$ holds.

Thus, by induction, if $n \in \mathbb{Z}^+$ then there exists m in $\mathbb{Z}^+$ such that $ns^n(f) = rs^m(f)$.

(7)

Let f be a formula and F be a finite set of formulas. Let $n = N(\{\wedge, \vee\}, f)$. By part (6.1), $ns(f) = rs^n(f)$.

(7.1)

By part (4.1), $N(\wedge, f) \geq N(\wedge, rs(f)) \geq N(\wedge, rs^n(f)) = N(\wedge, ns(f))$. So $N(\wedge, F) = \sum(N(\wedge, f) : f \in F) \geq \sum(N(\wedge, ns(f)) : f \in F) = N(\wedge, \{ns(f) : f \in F\}) = N(\wedge, ns(F))$.

(7.2)

By part (4.2), $N(\vee, f) \geq N(\vee, rs(f)) \geq N(\vee, rs^n(f)) = N(\vee, ns(f))$. So $N(\vee, F) = \sum(N(\vee, f) : f \in F) \geq \sum(N(\vee, ns(f)) : f \in F) = N(\vee, \{ns(f) : f \in F\}) = N(\vee, ns(F))$.

(7.3)

By part (4.3), $N(\{\wedge, \vee\}, f) \geq N(\{\wedge, \vee\}, rs(f)) \geq N(\{\wedge, \vee\}, rs^n(f)) = N(\{\wedge, \vee\}, ns(f))$.

(7.4)

Suppose $N(s, f) > 0$. By parts (4.4, 4.3) we have $N(\{\wedge, \vee\}, f) > N(\{\wedge, \vee\}, rs(f)) \geq N(\{\wedge, \vee\}, rs^n(f)) = N(\{\wedge, \vee\}, ns(f))$.

(8)

Suppose f is any quasi-formula, F is any set of quasi-formulas, and $i \in \mathbb{Z}^+$. The result is proved by induction on i. Let $X(i)$ be the following statement. "If f is any quasi-formula then $ns^i(f) = ns(f)$." Before we start the induction we shall need $X(2)$, which we prove now.

Case 1: f is any formula.

Let $n = N(\{\wedge, \vee\}, f)$. By part (6.1), $ns(f) = rs^n(f)$. By part (3), $ns(rs^n(f)) = ns(f)$. Therefore $ns(ns(f)) = ns(f)$.

Case 2: f is any quasi-formula that is not a formula.

Then $f = \neg^n \lozenge G$, where $n \in \mathbb{N}$, $\lozenge \in \{\wedge, \vee\}$, and G is an infinite set of formulas. By Case 1 we have $ns(ns(f)) = ns(ns(\neg^n \lozenge G)) = ns(\neg^n \lozenge ns(G)) = \neg^n \lozenge ns(ns(G)) = \neg^n \lozenge ns(\{ns(g) : g \in G\}) = \neg^n \lozenge \{ns(ns(g)) : g \in G\} = \neg^n \lozenge \{ns(g) : g \in G\} = \neg^n \lozenge ns(G) = ns(\neg^n \lozenge G) = ns(f)$.

Thus if f is any quasi-formula then $ns(ns(f)) = ns(f)$; and so $X(2)$ is proved.

We start the induction by noting that $X(1)$ is immediate.

Take any k in $\mathbb{Z}^+$ and suppose that $X(k)$ holds. We show $X(k+1)$ holds. By $X(k)$ and $X(2)$, $ns^{k+1}(f) = ns(ns^k(f)) = ns(ns(f)) = ns(f)$. So $X(k+1)$ holds.

Thus, by induction, if $i \in \mathbb{Z}^+$ and f is any quasi-formula then $ns^i(f) = ns(f)$.

Let F be any set of quasi-formulas. Then using this result for quasi-formulas we have the following. $ns^i(F) = \{ns^i(f) : f \in F\} = \{ns(f) : f \in F\} = ns(F)$.

(9)

Let F be any set of formulas.

(9.1)

Suppose that for all f in F, $rs(f) = f$. Then $rs(F) = \{rs(f) : f \in F\} = \{f : f \in F\} = F$.

Now suppose F is finite and $rs(F) = F$. Let $G = \{f \in F : rs(f) = f\}$. Then $G \subseteq F$. Assume $F \neq G$. Then $\{N(\{\wedge, \vee\}, f) : f \in F - G\} \neq \{\}$. Let $m = \max\{N(\{\wedge, \vee\}, f) : f \in F - G\}$. Then there exists f_m in $F - G$ such that $m = N(\{\wedge, \vee\}, f_m)$. Since $f_m \notin G$,

$rs(f_m) \neq f_m$. But $f_m \in F = rs(F)$. So there exists h in F such that $rs(h) = f_m$. If $h = rs(h)$ then $h = f_m$ and so $rs(f_m) = rs(h) = f_m$. This contradiction shows that $h \neq rs(h)$ hence $h \notin G$ and so $h \in F-G$. By parts (2, 5), $N(s,h) > 0$. By part (4.4), $N(\{\wedge,\vee\},h) > N(\{\wedge,\vee\},rs(h)) = N(\{\wedge,\vee\},f_m) = m$. Since $h \in F-G$, this contradicts the maximality of m.

Therefore $F = G$. Hence for all f in F, $rs(f) = f$.

Thus if F is any finite set of formulas then $rs(F) = F$ iff for all f in F, $rs(f) = f$.

(9.2)

Suppose that for all f in F, $ns(f) = f$. Then $ns(F) = \{ns(f) : f \in F\} = \{f : f \in F\} = F$.

Conversely, suppose $ns(F) = F$. Let $G = \{f \in F : ns(f) = f\}$. Then $G \subseteq F$. Assume $F \neq G$. Then there exists g_0 in $F-G$. So $ns(g_0) \neq g_0$. But $g_0 \in F = ns(F)$. So there exists g_1 in F such that $ns(g_1) = g_0$. Let $n = N(\{\wedge,\vee\},g_1)$. By part (8), $ns(ns(g_1)) = ns(g_1)$. Therefore $g_0 \neq ns(g_0) = ns(ns(g_1)) = ns(g_1) = g_0$. This contradiction shows that $F = G$. Hence for all f in F, $ns(f) = f$.

Thus if F is any set of formulas then $ns(F) = F$ iff for all f in F, $ns(f) = f$.

(10)

Let f be any quasi-formula.

(10.1) is equivalent to (10.2)

For any quasi-formula g let $S(g)$ be the following statement. "g has no singletons iff $N(s,g) = 0$." We prove $S(g)$ by induction on g. Let a be an atom. Then a has no singletons and $N(s,a) = 0$. Take any quasi-formula g and suppose that for all strict subformulas h of g, $S(h)$ holds. We show $S(g)$ holds.

Suppose $g = \neg h$ where h is a formula. By $S(h)$, g has no singletons iff h has no singletons iff $N(s,h) = 0$ iff $N(s,\neg h) = 0$ iff $N(s,g) = 0$.

Suppose $g \in \{\wedge\{h\}, \vee\{h\}\}$ where h is a formula. Then g has a singleton and $N(s,g) = 1 + N(s,h)$.

Suppose $g \in \{\wedge H, \vee H\}$ where H is a finite set of formulas such that $|H| \neq 1$. By $S(h)$, g has no singletons iff for all h in H, h has no singletons iff for all h in H, $N(s,h) = 0$ iff $\sum(N(s,h) : h \in H) = 0$ iff $N(s,H) = 0$ iff $N(s,g) = 0$.

Suppose $g = \neg^n \lozenge H$ where $n \in \mathbb{N}$, $\lozenge \in \{\wedge,\vee\}$, and H is an infinite set of formulas. By $S(h)$, g has no singletons iff for all h in H, h has no singletons iff for all h in H, $N(s,h) = 0$ iff $N(s,\neg^n \lozenge H) = 0$ iff $N(s,g) = 0$.

Thus, by induction, $S(g)$ holds.

(10.2) implies (10.3)

Suppose $N(s,f) = 0$.

Case 1: f is a formula.

By part (5), $rs(f) = f$ and so $ns(f) = f$.

Case 2: f is not a formula.

Then $f = \neg^n \lozenge G$, where $n \in \mathbb{N}$, $\lozenge \in \{\wedge,\vee\}$, and G is an infinite set of formulas. So for each g in G, $N(s,g) = 0$. By Case 1, for each g in G, $ns(g) = g$. So $ns(f) = ns(\neg^n \lozenge G) = \neg^n \lozenge ns(G) = \neg^n \lozenge \{ns(g) : g \in G\} = \neg^n \lozenge \{g : g \in G\} = \neg^n \lozenge G = f$.

(10.3) implies (10.4)

Suppose $ns(f) = f$. The proof is by induction on i. If $i = 1$ then $ns^i(f) = ns(f) = f$.

2.5 Singletons

Take any k in $\mathbb{Z}^+$ and suppose $ns^k(f) = f$. Then $ns^{k+1}(f) = ns(ns^k(f)) = ns(f) = f$. So by induction for each i in $\mathbb{Z}^+$, $ns^i(f) = f$.

(10.4) implies (10.5)
This follows because $\mathbb{Z}^+$ is not empty.

(10.5) implies (10.2)
Suppose $i \in \mathbb{Z}^+$ and $ns^i(f) = f$. Then by part (8), $f = ns^i(f) = ns(f)$.

Case 1: f is a formula.
Let $n = N(\{\wedge, \vee\}, f)$. Then by part (6.1), $ns(f) = rs^n(f)$. So $rs^n(f) = ns(f) = f$. By part (5), $N(s, f) = 0$.

Case 2: f is any quasi-formula that is not a formula.
Then $f = \neg^n \lozenge G$, where $n \in \mathbb{N}$, $\lozenge \in \{\wedge, \vee\}$, and G is an infinite set of formulas. So $\neg^n \lozenge G = f = ns(f) = ns(\neg^n \lozenge G) = \neg^n \lozenge ns(G) = \neg^n \lozenge \{ns(g) : g \in G\}$. Hence $G = ns(G)$. By part (9.2), for each g in G, $ns(g) = g$. So by Case 1, for each g in G, $N(s, g) = 0$. Hence $N(s, \neg^n \lozenge G) = 0$. Therefore $N(s, f) = 0$.

(11)
Case 1: f is any formula.
Suppose $n = N(\{\wedge, \vee\}, f)$. By part (6.1), $rs^{n+1}(f) = ns(f) = rs^n(f)$. So $rs(ns(f)) = ns(f)$. By part (5), $N(s, ns(f)) = 0$. Thus if f is any formula then $N(s, ns(f)) = 0$.

Case 2: f is any quasi-formula that is not a formula.
Then $f = \neg^n \lozenge G$, where $n \in \mathbb{N}$, $\lozenge \in \{\wedge, \vee\}$, and G is an infinite set of formulas. So $ns(f) = ns(\neg^n \lozenge G) = \neg^n \lozenge ns(G) = \neg^n \lozenge \{ns(g) : g \in G\}$. By Case 1, for all g in G, $N(s, ns(g)) = 0$. But $N(s, ns(f)) = 0$ iff $N(s, \neg^n \lozenge \{ns(g) : g \in G\}) = 0$ iff for each g in G, $N(s, ns(g)) = 0$. Therefore $N(s, ns(f)) = 0$.

Thus if f is any quasi-formula then $N(s, ns(f)) = 0$.

(12)
Let f be any quasi-formula such that $N(s, f) = 0$. If f is a formula then by part (5), $rs(f) = f$.

So suppose f is not a formula. Then $f = \neg^n \lozenge G$, where $n \in \mathbb{N}$, $\lozenge \in \{\wedge, \vee\}$, and G is an infinite set of formulas. By part (5), $N(s, f) = 0$
iff $N(s, \neg^n \lozenge G) = 0$
iff for each g in G, $N(s, g) = 0$
iff for each g in G, $rs(g) = g$
implies $rs(G) = G$
iff $\neg^n \lozenge rs(G) = \neg^n \lozenge G$
iff $rs(\neg^n \lozenge G) = \neg^n \lozenge G$
iff $rs(f) = f$.

Thus if f is any quasi-formula and $N(s, f) = 0$ then $rs(f) = f$.

(13)
Let f be any quasi-formula. By part (11), $N(s, ns(f)) = 0$. Let $g = ns(f)$. Then $N(s, g) = 0$. So by part (12), $rs(g) = g$. Hence $rs(ns(f)) = rs(g) = g = ns(f)$.

(14)
We shall use induction on the structure of quasi-formulas. If a is an atom then $rs(a) = a \equiv a$.

Let f be a quasi-formula such that $rs(f) \equiv f$. Then $rs(\neg f) = \neg rs(f) \equiv \neg f$. Also $rs(\wedge\{f\}) = rs(f) \equiv f$. If v is any valuation then $v(\wedge\{f\}) = \text{AND}\{v(f)\} = v(f)$. So $f \equiv \wedge\{f\}$. Similarly $rs(\vee\{f\}) = rs(f) \equiv f$. If v is any valuation then $v(\vee\{f\}) = \text{OR}\{v(f)\} = v(f)$. So $f \equiv \vee\{f\}$.

Let F be any set of formulas such that $|F| \neq 1$. Suppose that for all f in F, $rs(f) \equiv f$. Then $rs(\wedge F) = \wedge\{rs(f) : f \in F\} \equiv \wedge\{f : f \in F\} = \wedge F$. Similarly, $rs(\vee F) = \vee\{rs(f) : f \in F\} \equiv \vee\{f : f \in F\} = \vee F$.

Thus if f is any quasi-formula then $rs(f) \equiv f$.

We show $ns(f) \equiv f$ by considering two cases.

Case 1: f is any formula.

Let $n = N(\{\wedge, \vee\}, f)$. Since $rs(f) \equiv f$ we have $rs^n(f) \equiv f$. By part (6.1), $ns(f) = rs^n(f)$. Hence $ns(f) \equiv f$.

Case 2: f is any quasi-formula that is not a formula.

Then $f = \neg^n \Diamond G$, where $n \in \mathbb{N}$, $\Diamond \in \{\wedge, \vee\}$, and G is an infinite set of formulas. By Case 1, $ns(f) = ns(\neg^n \Diamond G) = \neg^n \Diamond ns(G) = \neg^n \Diamond \{ns(g) : g \in G\} \equiv \neg^n \Diamond \{g : g \in G\} = \neg^n \Diamond G = f$.

(15)

Suppose F is any set of formulas and $\Diamond \in \{\wedge, \vee\}$.

By part (14), $\Diamond ns(F) = \Diamond \{ns(f) : f \in F\} \equiv \Diamond \{f : f \in F\} = \Diamond F$.

By part (14), $\Diamond rs(F) = \Diamond \{rs(f) : f \in F\} \equiv \Diamond \{f : f \in F\} = \Diamond F$.

Also from the definitions, $\wedge F \equiv F$ and $ns(F) \equiv \wedge ns(F)$ and $rs(F) \equiv \wedge rs(F)$.

EndProofLemma2.5.6

Mostly we want to remove singletons from a clause or a meet. In these cases rs and ns give the same result. This and other results concerning clauses and meets are in our next lemma.

Lemma 2.5.7. Let C be a set of clauses, M be a set of meets, G be either a set of clauses or a set of meets, l be a literal, and g be either a clause or a meet. If g is a clause or G is a set of clauses then let $\Diamond = \vee$. If g is a meet or G is a set of meets then let $\Diamond = \wedge$.

1) $rs(l) = l = ns(l)$.
   If $|Lit(g)| = 1$ then $g \in \{l, \Diamond\{l\}\}$ and $rs(g) = l = ns(g)$.
   If $|Lit(g)| \neq 1$ then $rs(g) = g = ns(g)$.
   Therefore $ns(g) = rs(g)$; $rs(rs(g)) = rs(g)$; and $ns(rs(g)) = rs(g)$.
2) $ns(G) = rs(G) = \{\Diamond L \in G : |L| \neq 1\} \cup \{l : \Diamond\{l\} \in G\} \cup \{l : l \in G\}$.
3) $ns(G) = \{\}$ iff $G = \{\}$.
4) $ns(\wedge C) = rs(rs(\wedge C))$ and $ns(\vee M) = rs(rs(\vee M))$.
5) $Lit(ns(g)) = Lit(g)$, and $Lit(ns(G)) = Lit(G)$.
6) $Res(ns(G)) = Res(G)$.
7) $ns(G) \subseteq ns(Res(G))$.
8) $\sim ns(g) = ns(\sim g)$. So $ns(g) = \sim ns(\sim g)$.
9) $\sim ns(G) = ns(\sim G)$. So $ns(G) = \sim ns(\sim G)$.
10) $ns(\Diamond Lit(g)) = ns(g)$ and $ns(\Diamond .Lit(G)) = ns(G)$.

2.5 Singletons

Proof

Let C be a set of clauses, M be a set of meets, G be either a set of clauses or a set of meets, l be a literal, and g be either a clause or a meet. If g is a clause or G is a set of clauses then let $\Diamond = \vee$. If g is a meet or G is a set of meets then let $\Diamond = \wedge$.

(1) Let a be an atom. Then $rs(a) = a$ and $rs(\neg a) = \neg rs(a) = \neg a$. So if l is a literal then $rs(l) = l$. Hence $ns(l) = l$.

Suppose $|Lit(g)| = 1$. Then $g \in \{l, \Diamond\{l\}\}$. The case $g = l$ has already been considered. So suppose $g = \Diamond\{l\}$. Then $rs(g) = rs(\Diamond\{l\}) = rs(l) = l$. Hence $ns(g) = ns(\Diamond\{l\}) = ns(rs(\Diamond\{l\})) = ns(l) = l$.

Suppose $|Lit(g)| \neq 1$. Then $g = \Diamond L$, where L is a set of literals such that $|L| \neq 1$. So $rs(g) = rs(\Diamond L) = \Diamond\{rs(l) : l \in L\} = \Diamond\{l : l \in L\} = \Diamond L = g$. Hence $ns(g) = g$.

By the above and Definition 2.5.3 we have $ns(g) = rs(g)$; $rs(rs(g)) = rs(g)$; and $ns(rs(g)) = rs(g)$.

(2) This follows directly from part (1).

(3) This follows directly from part (2).

(4) Suppose that F is either a set of clauses or a set of meets. If F is a set of clauses let $\Diamond$ denote $\wedge$. If F is a set of meets let $\Diamond$ denote $\vee$. We show that $ns(\Diamond F) = rs(rs(\Diamond F))$.

First we make the following observations. If f is a clause or a meet then by part (1) we have (A), (B), and (C) below.

(A) $ns(f) = rs(f)$.   (B) $rs(rs(f)) = rs(f)$.   (C) $ns(rs(f)) = rs(f)$.

Case 1: F is finite.

Then $\Diamond F$ is a formula. If $rs(\Diamond F) = \Diamond F$ then $ns(\Diamond F) = \Diamond F = rs(\Diamond F) = rs(rs(\Diamond F))$. So suppose $rs(\Diamond F) \neq \Diamond F$. Then $ns(\Diamond F) = ns(rs(\Diamond F))$.

If $|F| = 1$ let $F = \{f\}$. Then f is either a clause or a meet. So by (C) and (B), $ns(\Diamond F) = ns(rs(\Diamond F)) = ns(rs(\Diamond\{f\})) = ns(rs(f)) = rs(f)$. Also, by (B), $rs(rs(\Diamond F)) = rs(rs(\Diamond\{f\})) = rs(rs(f)) = rs(f)$. Hence $ns(\Diamond F) = rs(rs(\Diamond F))$.

So suppose $|F| \neq 1$. By (B), $rs(rs(\Diamond F)) = \Diamond rs(rs(F)) = \Diamond rs(\{rs(f) : f \in F\}) = \Diamond\{rs(rs(f)) : f \in F\} = \Diamond\{rs(f) : f \in F\} = \Diamond rs(F) = rs(\Diamond F)$. So by Definition 2.5.3, $ns(\Diamond F) = ns(rs(\Diamond F)) = rs(\Diamond F)$. Hence $ns(\Diamond F) = rs(rs(\Diamond F))$.

Thus if F is finite then $ns(\Diamond F) = rs(rs(\Diamond F))$.

Case 2: F is infinite.

Then by (A) and (B), $ns(\Diamond F) = \Diamond ns(F)) = \Diamond\{ns(f) : f \in F\} = \Diamond\{rs(f) : f \in F\} = \Diamond\{rs(rs(f)) : f \in F\} = \Diamond rs(\{rs(f) : f \in F\}) = \Diamond rs(rs(F)) = rs(rs(\Diamond F))$.

(5) Case 1: $|Lit(g)| \neq 1$.

Then $g = \Diamond L$, where L is a set of literals such that $|L| \neq 1$. By part (1), $ns(g) = g$. Hence $Lit(ns(g)) = Lit(g)$.

Case 2: $|Lit(g)| = 1$.

Then there is a literal l such that $Lit(g) = \{l\} = Lit(l)$. By part (1), $ns(g) = l$. Hence $Lit(ns(g)) = Lit(g)$.

From this result we get $Lit(ns(G)) = Lit(\{ns(g) : g \in G\}) = \{Lit(ns(g)) : g \in G\} = \{Lit(g) : g \in G\} = Lit(G)$.

(6) By part (5), $Res(ns(G)) = \Diamond.ResLit(ns(G)) = \Diamond.ResLit(G) = Res(G)$.

(7) By Definitions 2.4.6 and 2.4.8, $Res(G) = \Diamond.ResLit(G)$. Suppose l is a literal and L is a finite set of literals such that $|L| \neq 1$.

If $\Diamond L \in ns(G)$ then $\Diamond L \in G$. Hence $\Diamond L \in Res(G)$ and so $\Diamond L \in ns(Res(G))$.

If $l \in ns(G)$ then either $l \in G$ or $\Diamond\{l\} \in G$. Hence $\Diamond\{l\} \in Res(G)$ and therefore $l \in ns(Res(G))$.

So by part (2), $ns(G) \subseteq ns(Res(G))$.

(8) By Lemma 2.3.10(12,13,14), $\sim Lit(g) = Lit(\sim g)$, and so $|Lit(g)| = |\sim Lit(g)| = |Lit(\sim g)|$.

If $|Lit(g)| \neq 1$ then by part (1), $ns(g) = g$. Also $|Lit(\sim g)| \neq 1$, and so by part (1), $ns(\sim g) = \sim g$. Therefore $\sim ns(g) = ns(\sim g)$ and so $\sim ns(\sim g) = \sim\sim ns(g) = ns(g)$.

If $|Lit(g)| = 1$ then there is a literal l such that $Lit(g) = \{l\}$. By part (1), $ns(g) = l$ and so $\sim ns(g) = \sim l$. Also $Lit(\sim g) = \sim Lit(g) = \sim\{l\} = \{\sim l\}$. By part (1), $ns(\sim g) = \sim l$. Thus $\sim ns(g) = ns(\sim g)$, and so $ns(g) = \sim ns(\sim g)$.

(9) By part (8), $ns(\sim G) = \{ns(h) : h \in \sim G\} = \{ns(\sim g) : g \in G\} = \{\sim ns(g) : g \in G\} = \sim\{ns(g) : g \in G\} = \sim ns(G)$. Hence $\sim ns(\sim G) = ns(\sim\sim G) = ns(G)$.

(10) Case 1: $|Lit(g)| \neq 1$.

Then $g = \Diamond L$, where L is a set of literals such that $|L| \neq 1$. Hence $Lit(g) = L$ and so $\Diamond Lit(g) = \Diamond L = g$. Therefore $ns(\Diamond Lit(g)) = ns(g)$.

Case 2: $|Lit(g)| = 1$.

Then there is a literal l such that $Lit(g) = \{l\}$. By part (1), $ns(g) = l$. Also by part (1), $ns(\Diamond Lit(g)) = ns(\Diamond\{l\}) = ns(l) = l = ns(g)$.

Thus $ns(\Diamond Lit(g)) = ns(g)$.

From this result we have $ns(\Diamond.Lit(G)) = ns(\Diamond.\{Lit(g) : g \in G\}) = ns(\{\Diamond Lit(g) : g \in G\}) = \{ns(\Diamond Lit(g)) : g \in G\} = \{ns(g) : g \in G\} = ns(G)$.

EndProofLemma2.5.7

2.6 Core

Let C be any set of clauses. By Lemma 2.2.19(1), removing from C all tautologies and all clauses that have a strict subclause in C leaves a subset C' of C such that $C' \equiv C$. We also want to replace any clause $\vee\{l\}$ in C by l. The result will be the core of C.

Dually, let M be any set of meets. By Lemma 2.2.19(2), removing from M all contradictions and all meets that have a strict submeet in M leaves a subset M' of M such that $\vee M' \equiv \vee M$. We also want to replace any meet $\wedge\{l\}$ in M by l. The result will be the core of M.

These ideas are formalised in the following definition.

Definition 2.6.1. Let G be either a set of clauses or a set of meets.
1) The set of contingent or empty elements of G, $Ctge(G)$, is defined as follows.
$Ctge(G) = \{g \in G : g \text{ is contingent or empty}\}$.
2) The set of minimal elements of G, $Min(G)$, is defined by the following equation.
$Min(G) = \{g \in G : \text{if } g' \in G \text{ then } Lit(g') \not\subset Lit(g)\} = G - Sup(G)$.

2.6 Core

3) The *core* of G, $Cor(G)$, is defined by $Cor(G) = ns(Min(Ctge(G)))$.
4) The following notation is convenient.
   $nsMin(G) = ns(Min(G))$,
   $MinCtge(G) = Min(Ctge(G))$,
   $nsMinCtge(G) = ns(Min(Ctge(G)))$.

The next three lemmas concern $Ctge(.)$, $Min(.)$, and $Cor(.)$.

Lemma 2.6.2. Let each of C, C_1, and C_2 be a set of clauses, M be a set of meets, and G be either a set of clauses or a set of meets.
1) $Ctge(G) \subseteq G$.
2) $Ctge(C) = C - Taut$.
3) $Ctge(Ctge(C)) = Ctge(C)$.
4) $Ctge(M) = M - Contrad$.
5) $Ctge(Ctge(M)) = Ctge(M)$.
6) $Ctge(C) \equiv C$ and $\lor Ctge(M) \equiv \lor M$.
7) $\sim Ctge(G) = Ctge(\sim G)$. So $Ctge(G) = \sim Ctge(\sim G)$.
8) $Ctge(C_1 \cup C_2) = Ctge(C_1) \cup Ctge(C_2)$.

Proof

Let each of C, C_1, and C_2 be a set of clauses, M be a set of meets, and G be either a set of clauses or a set of meets.

(1) This follows directly from Definition 2.6.1.

(2) A clause $\lor L$ is not contingent and not empty iff L contains an atom and its negation iff $\lor L$ is a tautology.

(3) $Ctge(Ctge(C)) = Ctge(C - Taut)$ by part (2),
$= (C - Taut) - Taut$ by part (2),
$= (C - Taut)$
$= Ctge(C)$ by part (2).

(4) A meet $\land L$ is not contingent and not empty iff L contains an atom and its negation iff $\land L$ is a contradiction.

(5) $Ctge(Ctge(M)) = Ctge(M - Contrad)$ by part (4),
$= (M - Contrad) - Contrad$ by part (4),
$= (M - Contrad)$
$= Ctge(M)$ by part (4).

(6) Let v be any valuation. By Lemma 2.6.2(2), $v(Ctge(C)) = v(C - Taut) = v(C)$. Thus $Ctge(C) \equiv C$.

By Lemma 2.6.2(4), $v(\lor Ctge(M)) = v(\lor[M - Contrad]) = v(\lor M)$. Thus $\lor Ctge(M) \equiv \lor M$.

(7) Suppose $g \in G$. By Lemma 2.3.10(5), g is contingent iff $\sim g$ is contingent. By Lemma 2.3.10(14), $g \in G$ iff $\sim g \in \sim G$. Also g is empty iff $\sim g$ is empty. So $\sim Ctge(G)$
$= \sim \{g \in G : g \text{ is contingent or empty}\}$
$= \{\sim g : g \in G \text{ and } g \text{ is contingent or empty}\}$
$= \{\sim g : \sim g \in \sim G \text{ and } \sim g \text{ is contingent or empty}\}$
$= \{h : h \in \sim G \text{ and } h \text{ is contingent or empty}\}$

$= Ctge(\sim G)$.
Hence by Lemma 2.3.10(4), $\sim Ctge(\sim G) = Ctge(\sim\sim G) = Ctge(G)$.
 (8) $Ctge(C_1 \cup C_2) = (C_1 \cup C_2) - Taut$ by Lemma 2.6.2(2),
$= (C_1 - Taut) \cup (C_2 - Taut)$
$= Ctge(C_1) \cup Ctge(C_2)$ by Lemma 2.6.2(2).
EndProofLemma2.6.2

Lemma 2.6.3. Suppose each of C, C_1, and C_2 is a set of clauses. Let M be any set of meets, and G be either a set of clauses or a set of meets.
1) $Min(G) \subseteq G$.
2) $Min(G) = \{\}$ iff $G = \{\}$.
3) $Min(Min(G)) = Min(G)$.
4) $Sup(G) = G - Min(G) = \{g \in G : \text{there exists } g' \text{ in } Min(G) \text{ such that } g' < g\}$.
5) If $g \in G$ then there exists g' in $Min(G)$ such that $Lit(g') \subseteq Lit(g)$.
6) $\sim Min(G) = Min(\sim G)$. So $Min(G) = \sim Min(\sim G)$.
7) $Min(C) \equiv C$ and $\vee Min(M) \equiv \vee M$.
8) $nsMin(G) = Min(ns(G))$.
9) $Min(C_1 \cup C_2) \subseteq Min(C_1) \cup Min(C_2)$.
Proof
 Suppose each of C, C_1, and C_2 is a set of clauses. Let M be any set of meets, and G be either a set of clauses or a set of meets.
 (1) This follows directly from Definition 2.6.1.
 (2) This follows directly from Definition 2.6.1.
 (3) Recall $Min(G) = \{g \in G : \text{if } g' \in G \text{ then } Lit(g') \not\subset Lit(g)\}$. So $Min(Min(G)) = \{g \in Min(G) : \text{if } g' \in Min(G) \text{ then } Lit(g') \not\subset Lit(g)\}$. Clearly $Min(Min(G)) \subseteq Min(G) \subseteq G$.
 Conversely, take any g in $Min(G)$. Then $g \in Min(G)$ and if $g' \in G$ then $Lit(g') \not\subset Lit(g)$. So if $g' \in Min(G)$ then $Lit(g') \not\subset Lit(g)$. Hence $g \in Min(Min(G))$. But g was arbitrary, so $Min(G) \subseteq Min(Min(G))$.
 Thus $Min(Min(G)) = Min(G)$.
 (4) $Sup(G) = G - Min(G)$ follows directly from the definitions.
 Suppose $g \in G - Min(G)$. Then there exists g_1 in G such that $g_1 < g$. If $g_1 \in G - Min(G)$ then there exists g_2 in G such that $g_2 < g_1$. If $g_2 \in G - Min(G)$ then there exists g_3 in G such that $g_3 < g_2$. We can continue this reasoning indefinitely. But $Lit(g)$ is finite, so there is an n such that $g_n \in Min(G)$ and $g_n < g_{n-1} < ... < g_1 < g$. Hence $g_n < g$. Therefore if $g \in G - Min(G)$ then there exists g' in $Min(G)$ such that $g' < g$. Thus $G - Min(G) = \{g \in G : \text{there exists } g' \text{ in } Min(G) \text{ such that } g' < g\}$.
 (5) Take any g in G. If $g \in Min(G)$ then let $g' = g$. So suppose $g \in G - Min(G)$. By Lemma 2.6.3(4), there exists g' in $Min(G)$ such that $Lit(g') \subset Lit(g)$. Thus if $g \in G$ then there exists g' in $Min(G)$ such that $Lit(g') \subseteq Lit(g)$.
 (6) If $\{f,g\} \subseteq G$ then $f < g$ iff $\sim f < \sim g$. So
$\sim Min(G) = \sim \{g \in G : \text{if } g' < g \text{ then } g' \notin G\}$
$= \{\sim g : g \in G \text{ and if } g' < g \text{ then } g' \notin G\}$
$= \{\sim g : \sim g \in \sim G \text{ and if } \sim g' < \sim g \text{ then } \sim g' \notin \sim G\}$

2.6 Core

$= \{h : h \in \sim G \text{ and if } h' < h \text{ then } h' \notin \sim G\}$
$= Min(\sim G)$.
Hence by Lemma 2.3.10(4), $\sim Min(\sim G) = Min(\sim \sim G) = Min(G)$.

(7) Let v be any valuation. Let L be a finite set of literals and suppose $L' \subset L$. If $v(\vee L') = \mathbf{T}$ then $v(\vee L) = \mathbf{T}$ and so if $v(\vee L) = \mathbf{F}$ then $v(\vee L') = \mathbf{F}$. Suppose $c \in C - Min(C)$. By Lemma 2.6.3(4), there exists c' in $Min(C)$ such that $c' < c$. We shall show that $v(C) = v(C - \{c\})$.

If $v(c) = \mathbf{T}$ then $v(C) = v(C - \{c\})$. So suppose $v(c) = \mathbf{F}$. Then $v(c') = \mathbf{F}$. But $c' \in Min(C) \subseteq (C - \{c\})$. So $v(C) = \mathbf{F} = v(C - \{c\})$.

So the clauses removed from C to get $Min(C)$ do not change the valuation of C. Thus $Min(C) \equiv C$.

By Lemma 2.6.3(6, and the first part of 7), and Lemma 2.3.10(6,3), $\vee Min(M) \equiv \sim \sim \vee Min(M) = \sim[\wedge \sim Min(M)] = \sim[\wedge Min(\sim M)] \equiv \sim[\wedge(\sim M)] = \sim[\sim \vee (M)] \equiv \vee M$.

(8) $nsMin(G) = ns(\{g \in G : \text{if } g' \in G \text{ then } Lit(g') \not\subset Lit(g)\})$
$= \{ns(g) \in G : \text{if } g' \in G \text{ then } Lit(g') \not\subset Lit(g)\}$
$= \{g \in ns(G) : \text{if } g' \in ns(G) \text{ then } Lit(g') \not\subset Lit(g)\}$ by Lemma 2.5.7(5),
$= Min(ns(G))$.

(9) $Min(C_1 \cup C_2) = \{c \in C_1 \cup C_2 : \text{if } c' \in C_1 \cup C_2 \text{ then } Lit(c') \not\subset Lit(c)\}$
$\subseteq \{c \in C_1 : \text{if } c' \in C_1 \text{ then } Lit(c') \not\subset Lit(c)\} \cup \{c \in C_2 : \text{if } c' \in C_2 \text{ then } Lit(c') \not\subset Lit(c)\}$
$= Min(C_1) \cup Min(C_2)$.

EndProofLemma2.6.3

It is convenient to have the following notation.

Definition 2.6.4. Let F any set of formulas. Define $CorClsC_\models(F) = Cor(Cls(C_\models(F)))$, and $CorMtsImp(\vee F) = Cor(Mts(Imp(\vee F)))$.

Lemma 2.6.5. Suppose each of C, C_1, and C_2 is a set of clauses. Let M be a set of meets, F be a set of formulas, and G be either a set of clauses or a set of meets.
1) If $g \in Cor(G)$ and $|Lit(g)| \neq 1$ then $g \in G$.
2) If $g \in Cor(G)$ and $|Lit(g)| = 1$ then g is a literal and either $g \in G$ or $\vee\{g\} \in G$ or $\wedge\{g\} \in G$.
3) $\wedge Cor(C) \equiv Cor(C) \equiv C \equiv \wedge C$, and $\vee Cor(M) \equiv \vee M$.
4) $\sim Cor(G) = Cor(\sim G)$. So $Cor(G) = \sim Cor(\sim G)$.
5) $\sim CorClsC_\models(F) = CorMtsImp(\vee \sim F)$. So $CorClsC_\models(F) = \sim CorMtsImp(\vee \sim F)$.
6) $Cor(C) \subseteq ns(C) - Taut$.
So there are no tautologies in $Cor(C)$. That is, $Cor(C) - Taut = Cor(C)$.
7) If $c \in C - Taut$ then there exists c' in $Cor(C)$ such that $Lit(c') \subseteq Lit(c)$.
8) $Cor(Cor(G)) = Cor(G)$.
9) $Cor(C_1 \cup C_2) \subseteq Cor(C_1) \cup Cor(C_2)$.

Proof

Suppose each of C, C_1, and C_2 is a set of clauses. Let M be a set of meets, F be a set of formulas, and G be either a set of clauses or a set of meets.

(1,2) Suppose $g \in Cor(G)$. Then $g \in nsMinCtge(G)$. By Lemma 2.6.2(1) and Lemma 2.6.3(1), $MinCtge(G) \subseteq G$. So by Lemma 2.5.6(1), we get $g \in ns(G)$. By

Lemma 2.5.7(2), if $|Lit(g)| \neq 1$ then $g \in G$; and if $|Lit(g)| = 1$ then g is a literal, and so either $g \in G$ or $\vee\{g\} \in G$ or $\wedge\{g\} \in G$.

(3) By Lemma 2.5.6(15), Lemma 2.6.3(7), and Lemma 2.6.2(6), $\vee Cor(M) = \vee ns(Min(Ctge(M))) \equiv \vee Min(Ctge(M)) \equiv \vee Ctge(M) \equiv \vee M$, and also $Cor(C) = ns(Min(Ctge(C))) \equiv Min(Ctge(C)) \equiv Ctge(C) \equiv C$. So by Lemma 2.2.18(8), $\wedge Cor(C) \equiv Cor(C) \equiv C \equiv \wedge C$.

(4) By Lemma 2.5.7(9), Lemma 2.6.3(6), and Lemma 2.6.2(7), we get the following. $\sim Cor(G) = \sim ns(Min(Ctge(G))) = ns(\sim Min(Ctge(G))) = ns(Min(\sim Ctge(G))) = ns(Min(Ctge(\sim G))) = Cor(\sim G)$. Hence by Lemma 2.3.10(4), $\sim Cor(\sim G) = Cor(\sim\sim G) = Cor(G)$.

(5) By Lemma 2.6.5(4) and Lemma 2.3.10(11), we obtain $\sim CorClsC_\models(F) = \sim Cor[ClsC_\models(F)] = Cor[\sim ClsC_\models(F)] = Cor[MtsImp(\vee \sim F)] = CorMtsImp(\vee \sim F)$. By Lemma 2.3.10(4), $CorClsC_\models(F) = \sim\sim CorClsC_\models(F) = \sim CorMtsImp(\vee \sim F)$.

(6) If C is any set of clauses then $Cor(C) = nsMinCtge(C)$. By Lemma 2.6.2(2), $Ctge(C) = C - Taut$. So $Cor(C) = nsMin(C - Taut) \subseteq ns(C - Taut) = ns(C) - Taut$. Hence there are no tautologies in $Cor(C)$, and so $Cor(C) - Taut = Cor(C)$.

(7) Take any c in $C - Taut$. $Cor(C) = ns(Min(Ctge(C)))$. By Lemma 2.6.2(2), $c \in C - Taut = Ctge(C)$. By Lemma 2.6.3(5), there exists c_1 in $Min(Ctge(C))$ such that $Lit(c_1) \subseteq Lit(c)$. Hence $ns(c_1) \in ns(Min(Ctge(C)))$. By Lemma 2.5.7(5), $Lit(ns(c_1)) = Lit(c_1)$. Finally let $c' = ns(c_1)$. Then $c' \in Cor(C)$ and $Lit(c') = Lit(c_1) \subseteq Lit(c)$.

(8) Recall $Cor(G) = nsMinCtge(G)$. So $Cor(Cor(G)) = nsMinCtge(Cor(G))$
$= nsMin(Cor(G))$ by part (6),
$= nsMin(nsMinCtge(G))$
$= ns(nsMin(MinCtge(G)))$ by Lemma 2.6.3(8)
$= nsMin(MinCtge(G))$ by Lemma 2.5.6(8)
$= ns(MinCtge(G))$ by Lemma 2.6.3(3)
$= Cor(G)$.

(9) $Cor(C_1 \cup C_2) = nsMinCtge(C_1 \cup C_2)$
$= nsMin(Ctge(C_1) \cup Ctge(C_2))$ by Lemma 2.6.2(8),
$\subseteq ns(MinCtge(C_1) \cup MinCtge(C_2))$ by Lemma 2.6.3(9),
$= nsMinCtge(C_1) \cup nsMinCtge(C_2)$
$= Cor(C_1) \cup Cor(C_2)$.

EndProofLemma2.6.5

If f is a formula then $nnf(f)$ is a formula equivalent to f, and in a standard or normal form, namely negation normal form. If C is a set of clauses then $Cor(C)$ is a set of clauses equivalent to C, and we could regard $Cor(C)$ as a standard or normal form, perhaps core normal form.

Unfortunately both these normal forms are not unique. That is, if $f \equiv g$ then $nnf(f)$ and $nnf(g)$ may be different formulas. Also if $C \equiv C'$ then $Cor(C)$ and $Cor(C')$ may be different sets of clauses. As a step to getting unique normal forms for formulas, we shall now define what will turn out to be a unique normal form for a set of clauses (see Theorem 2.8.1).

2.6 Core

Definition 2.6.6. Let G be either a set of clauses or a set of meets. The following notation is convenient:
$CorRes(G) = Cor(Res(G))$.
$nsMinCtgeRes(G) = ns(Min(Ctge(Res(G))))$.
Define G to be *stable* iff $CorRes(G) = G$.

Some examples of simple stable sets are given in the following lemma.

Lemma 2.6.7. Let L be a set of literals.
1) $CorRes(\{\}) = \{\}$ and so $\{\}$ is stable.
2) $CorRes(\{\vee\{\}\}) = \{\vee\{\}\}$, and so $\{\vee\{\}\}$ is stable.
3) $CorRes(\{\wedge\{\}\}) = \{\wedge\{\}\}$, and so $\{\wedge\{\}\}$ is stable.
4) L is stable iff L is contingent or empty.

Proof

Let L be a set of literals.

(1) $CorRes(\{\}) = Cor(\{\}) = nsMinCtge(\{\}) = nsMin(\{\}) = ns(\{\}) = \{\}$. So $\{\}$ is stable.

(2) $CorRes(\{\vee\{\}\}) = Cor(\{\vee\{\}\}) = nsMinCtge(\{\vee\{\}\}) = nsMin(\{\vee\{\}\}) = ns(\{\vee\{\}\}) = \{\vee\{\}\}$. So $\{\vee\{\}\}$ is stable.

(3) $CorRes(\{\wedge\{\}\}) = Cor(\{\wedge\{\}\}) = nsMinCtge(\{\wedge\{\}\}) = nsMin(\{\wedge\{\}\}) = ns(\{\wedge\{\}\}) = \{\wedge\{\}\}$. So $\{\wedge\{\}\}$ is stable.

(4) Suppose L is contingent or empty. If L is empty then by part (1), L is stable. So suppose L is contingent. By Lemma 2.2.16(4), L is not empty and for each atom a either $a \notin L$ or $\neg a \notin L$. Therefore $CorRes(L) = Cor(\vee.ResLit(L)) = Cor(\vee.Res(\{Lit(l) : l \in L\})) = Cor(\vee.Res(\{\{l\} : l \in L\})) = Cor(\vee.\{\{l\} : l \in L\}) = Cor(\{\vee\{l\} : l \in L\}) = nsMinCtge(\{\vee\{l\} : l \in L\}) = nsMin(\{\vee\{l\} : l \in L\}) = ns(\{\vee\{l\} : l \in L\}) = \{l : l \in L\} = L$. Thus L is stable.

Conversely suppose L is not contingent and not empty. By Lemma 2.2.16(4), there is an atom a such that $\{a, \neg a\} \subseteq L$. Therefore $CorRes(L) = Cor(\vee.ResLit(L)) = Cor(\vee.Res(\{Lit(l) : l \in L\})) = Cor(\vee.Res(\{\{l\} : l \in L\})) = Cor(\vee.\{\{l\} : l \in L\} \cup \{\vee\{\}\}) = Cor(\{\vee\{l\} : l \in L\} \cup \{\vee\{\}\}) = nsMinCtge(\{\vee\{l\} : l \in L\} \cup \{\vee\{\}\}) = nsMin(\{\vee\{l\} : l \in L\} \cup \{\vee\{\}\}) = ns(\{\vee\{\}\}) = \{\vee\{\}\} \neq L$. Thus L is not stable.
EndProofLemma2.6.7

Our next result shows that $CorRes(G)$ is stable.

Lemma 2.6.8. Let C be a set of clauses, M be a set of meets, and G be either a set of clauses or a set of meets.
1) $\wedge CorRes(C) \equiv CorRes(C) \equiv C \equiv \wedge C$, and $\vee CorRes(M) \equiv \vee M$.
2) $\sim CorRes(G) = CorRes(\sim G)$. So $CorRes(G) = \sim CorRes(\sim G)$.
3) $MinCtgeRes(G)) \subseteq Res(MinCtgeRes(G)) \subseteq Res(G)$.
4) $MinCtgeRes(MinCtgeRes(G)) = MinCtgeRes(G)$.
5) $CorRes(CorRes(G)) = CorRes(G)$. So $CorRes(G)$ is stable.

Proof

Let C be a set of clauses, M be a set of meets, and G be either a set of clauses or a set of meets.

(1) By Lemma 2.2.18(8), Lemma 2.6.5(3), and Lemma 2.4.12(2), $\wedge CorRes(C) \equiv CorRes(C) = Cor(Res(C)) \equiv Res(C) \equiv C \equiv \wedge C$. By using Lemma 2.6.5(3) and Lemma 2.4.12(3), $\vee CorRes(M) = \vee Cor(Res(M)) \equiv \vee Res(M) \equiv \vee M$.

(2) By Lemma 2.6.5(4) and Lemma 2.4.11(16), $\sim CorRes(G) = Cor(\sim Res(G)) = Cor(Res(\sim G)) = CorRes(\sim G)$. By Lemma 2.3.10(4), $CorRes(G) = \sim\sim CorRes(G) = \sim CorRes(\sim G)$.

(3) By Lemma 2.4.11(4), Lemma 2.6.2(1), Lemma 2.6.3(1), and Lemma 2.4.11(3), $MinCtgeRes(G) \subseteq Res(MinCtgeRes(G))$. By Lemma 2.6.2(1) and Lemma 2.6.3(1), $MinCtgeRes(G) \subseteq Res(G)$. So by Lemma 2.4.11(5,6), we get $Res(MinCtgeRes(G)) \subseteq Res(Res(G)) = Res(G)$.

(4) We show $MinCtgeRes(G) \subseteq MinCtgeRes(MinCtgeRes(G))$. Take any g in $MinCtgeRes(G)$. So by Lemma 2.6.3(1), $g \in CtgeRes(G)$, and hence g is either contingent or empty. By part (3) we have, $g \in Res(MinCtgeRes(G))$ and so $g \in CtgeRes(MinCtgeRes(G))$. Assume $g \notin MinCtgeRes(MinCtgeRes(G))$. Then there exists f in $CtgeRes(MinCtgeRes(G))$ such that $f < g$. Therefore f is contingent or empty. By Lemma 2.6.2(1), $f \in Res(MinCtgeRes(G))$ and so by part (3), $f \in Res(G)$. Since f is contingent or empty, $f \in Ctge(Res(G))$. But $g \in MinCtgeRes(G)$ and $f < g$ hence $f \notin Ctge(Res(G))$. This contradiction shows that the assumption is wrong and therefore $g \in MinCtgeRes(MinCtgeRes(G))$. Thus $MinCtgeRes(G) \subseteq MinCtgeRes(MinCtgeRes(G))$.

Conversely to show $MinCtgeRes(MinCtgeRes(G)) \subseteq MinCtgeRes(G)$ take any g in $MinCtgeRes(MinCtgeRes(G))$. By Lemma 2.6.3(1), $g \in CtgeRes(MinCtgeRes(G))$ and hence g is either contingent or empty. By Lemma 2.6.2(1) and part (3), $g \in CtgeRes(MinCtgeRes(G)) \subseteq Res(MinCtgeRes(G)) \subseteq Res(G)$. Since g is either contingent or empty, $g \in CtgeRes(G)$. Assume $g \notin MinCtgeRes(G)$. Then there exists f in $CtgeRes(G)$ such that $f < g$. Therefore f is contingent or empty. Moreover we can choose f so that $f \in MinCtgeRes(G)$. But $g \in MinCtgeRes(MinCtgeRes(G))$ and $f < g$ so $f \notin CtgeRes(MinCtgeRes(G))$. Since f is contingent or empty, we have $f \notin Res(MinCtgeRes(G))$. By part (3), $MinCtgeRes(G) \subseteq Res(MinCtgeRes(G))$ and so $f \notin MinCtgeRes(G)$. This contradiction shows that $g \in MinCtgeRes(G)$, and hence $MinCtgeRes(MinCtgeRes(G)) \subseteq MinCtgeRes(G)$.

Thus $MinCtgeRes(MinCtgeRes(G)) = MinCtgeRes(G)$.

(5) By part (4), we have $MinCtgeRes(MinCtgeRes(G)) = MinCtgeRes(G)$. Hence $ns(MinCtgeRes(MinCtgeRes(G))) = ns(MinCtgeRes(G))$. This can be rewritten as, $CorRes(MinCtgeRes(G)) = CorRes(G)$. By Lemma 2.5.7(6) we have the following, $Res(ns(MinCtgeRes(G))) = Res(MinCtgeRes(G))$. That is, $Res(CorRes(G)) = Res(MinCtgeRes(G))$. Therefore $CorRes(CorRes(G)) = CorRes(MinCtgeRes(G)) = CorRes(G)$.

EndProofLemma2.6.8

Resolution can be regarded as a method for deriving a clause from a set of clauses. Two important properties that a derivation method may have are soundness and completeness. Roughly soundness means that every derived formula is true; and completeness means that every true formula is derivable. A more general and precise definition follows.

2.6 Core

Definition 2.6.9. A method for deriving a formula from a set of formulas is *sound* (with respect to $\models$) iff for every set F of formulas, if a formula f can be derived from F then $F \models f$.

A method for deriving a formula from a set of formulas is *complete* (with respect to $\models$) iff for every set F of formulas, if f is a formula such that $F \models f$ then f can be derived from F.

The following theorem gives a restricted soundness and completeness result for resolution.

Theorem 2.6.10. Let C be any set of clauses, M be any set of meets, and L be any finite set of literals.
1) $MinCtgeRes(C) \subseteq Res(C) \subseteq ClsC_{\models}(C)$. (soundness)
2) If $\vee L \in MinCtgeClsC_{\models}(C)$ then $\vee L \in MinCtgeRes(C)$.
3) $MinCtgeRes(C) \subseteq MinCtgeClsC_{\models}(C)$.
4) $CorRes(C) = CorClsC_{\models}(C)$. (soundness and core-completeness)
5) $CorRes(M) = CorMtsImp(\vee M)$. (soundness and core-completeness)

Proof

Let C be any set of clauses, and M be any set of meets.

(1) By Lemma 2.6.2(1) and Lemma 2.6.3(1), $MinCtgeRes(C) \subseteq Res(C)$. Take any c in $Res(C)$. By Lemma 2.4.12(2), $C \models c$. So $c \in ClsC_{\models}(C)$. Thus $Res(C) \subseteq ClsC_{\models}(C)$.

(2) Take any clause $\vee L$ in $MinCtgeClsC_{\models}(C)$. By using Lemma 2.6.3(1) and Lemma 2.6.2(1), $\vee L \in CtgeClsC_{\models}(C)$ and $\vee L \in ClsC_{\models}(C)$. So $\vee L$ is contingent or empty, and hence L is contingent or empty. So $C \models \vee L$. Hence $C \cup \sim L$ is unsatisfiable and so $C \cup \{\vee\{\sim l\} : l \in L\}$ is unsatisfiable. So by Lemma 2.4.12(5), $\vee\{\} \in Res(C \cup \{\vee\{\sim l\} : l \in L\})$. By Lemma 2.4.11(14), there is a finite subset L_1 of L such that $\vee L_1 \in Res(C)$. So $\vee L_1$ is contingent or empty. Therefore $\vee L_1 \in CtgeRes(C)$. So there exists a subset L_2 of L_1 such that L_2 is contingent or empty and $\vee L_2 \in MinCtgeRes(C)$. By part (1), $\vee L_2 \in ClsC_{\models}(C)$. Since L_2 is contingent or empty, $\vee L_2 \in CtgeClsC_{\models}(C)$. So there exists a subset L_3 of L_2 such that L_3 is contingent or empty and $\vee L_3 \in MinCtgeClsC_{\models}(C)$. But $L_3 \subseteq L$ and $\vee L \in MinCtgeClsC_{\models}(C)$, hence $L_3 = L$. Therefore $L_3 = L_2 = L_1 = L$ and so $\vee L \in MinCtgeRes(C)$.

(3) Take any clause $\vee L$ in $MinCtgeRes(C)$. By Lemma 2.6.3(1), $\vee L \in CtgeRes(C)$. So $\vee L$ is contingent or empty. By part (1), $\vee L \in ClsC_{\models}(C)$. Since $\vee L$ is contingent or empty, $\vee L \in CtgeClsC_{\models}(C)$. So there is a subset L_1 of L such that $\vee L_1$ is contingent or empty and $\vee L_1 \in MinCtgeClsC_{\models}(C)$. By the previous paragraph, $\vee L_1 \in MinCtgeRes(C)$. But $L_1 \subseteq L$ and $\vee L \in MinCtgeRes(C)$, hence $L_1 = L$. Hence $\vee L \in MinCtgeClsC_{\models}(C)$. Therefore $MinCtgeRes(C) \subseteq MinCtgeClsC_{\models}(C)$.

(4) By parts (2) and (3), $ns(MinCtgeRes(C)) = ns(MinCtgeClsC_{\models}(C))$. That is, $CorRes(C) = CorClsC_{\models}(C)$.

(5) By using Lemma 2.3.10(4), Lemma 2.6.8(2), part (4), Lemma 2.6.5(5), and Lemma 2.3.10(4) we have, $CorRes(M) = \sim\sim CorRes(M) = \sim[CorRes(\sim M)] = \sim[CorClsC_{\models}(\sim M)] = \sim CorClsC_{\models}(\sim M) = CorMtsImp(\vee\sim\sim M) = CorMtsImp(\vee M)$.

EndProofTheorem2.6.10

2.7 Normal Forms

We have already seen that formulas can be put into an equivalent standard or normal form, namely negation normal form. In this section we shall introduce two normal forms that are more constrained than negation normal form, namely conjunctive normal form and disjunctive normal form.

Definition 2.7.1. A *cnf-formula* is either a clause or the conjunction, $\wedge C$, of a finite set, C, of clauses. A formula is in *conjunctive normal form* iff it is a cnf-formula. The set of all cnf-formulas is denoted by *CnfFml*.

A *dnf-formula* is either a meet or the disjunction, $\vee M$, of a finite set, M, of meets. A formula is in *disjunctive normal form* iff it is a dnf-formula. The set of all dnf-formulas is denoted by *DnfFml*.

Hence every cnf-formula is an nnf-formula and every dnf-formula is an nnf-formula.

Lemma 2.7.2.
1) If f is a cnf-formula then $f = \sim\sim f$.
2) If f is a dnf-formula then $f = \sim\sim f$.
3) If $F \subseteq CnfFml \cup DnfFml$ then $F = \sim\sim F$.

Proof

(1) Let f be a cnf-formula. If f is a clause then by Lemma 2.3.10(4), $f = \sim\sim f$. So suppose $f = \wedge C$ where C is a finite set of clauses. By Lemma 2.3.10(3,4), $\sim\sim f = \sim\sim \wedge C = \wedge \sim\sim C = \wedge C = f$.

(2) Let f be a dnf-formula. If f is a meet then by Lemma 2.3.10(4), $f = \sim\sim f$. So suppose $f = \vee M$ where M is a finite set of meets. By Lemma 2.3.10(3,4), $\sim\sim f = \sim\sim \vee M = \vee \sim\sim M = \vee M = f$.

(3) Suppose $F \subseteq CnfFml \cup DnfFml$. Then by parts (1) and (2) we have, $\sim\sim F = \sim\sim \{f : f \in F\} = \{\sim\sim f : f \in F\} = \{f : f \in F\} = F$.
EndProofLemma2.7.2

The rest of this section concerns the transformation of a formula into an equivalent formula in conjunctive normal form, and also the transformation of a formula into an equivalent formula in disjunctive normal form. The first step in the transformation is to convert a given formula f into an equivalent formula in negation normal form. This is accomplished by applying *nnf* to f to get $nnf(f)$.

2.7.1 Rewrite Functions

Formulas in negation normal form are then transformed into either conjunctive normal form or disjunctive normal form by applying the following rewrite functions. Recall that *NnfFml* denotes the set of all nnf-formulas.

2.7 Normal Forms

Definition 2.7.3. The following notation is convenient. Let $f_{1,k}$ denote $f_1,\ldots,f_k$; $h_{1,n}$ denote $h_1,\ldots,h_n$; $l_{1,k}$ denote $l_1,\ldots,l_k$; $c_{1,j}$ denote $c_1,\ldots,c_j$; and $m_{1,j}$ denote $m_1,\ldots,m_j$.

The names of the 6 rewrite functions are:
dc for distributivity used to get conjunctive normal form (cnf),
ndc for nested disjunction used to get cnf,
ncc for nested conjunction used to get cnf,
dd for distributivity used to get disjunctive normal form (dnf),
ncd for nested conjunction used to get dnf, and
ndd for nested disjunction used to get dnf.

If $r \in \{dc, ndc, ncc, dd, ncd, ndd\}$ then the domain of r is denoted by Dr.
The rewrite functions are defined as follows.

1) $Ddc = \{\vee\{\wedge C_1,\ldots,\wedge C_j, l_{1,k}\} \in NnfFml : j \in \mathbb{Z}^+, k \in \mathbb{N}$, for each c in $C_1 \cup \ldots \cup C_j$, c is a clause; and for each i in $[1..k]$, l_i is a literal$\}$.

2) If there exist i in $[1..j]$ such that $C_i = \{\}$ then $dc(\vee\{\wedge C_1,\ldots,\wedge C_j, l_{1,k}\}) = \wedge\{\}$; else $dc(\vee\{\wedge C_1,\ldots,\wedge C_j, l_{1,k}\}) = \wedge\{\vee\{c_{1,j}, l_{1,k}\} :$ for each i in $[1..j]$, $c_i \in C_i\}$.

3) $Dndc = \{\vee\{\vee H_1,\ldots,\vee H_j, f_{1,k}\} \in NnfFml : j \in \mathbb{Z}^+, k \in \mathbb{N}$, for each i in $[1..k]$, f_i does not start with the $\vee$ symbol; and if g is a strict subformula of $\vee\{\vee H_1,\ldots,\vee H_j, f_{1,k}\}$ then $g \notin Ddc \cup Dndc \cup Dncc\}$.

4) $ndc(\vee\{\vee H_1,\ldots,\vee H_j, f_{1,k}\}) = \vee\{h_{1,n}, f_{1,k}\}$, where $\{h_{1,n}\} = H_1 \cup \ldots \cup H_j$.

5) $Dncc = \{\wedge\{\wedge H_1,\ldots,\wedge H_j, f_{1,k}\} \in NnfFml : j \in \mathbb{Z}^+, k \in \mathbb{N}$, for each i in $[1..k]$, f_i does not start with the $\wedge$ symbol; and if g is a strict subformula of $\wedge\{\wedge H_1,\ldots,\wedge H_j, f_{1,k}\}$ then $g \notin Ddc \cup Dndc \cup Dncc\}$.

6) $ncc(\wedge\{\wedge H_1,\ldots,\wedge H_j, f_{1,k}\}) = \wedge\{h_{1,n}, f_{1,k}\}$, where $\{h_{1,n}\} = H_1 \cup \ldots \cup H_j$.

7) $Ddd = \{\wedge\{\vee M_1,\ldots,\vee M_j, l_{1,k}\} \in NnfFml : j \in \mathbb{Z}^+, k \in \mathbb{N}$, for each m in $M_1 \cup \ldots \cup M_j$, m is a meet; and for each i in $[1..k]$, l_i is a literal$\}$.

8) If there exist i in $[1..j]$ such that $M_i = \{\}$ then $dd(\wedge\{\vee M_1,\ldots,\vee M_j, l_{1,k}\}) = \vee\{\}$; else $dd(\wedge\{\vee M_1,\ldots,\vee M_j, l_{1,k}\}) = \vee\{\wedge\{m_{1,j}, l_{1,k}\} :$ for each i in $[1..j]$, $m_i \in M_i\}$.

9) $Dncd = \{\wedge\{\wedge H_1,\ldots,\wedge H_j, f_{1,k}\} \in NnfFml : j \in \mathbb{Z}^+, k \in \mathbb{N}$, for each i in $[1..k]$, f_i does not start with the $\wedge$ symbol; and if g is a strict subformula of $\wedge\{\wedge H_1,\ldots,\wedge H_j, f_{1,k}\}$ then $g \notin Ddd \cup Dncd \cup Dndd\}$.

10) $ncd(\wedge\{\wedge H_1,\ldots,\wedge H_j, f_{1,k}\}) = \wedge\{h_{1,n}, f_{1,k}\}$, where $\{h_{1,n}\} = H_1 \cup \ldots \cup H_j$.

11) $Dndd = \{\vee\{\vee H_1,\ldots,\vee H_j, f_{1,k}\} \in NnfFml : j \in \mathbb{Z}^+, k \in \mathbb{N}$, for each i in $[1..k]$, f_i does not start with the $\vee$ symbol; and if g is a strict subformula of $\vee\{\vee H_1,\ldots,\vee H_j, f_{1,k}\}$ then $g \notin Ddd \cup Dncd \cup Dndd\}$.

12) $ndd(\vee\{\vee H_1,\ldots,\vee H_j, f_{1,k}\}) = \vee\{h_{1,n}, f_{1,k}\}$, where $\{h_{1,n}\} = H_1 \cup \ldots \cup H_j$.

Suppose $r \in \{dc, ndc, ncc, dd, ncd, ndd\}$ and $f \in NnfFml$. Then r is applicable to f iff there is a subformula g of f such that $g \in Dr$.

To convert an nnf-formula to conjunctive normal form we use only the rewrite functions in $\{dc, ndc, ncc\}$. To convert an nnf-formula into disjunctive normal form we use only the rewrite functions in $\{dd, ncd, ndd\}$. We keep applying the relevant rewrite functions until they cannot be applied anymore. It may be noticed that the rewrite functions are only applied to the innermost subformulas.

The next lemma shows that no element of $\{dc, ndc, ncc\}$ can be applied to a cnf-formula, and that no element of $\{dd, ncd, ndd\}$ can be applied to a dnf-formula. It also shows that the elements of $\{Ddc, Dndc, Dncc\}$ are pairwise disjoint, and that the elements of $\{Ddd, Dncd, Dndd\}$ are pairwise disjoint.

Lemma 2.7.4.
1) $CnfFml \cap (Ddc \cup Dndc \cup Dncc) = \{\}$.
2) $DnfFml \cap (Ddd \cup Dncd \cup Dndd) = \{\}$.
3) Suppose that either $\{r_1, r_2\} \subseteq \{dc, ndc, ncc\}$ or $\{r_1, r_2\} \subseteq \{dd, ncd, ndd\}$. Then $r_1 \neq r_2$ iff $Dr_1 \cap Dr_2 = \{\}$.

Proof

All three parts follow immediately from the definitions.
EndProofLemma2.7.4

The following arrow notation is useful.

Definition 2.7.5. The notation $f \xrightarrowtail{r} f'$ means the following.
1) Both f and f' are nnf-formulas.
2) $r \in \{dc, ndc, ncc, dd, ncd, ndd\}$.
3) There is a subformula g of f such that $g \in Dr$.
4) f' is the result of replacing exactly one occurrence of g in f by $r(g)$.

A sequence of steps that converts a formula into either a cnf-formula or a dnf-formula is called a conversion, as we now define.

Definition 2.7.6.
1) If f is an nnf-formula then $cnfNext(f) = \{f' : r \in \{dc, ndc, ncc\} \text{ and } f \xrightarrowtail{r} f'\}$.
2) A sequence $(f_1, f_2, \ldots)$ of formulas is a *cnf-conversion* of f_1 iff
   2.1) Every formula in $(f_1, f_2, \ldots)$ is an nnf-formula.
   2.2) If f_i is in $(f_1, f_2, \ldots)$ and $cnfNext(f_i) \neq \{\}$ then there exists f_{i+1} in $(f_1, f_2, \ldots)$ such that $f_{i+1} \in cnfNext(f_i)$.
   2.3) If f_n is in $(f_1, f_2, \ldots)$ and $cnfNext(f_n) = \{\}$ then f_n is the last element of $(f_1, f_2, \ldots)$.
3) If f is an nnf-formula then $dnfNext(f) = \{f' : r \in \{dd, ncd, ndd\} \text{ and } f \xrightarrowtail{r} f'\}$.
4) A sequence $(f_1, f_2, \ldots)$ of formulas is a *dnf-conversion* of f_1 iff
   4.1) Every formula in $(f_1, f_2, \ldots)$ is an nnf-formula.
   4.2) If f_i is in $(f_1, f_2, \ldots)$ and $dnfNext(f_i) \neq \{\}$ then there exists f_{i+1} in $(f_1, f_2, \ldots)$ such that $f_{i+1} \in dnfNext(f_i)$.
   4.3) If f_n is in $(f_1, f_2, \ldots)$ and $dnfNext(f_n) = \{\}$ then f_n is the last element of $(f_1, f_2, \ldots)$.
5) A sequence $(f_1, f_2, \ldots)$ of formulas is a *conversion* of f_1 iff it is either a cnf-conversion of f_1 or a dnf-conversion of f_1.

2.7 Normal Forms

It is easy to check that if f is an nnf-formula then every element of $cnfNext(f) \cup dnfNext(f)$ is an nnf-formula.

Example 2.7.7 (A conjunctive normal form conversion).
Let a, b, c, and d be four different atoms. The following is a common representation of a cnf-conversion of $\vee\{\wedge\{a,b\},\wedge\{c,d\},\vee\{\neg a,\neg b\}\}$.

$$\vee\{\wedge\{a,b\},\wedge\{c,d\},\vee\{\neg a,\neg b\}\} \xmapsto{ndc} \vee\{\wedge\{a,b\},\wedge\{c,d\},\neg a,\neg b\} \xmapsto{dc}$$
$$\wedge\{\vee\{a,c,\neg a,\neg b\},\vee\{a,d,\neg a,\neg b\},\vee\{b,c,\neg a,\neg b\},\vee\{b,d,\neg a,\neg b\}\}.$$

Several questions now arise. Is every formula in a conversion equivalent to every other formula in the conversion? Is each conversion finite? If a cnf-conversion is finite, is the last formula in the sequence a cnf-formula? Similarly, if a dnf-conversion is finite, is the last formula in the sequence a dnf-formula? We shall answer these questions in the following subsections.

2.7.2 Equivalence Is Preserved

The next lemma shows that every formula in a conversion is equivalent to every other formula in the conversion; that is, the rewrite functions preserve equivalence.

Lemma 2.7.8.
1) If $r \in \{dc, ndc, ncc, dd, ncd, ndd\}$ and $f \in Dr$ then $f \equiv r(f)$.
2) If $f \xmapsto{r} f'$ then $f \equiv f'$.
3) Any two formulas in a conversion are equivalent.

Proof
(1) Suppose $r \in \{dc, ndc, ncc, dd, ncd, ndd\}$ and $f \in Dr$.
Case 1: $r = dc$.
Then $f = \vee\{\wedge C_1,...,\wedge C_j, l_{1,k}\}$ where $j \in \mathbb{Z}^+$, $k \in \mathbb{N}$, for each i in $[1..j]$, C_i is a set of clauses, and for each i in $[1..k]$, l_i is a literal. If there exists i in $[1..j]$ such that $C_i = \{\}$, then $r(f) = dc(\vee\{\wedge C_1,...,\wedge C_j, l_{1,k}\}) = \wedge\{\}$; and so $f \equiv r(f)$. So suppose for all i in $[1..j]$, $C_i \neq \{\}$. Then $dc(f) = \wedge\{\vee\{c_{1,j}, l_{1,k}\} : \text{for each } i \text{ in } [1..j],\ c_i \in C_i\}$. Let $D = \{\vee\{c_{1,j}, l_{1,k}\} : \text{for each } i \text{ in } [1..j],\ c_i \in C_i\}$. Then $dc(f) = \wedge D$.
Let v be a valuation that makes f true; that is, $v(f) = \mathbf{T}$. Then either there is an i in $[1..k]$ such that $v(l_i) = \mathbf{T}$, or there is an i in $[1..j]$ such that $v(\wedge C_i) = \mathbf{T}$. In the first case, for all d in D, $v(d) = \mathbf{T}$; and so $v(dc(f)) = v(\wedge D) = \mathbf{T}$. In the second case, for all c in C_i, $v(c) = \mathbf{T}$. Hence for all d in D, $v(d) = \mathbf{T}$; and so $v(dc(f)) = v(\wedge D) = \mathbf{T}$. Therefore $f \models dc(f)$.
Conversely, let v be a valuation that makes f false; that is, $v(f) = \mathbf{F}$. Then for all i in $[1..k]$, $v(l_i) = \mathbf{F}$; and for all i in $[1..j]$, $v(\wedge C_i) = \mathbf{F}$. So for all i in $[1..j]$ there exists c'_i in C_i such that $v(c'_i) = \mathbf{F}$. Let $d' = \vee\{c'_1,...,c'_j, l_1,...l_k\}$. Then $d' \in D$ and $v(d') = \mathbf{F}$. Hence $v(dc(f)) = v(\wedge D) = \mathbf{F}$. Therefore $dc(f) \models f$.
Thus $f \equiv dc(f) = r(f)$.

Case 2: $r \in \{ndc, ndd\}$.
Then $f = \vee\{\vee H_1, ..., \vee H_j, f_{1,k}\}$ where $\{h_{1,n}\} = H_1 \cup ... \cup H_j$. So $f \equiv \vee\{h_{1,n}, f_{1,k}\} = r(f)$.

Case 3: $r \in \{ncc, ncd\}$.
Then $f = \wedge\{\wedge H_1, ..., \wedge H_j, f_{1,k}\}$ where $\{h_{1,n}\} = H_1 \cup ... \cup H_j$. So $f \equiv \wedge\{h_{1,n}, f_{1,k}\} = r(f)$.

Case 4: $r = dd$.
Then $f = \wedge\{\vee M_1, ..., \vee M_j, l_{1,k}\}$ where $j \in \mathbb{Z}^+$, $k \in \mathbb{N}$, for each i in $[1..j]$, M_i is a set of meets, and for each i in $[1..k]$, l_i is a literal. If there exists i in $[1..j]$ such that $M_i = \{\}$, then $r(f) = dd(\wedge\{\vee M_1, ..., \vee M_j, l_{1,k}\}) = \vee\{\}$; and so $f \equiv r(f)$. So suppose for all i in $[1..j]$, $M_i \neq \{\}$. Then $dd(f) = \vee\{\wedge\{m_{1,j}, l_{1,k}\} :$ for each i in $[1..j]$, $m_i \in M_i\}$. Let $C = \{\wedge\{m_{1,j}, l_{1,k}\} :$ for each i in $[1..j]$, $m_i \in M_i\}$. Then $dd(f) = \vee C$.

Let v be a valuation that makes f false; that is, $v(f) = \mathbf{F}$. Then either there is an i in $[1..k]$ such that $v(l_i) = \mathbf{F}$, or there is an i in $[1..j]$ such that $v(\vee M_i) = \mathbf{F}$. In the first case, for all c in C, $v(c) = \mathbf{F}$; and so $v(dd(f)) = v(\vee C) = \mathbf{F}$. In the second case, for all m in M_i, $v(m) = \mathbf{F}$. Hence for all c in C, $v(c) = \mathbf{F}$; and so $v(dd(f)) = v(\vee C) = \mathbf{F}$. Therefore $dd(f) \models f$.

Conversely, let v be a valuation that makes f true; that is, $v(f) = \mathbf{T}$. Then for all i in $[1..k]$, $v(l_i) = \mathbf{T}$; and for all i in $[1..j]$, $v(\vee M_i) = \mathbf{T}$. So for all i in $[1..j]$ there exists m'_i in M_i such that $v(m'_i) = \mathbf{T}$. Let $c' = \wedge\{m'_1, ..., m'_j, l_1, ...l_k\}$. Then $c' \in C$ and $v(c') = \mathbf{T}$. Hence $v(dd(f)) = v(\vee C) = \mathbf{T}$. Therefore $f \models dd(f)$.

Thus $f \equiv dd(f) = r(f)$.

(2) This follows from part (1) and Lemma 2.2.18(9).

(3) So by part (2), every formula in $cnfNext(f)$ is equivalent to f, and every formula in $dnfNext(f)$ is equivalent to f. By the transitivity of $\equiv$, any two formulas in a conversion are equivalent.
EndProofLemma2.7.8

2.7.3 Termination

In this subsection we show that each cnf-conversion and each dnf-conversion is finite. A standard way to do this is to define a function, say R (for rank), that maps a formula to a natural number such that if f_i and f_{i+1} are adjacent in a conversion then $R(f) > R(f')$. In our case we shall need triples of natural numbers, rather than natural numbers, and so the definitions will be more complicated. Also we will need different ranks depending on whether we are converting to conjunctive normal form or disjunctive normal form.

Definition 2.7.9. Let l be a literal, f be an nnf-formula, and F be a finite set of nnf-formulas.
1) $OrAnd(l) = 0$.
   $OrAnd(\wedge F) = \sum(OrAnd(f) : f \in F)$.
   $OrAnd(\vee F) = N(\wedge, F) + \sum(OrAnd(f) : f \in F)$.

2.7 Normal Forms

2) $AndOr(l) = 0$.
   $AndOr(\wedge F) = N(\vee, F) + \sum(AndOr(f) : f \in F)$.
   $AndOr(\vee F) = \sum(AndOr(f) : f \in F)$.
3) $cnfRank(f) = (OrAnd(f), N(\wedge, f), N(\vee, f))$.
   $dnfRank(f) = (AndOr(f), N(\vee, f), N(\wedge, f))$.
   These triples are ordered as follows: $(j_1, j_2, j_3) > (k_1, k_2, k_3)$ iff one of the following three conditions hold.
   a) $j_1 > k_1$ and $j_2 \geq k_2$.
   b) $j_1 = k_1$ and $j_2 > k_2$.
   c) $j_1 = k_1$ and $j_2 = k_2$ and $j_3 > k_3$.

A formula in conjunctive normal form has no $\wedge$ signs inside an $\vee$ sign. $OrAnd(f)$ is a way of counting the number of $\wedge$ signs inside $\vee$ signs in f. Similarly, a formula in disjunctive normal form has no $\vee$ signs inside an $\wedge$ sign. $AndOr(f)$ is a way of counting the number of $\vee$ signs inside $\wedge$ signs in f.

The rank we associate with a formula is given in Definition 2.7.9(3). $cnfRank(f)$ is used if we are transforming f into conjunctive normal form; and $dnfRank(f)$ is used if we are transforming f into disjunctive normal form.

It is easy to check that the ordering $>$ on the triples is
irreflexive $((j_1, j_2, j_3) \not> (j_1, j_2, j_3))$,
asymmetric (if $(j_1, j_2, j_3) > (k_1, k_2, k_3)$ then $(k_1, k_2, k_3) \not> (j_1, j_2, j_3)$), and
transitive (if $(j_1, j_2, j_3) > (k_1, k_2, k_3)$ and $(k_1, k_2, k_3) > (n_1, n_2, n_3)$
        then $(j_1, j_2, j_3) > (n_1, n_2, n_3)$).
However, $>$ on triples is not linear or total as $(2, 1, 0)$ is not related to $(1, 2, 3)$.

Parts (3) and (6) of the next lemma show that conversions are finite.

Lemma 2.7.10.
1) If $r \in \{dc, ndc, ncc\}$ and $f \in Dr$ then $cnfRank(f) > cnfRank(r(f))$.
2) If $r \in \{dc, ndc, ncc\}$ and $f \xmapsto{r} f'$ then $cnfRank(f) > cnfRank(f')$.
3) Every cnf-conversion of an nnf-formula is finite.
4) If $r \in \{dd, ncd, ndd\}$ and $f \in Dr$ then $dnfRank(f) > dnfRank(r(f))$.
5) If $r \in \{dd, ncd, ndd\}$ and $f \xmapsto{r} f'$ then $dnfRank(f) > dnfRank(f')$.
6) Every dnf-conversion of an nnf-formula is finite.

Proof

(1)
Suppose $r \in \{dc, ndc, ncc\}$ and $f \in Dr$.
    Case: $r = dc$.
Then $f = \vee\{\wedge C_1, ..., \wedge C_j, l_{1,k}\}$ where $j \geq 1$. So
$cnfRank(f) = cnfRank(\vee\{\wedge C_1, ..., \wedge C_j, l_{1,k}\})$
$= (OrAnd(\vee\{\wedge C_1, ..., \wedge C_j, l_{1,k}\}), N(\wedge, \vee\{\wedge C_1, ..., \wedge C_j, l_{1,k}\}), x)$
$= (N(\wedge, \{\wedge C_1, ..., \wedge C_j, l_{1,k}\}) + \sum(OrAnd(g) : g \in \{\wedge C_1, ..., \wedge C_j, l_{1,k}\}), j, x)$
$= (j, j, x)$.
    If there exists i in $[1..j]$ such that $C_i = \{\}$ then $cnfRank(dc(f)) = cnfRank(\wedge\{\})$
$= (OrAnd(\wedge\{\}), N(\wedge, \wedge\{\}), N(\vee, \wedge\{\})) = (0, 1, 0) < (j, j, x) = cnfRank(f)$.

If for all i in $[1..j]$, $C_i \neq \{\}$ then $cnfRank(dc(f))$
$= cnfRank(\wedge\{\vee\{c_{1,j}, l_{1,k}\} : \text{for each } i \text{ in } [1..j],\ c_i \in C_i\})$
$= (OrAnd(\wedge\{\vee\{c_{1,j}, l_{1,k}\} : \text{for each } i \text{ in } [1..j],\ c_i \in C_i\}),$
$\quad N(\wedge, \wedge\{\vee\{c_{1,j}, l_{1,k}\} : \text{for each } i \text{ in } [1..j],\ c_i \in C_i\}), y)$
$= (0, 1, y) < (j, j, x)$
$= cnfRank(f)$.

Case: $r = ndc$.

Then $f = \vee\{\vee H_1, \ldots, \vee H_j, f_{1,k}\}$ where $j \geq 1$ and for each i in $[1..k]$, f_i does not start with the $\vee$ symbol.

$cnfRank(f) = (OrAnd(f), N(\wedge, f), N(\vee, f))$
$= (OrAnd(f), N(\wedge, \vee\{h_{1,n}, f_{1,k}\}), j + N(\vee, \vee\{h_{1,n}, f_{1,k}\}))$ where
$\qquad \{h_{1,n}\} = H_1 \cup \ldots \cup H_j$.

Now $OrAnd(f) = OrAnd(\vee\{\vee H_1, \ldots, \vee H_j, f_{1,k}\})$
$= N(\wedge, \{\vee H_1, \ldots, \vee H_j, f_{1,k}\}) + \Sigma(OrAnd(g) : g \in \{\vee H_1, \ldots, \vee H_j, f_{1,k}\})$
$= N(\wedge, \{h_{1,n}, f_{1,k}\}) + N(\wedge, \{h_{1,n}\}) + \Sigma(OrAnd(g) : g \in \{h_{1,n}, f_{1,k}\})$
$\geq N(\wedge, \{h_{1,n}, f_{1,k}\}) + \Sigma(OrAnd(g) : g \in \{h_{1,n}, f_{1,k}\})$
$= OrAnd(\vee\{h_{1,n}, f_{1,k}\})$.

So $cnfRank(f)$
$\geq (OrAnd(\vee\{h_{1,n}, f_{1,k}\}), N(\wedge, \{h_{1,n}, f_{1,k}\}), j + N(\vee, \vee\{h_{1,n}, f_{1,k}\}))$
$> (OrAnd(\vee\{h_{1,n}, f_{1,k}\}), N(\wedge, \vee\{h_{1,n}, f_{1,k}\}), N(\vee, \vee\{h_{1,n}, f_{1,k}\}))$
$= cnfRank(\vee\{h_{1,n}, f_{1,k}\})$
$= cnfRank(ndc(f))$.

Case: $r = ncc$.

Then $f = \wedge\{\wedge H_1, \ldots, \wedge H_j, f_{1,k}\}$ where $j \geq 1$ and for each i in $[1..k]$, f_i does not start with the $\wedge$ symbol.

$cnfRank(f) = (OrAnd(f), N(\wedge, f), N(\vee, f))$
$= (OrAnd(f), N(\wedge, \wedge\{\wedge H_1, \ldots, \wedge H_j, f_{1,k}\}), N(\vee, \wedge\{\wedge H_1, \ldots, \wedge H_j, f_{1,k}\}))$
$= (OrAnd(f), j + N(\wedge, \wedge\{h_{1,n}, f_{1,k}\}), N(\vee, \wedge\{h_{1,n}, f_{1,k}\}))$ where
$\qquad \{h_{1,n}\} = H_1 \cup \ldots \cup H_j$.

Now $OrAnd(f) = OrAnd(\wedge\{\wedge H_1, \ldots, \wedge H_j, f_{1,k}\})$
$= \Sigma(OrAnd(g) : g \in \{\wedge H_1, \ldots, \wedge H_j, f_{1,k}\})$
$= \Sigma(OrAnd(g) : g \in \{h_{1,n}, f_{1,k}\})$
$= OrAnd(\wedge\{h_{1,n}, f_{1,k}\})$.

So $cnfRank(f)$
$= (OrAnd(\wedge\{h_{1,n}, f_{1,k}\}), j + N(\wedge, \wedge\{h_{1,n}, f_{1,k}\}), N(\vee, \wedge\{h_{1,n}, f_{1,k}\}))$
$> (OrAnd(\wedge\{h_{1,n}, f_{1,k}\}), N(\wedge, \wedge\{h_{1,n}, f_{1,k}\}), N(\vee, \wedge\{h_{1,n}, f_{1,k}\}))$
$= cnfRank(\wedge\{h_{1,n}, f_{1,k}\})$
$= cnfRank(ncc(f))$.

(2)

Suppose $r \in \{dc, ndc, ncc\}$ and $f \xmapsto{r} f'$. Let g^o denote the occurrence of the subformula g of f that is replaced by $r(g)$ to obtain f'. By part (1), $cnfRank(g) > cnfRank(r(g))$.

2.7 Normal Forms

We prove $\mathit{cnfRank}(f) > \mathit{cnfRank}(f')$ by induction on the structure of f. Suppose the result holds for all strict subformulas of f. We show it holds for f by considering the following cases.

Case 1: f is a literal.
By Lemma 2.7.4(1), g^o does not exist. Hence this case cannot occur.

Case 2: $f = \Diamond H$, where $\Diamond \in \{\wedge, \vee\}$ and H is a finite set of nnf-formulas.
If $g^o = f$ then $f' = r(g)$, and so $\mathit{cnfRank}(f) = \mathit{cnfRank}(g) > \mathit{cnfRank}(r(g)) = \mathit{cnfRank}(f')$. So suppose g^o is a strict subformula of f. Then for some h_0 in H, g^o is a subformula of h_0. Let h'_0 be the result of replacing g^o in h_0 by $r(g)$. By the induction hypothesis, $\mathit{cnfRank}(h_0) > \mathit{cnfRank}(h'_0)$. Let $H' = (H - \{h_0\}) \cup \{h'_0\}$. Then $f' = \Diamond H'$. There are now two subcases.

Subcase 2.1: $f = \wedge H$.
Then $f' = \wedge H'$. So
$\mathit{cnfRank}(f) = \mathit{cnfRank}(\wedge H)$
$= (\mathit{OrAnd}(\wedge H), N(\wedge, \wedge H), N(\vee, \wedge H))$
$= (\sum(\mathit{OrAnd}(h) : h \in H), 1 + N(\wedge, H), N(\vee, H))$
$= (\sum(\mathit{OrAnd}(h) : h \in H), 1 + \sum(N(\wedge, h) : h \in H), \sum(N(\vee, h) : h \in H))$
$= (\sum(\mathit{OrAnd}(h) : h \in H - \{h_0\}) + \mathit{OrAnd}(h_0),$
$\quad 1 + \sum(N(\wedge, h) : h \in H - \{h_0\}) + N(\wedge, h_0), \sum(N(\vee, h) : h \in H - \{h_0\}) + N(\vee, h_0))$
$> (\sum(\mathit{OrAnd}(h) : h \in H - \{h_0\}) + \mathit{OrAnd}(h'_0),$
$\quad 1 + \sum(N(\wedge, h) : h \in H - \{h_0\}) + N(\wedge, h'_0), \sum(N(\vee, h) : h \in H - \{h_0\}) + N(\vee, h'_0))$
$= (\sum(\mathit{OrAnd}(h) : h \in H'), 1 + \sum(N(\wedge, h) : h \in H'), \sum(N(\vee, h) : h \in H'))$
$= (\mathit{OrAnd}(\wedge H'), 1 + N(\wedge, H'), N(\vee, H'))$
$= (\mathit{OrAnd}(\wedge H'), N(\wedge, \wedge H'), N(\vee, \wedge H'))$
$= \mathit{cnfRank}(\wedge H')$
$= \mathit{cnfRank}(f')$.

Subcase 2.2: $f = \vee H$.
Then $f' = \vee H'$. So
$\mathit{cnfRank}(f) = \mathit{cnfRank}(\vee H)$
$= (\mathit{OrAnd}(\vee H), N(\wedge, \vee H), N(\vee, \vee H))$
$= (N(\wedge, H) + \sum(\mathit{OrAnd}(h) : h \in H), N(\wedge, H), 1 + N(\vee, H))$
$= (\sum(N(\wedge, h) : h \in H) + \sum(\mathit{OrAnd}(h) : h \in H),$
$\quad \sum(N(\wedge, h) : h \in H), 1 + \sum(N(\vee, h) : h \in H))$
$= (\sum(N(\wedge, h) : h \in H - \{h_0\}) + N(\wedge, h_0) + \sum(\mathit{OrAnd}(h) : h \in H - \{h_0\}) + \mathit{OrAnd}(h_0),$
$\quad \sum(N(\wedge, h) : h \in H - \{h_0\}) + N(\wedge, h_0), 1 + \sum(N(\vee, h) : h \in H - \{h_0\}) + N(\vee, h_0))$
$> (\sum(N(\wedge, h) : h \in H - \{h_0\}) + N(\wedge, h'_0) + \sum(\mathit{OrAnd}(h) : h \in H - \{h_0\}) + \mathit{OrAnd}(h'_0),$
$\quad \sum(N(\wedge, h) : h \in H - \{h_0\}) + N(\wedge, h'_0), 1 + \sum(N(\vee, h) : h \in H - \{h_0\}) + N(\vee, h'_0))$
$= (\sum(N(\wedge, h) : h \in H') + \sum(\mathit{OrAnd}(h) : h \in H'),$
$\quad \sum(N(\wedge, h) : h \in H'), 1 + \sum(N(\vee, h) : h \in H'))$
$= (N(\wedge, H') + \sum(\mathit{OrAnd}(h) : h \in H'), N(\wedge, H'), 1 + N(\vee, H'))$
$= (\mathit{OrAnd}(\vee H'), N(\wedge, \vee H'), N(\vee, \vee H'))$
$= \mathit{cnfRank}(\vee H')$
$= \mathit{cnfRank}(f')$.

Thus by structural induction, for all nnf-formulas f, $\mathit{cnfRank}(f) > \mathit{cnfRank}(f')$.

(3)

Let $(f_1, f_2, ...)$ be a cnf-conversion of the nnf-formula f_1. If f_i and f_{i+1} are in $(f_1, f_2, ...)$ then there exists r in $\{dc, ndc, ncc\}$ such that $f_i \overset{r}{\rightarrowtail} f_{i+1}$. By part (2), $cnfRank(f_i) > cnfRank(f_{i+1})$.

But for any nnf-formula f, $cnfRank(f) \geq (0, 0, 0)$. Thus every cnf-conversion of an nnf-formula is finite.

(4)

Suppose $r \in \{dd, ncd, ndd\}$ and $f \in Dr$.

Case: $r = dd$.

Then $f = \wedge\{\vee M_1, ..., \vee M_j, l_{1,k}\}$ where $j \geq 1$.
$dnfRank(f) = dnfRank(\wedge\{\vee M_1, ..., \vee M_j, l_{1,k}\})$
$= (AndOr(\wedge\{\vee M_1, ..., \vee M_j, l_{1,k}\}), N(\vee, \wedge\{\vee M_1, ..., \vee M_j, l_{1,k}\}, x)$
$= (N(\vee, \{\vee M_1, ..., \vee M_j, l_{1,k}\}) + \Sigma(AndOr(g) : g \in \{\vee M_1, ..., \vee M_j, l_{1,k}\})), j, x)$
$= (j, j, x)$.

If there exists i in $[1..j]$ such that $M_i = \{\}$ then $dnfRank(dd(f)) = dnfRank(\vee\{\})$
$= (AndOr(\wedge\{\}), N(\vee, \wedge\{\}), N(\wedge, \wedge\{\})) = (0, 0, 1) < (j, j, x) = dnfRank(f)$.

If for all i in $[1..j]$, $M_i \neq \{\}$ then
$dnfRank(dd(f)) = dnfRank(\vee\{\wedge\{m_{1,j}, l_{1,k}\} : \text{for each } i \text{ in } [1..j], m_i \in M_i\})$
$= (AndOr(\vee\{\wedge\{m_{1,j}, l_{1,k}\} : \text{for each } i \text{ in } [1..j], m_i \in M_i\}),$
$\quad N(\vee, \vee\{\wedge\{m_{1,j}, l_{1,k}\} : \text{for each } i \text{ in } [1..j], m_i \in M_i\}, y)$
$= (0, 1, y)$
$< (j, j, x)$
$= dnfRank(f)$.

Case: $r = ncd$.

Then $f = \wedge\{\wedge H_1, ..., \wedge H_j, f_{1,k}\}$ where $j \geq 1$ and for each i in $[1..k]$, f_i does not start with the $\wedge$ symbol.
$dnfRank(f) = (AndOr(f), N(\vee, \wedge\{\wedge H_1, ..., \wedge H_j, f_{1,k}\}), N(\wedge, \wedge\{\wedge H_1, ..., \wedge H_j, f_{1,k}\}))$
$= (AndOr(f), N(\vee, \wedge\{h_{1,n}, f_{1,k}\}), j + N(\wedge, \wedge\{h_{1,n}, f_{1,k}\}))$ where
$$\{h_{1,n}\} = H_1 \cup ... \cup H_j.$$

Now $AndOr(f) = AndOr(\wedge\{\wedge H_1, ..., \wedge H_j, f_{1,k}\})$
$= N(\vee, \{\wedge H_1, ..., \wedge H_j, f_{1,k}\}) + \Sigma(AndOr(g) : g \in \{\wedge H_1, ..., \wedge H_j, f_{1,k}\})$
$= N(\vee, \{h_{1,n}, f_{1,k}\}) + N(\vee, \{h_{1,n}\}) + \Sigma(AndOr(g) : g \in \{h_{1,n}, f_{1,k}\})$
$\geq N(\vee, \{h_{1,n}, f_{1,k}\}) + \Sigma(AndOr(g) : g \in \{h_{1,n}, f_{1,k}\})$
$= AndOr(\wedge\{h_{1,n}, f_{1,k}\})$.

So $dnfRank(f)$
$\geq (AndOr(\wedge\{h_{1,n}, f_{1,k}\}), N(\vee, \wedge\{h_{1,n}, f_{1,k}\}), j + N(\wedge, \wedge\{h_{1,n}, f_{1,k}\}))$
$> (AndOr(\wedge\{h_{1,n}, f_{1,k}\}), N(\vee, \wedge\{h_{1,n}, f_{1,k}\}), N(\wedge, \wedge\{h_{1,n}, f_{1,k}\}))$
$= dnfRank(\wedge\{h_{1,n}, f_{1,k}\})$
$= dnfRank(ncd(f))$.

Case: $r = ndd$.

Then $f = \vee\{\vee H_1, ..., \vee H_j, f_{1,k}\}$ where $j \geq 1$ and for each i in $[1..k]$, f_i does not start with the $\vee$ symbol.
$dnfRank(f) = (AndOr(f), N(\vee, f), N(\wedge, f))$
$= (AndOr(f), N(\vee, \vee\{\vee H_1, ..., \vee H_j, f_{1,k}\}), N(\wedge, \vee\{\vee H_1, ..., \vee H_j, f_{1,k}\}))$

2.7 Normal Forms

$= (AndOr(f), j+N(\vee, \vee\{h_{1,n}, f_{1,k}\}), N(\wedge, \vee\{h_{1,n}, f_{1,k}\}))$ where
$\{h_{1,n}\} = H_1 \cup ... \cup H_j$.
Now $AndOr(f) = AndOr(\vee\{\vee H_1, ..., \vee H_j, f_{1,k}\})$
$= \sum(AndOr(g) : g \in \{\vee H_1, ..., \vee H_j, f_{1,k}\})$
$= \sum(AndOr(g) : g \in \{h_{1,n}, f_{1,k}\})$
$= AndOr(\vee\{h_{1,n}, f_{1,k}\})$.
So $dnfRank(f)$
$= (AndOr(\vee\{h_{1,n}, f_{1,k}\}), j+N(\vee, \vee\{h_{1,n}, f_{1,k}\}), N(\wedge, \vee\{h_{1,n}, f_{1,k}\}))$
$> (AndOr(\vee\{h_{1,n}, f_{1,k}\}), N(\vee, \vee\{h_{1,n}, f_{1,k}\}), N(\wedge, \vee\{h_{1,n}, f_{1,k}\}))$
$= dnfRank(\vee\{h_{1,n}, f_{1,k}\})$
$= dnfRank(ndd(f))$.

(5)

Suppose $r \in \{dd, ncd, ndd\}$ and $f \xmapsto{r} f'$. Let g^o denote the occurrence of the subformula g of f that is replaced by $r(g)$ to obtain f'. By part (4), $dnfRank(g) > dnfRank(r(g))$.

We prove $dnfRank(f) > dnfRank(f')$ by induction on the structure of f. Suppose the result holds for all strict subformulas of f. We show it holds for f by considering the following cases.

Case 1: f is a literal.
By Lemma 2.7.4(2), g^o does not exist. Hence this case cannot occur.

Case 2: $f = \Diamond H$, where $\Diamond \in \{\wedge, \vee\}$ and H is a finite set of nnf-formulas.
If $g^o = f$ then $f' = r(g)$, and so $dnfRank(f) = dnfRank(g) > dnfRank(r(g)) = dnfRank(f')$. So suppose g^o is a strict subformula of f. Then for some h_0 in H, g^o is a subformula of h_0. Let h'_0 be the result of replacing g^o in h_0 by $r(g)$. By the induction hypothesis, $dnfRank(h_0) > dnfRank(h'_0)$. Let $H' = (H - \{h_0\}) \cup \{h'_0\}$. Then $f' = \Diamond H'$. There are now two subcases.

Subcase 2.1: $f = \wedge H$.
Then $f' = \wedge H'$. So
$dnfRank(f) = dnfRank(\wedge H)$
$= (AndOr(\wedge H), N(\vee, \wedge H), N(\wedge, \wedge H))$
$= (N(\vee, H) + \sum(AndOr(h) : h \in H), N(\vee, H), 1 + N(\wedge, H))$
$= (\sum(N(\vee, h) : h \in H) + \sum(AndOr(h) : h \in H),$
$\quad \sum(N(\vee, h) : h \in H), 1 + \sum(N(\wedge, h) : h \in H))$
$= (\sum(N(\vee, h) : h \in H - \{h_0\}) + N(\vee, h_0) + \sum(AndOr(h) : h \in H - \{h_0\}) + AndOr(h_0),$
$\quad \sum(N(\vee, h) : h \in H - \{h_0\}) + N(\vee, h_0), 1 + \sum(N(\wedge, h) : h \in H - \{h_0\}) + N(\wedge, h_0))$
$> (\sum(N(\vee, h) : h \in H - \{h_0\}) + N(\vee, h'_0) + \sum(AndOr(h) : h \in H - \{h_0\}) + AndOr(h'_0),$
$\quad \sum(N(\vee, h) : h \in H - \{h_0\}) + N(\vee, h'_0), 1 + \sum(N(\wedge, h) : h \in H - \{h_0\}) + N(\wedge, h'_0))$
$= (\sum(N(\vee, h) : h \in H') + \sum(AndOr(h) : h \in H'),$
$\quad \sum(N(\vee, h) : h \in H'), 1 + \sum(N(\wedge, h) : h \in H'))$
$= (N(\vee, H') + \sum(AndOr(h) : h \in H'), N(\vee, H'), 1 + N(\wedge, H'))$
$= (AndOr(\wedge H'), N(\vee, \wedge H'), N(\wedge, \wedge H'))$
$= dnfRank(\wedge H')$
$= dnfRank(f')$.

Subcase 2.2: $f = \vee H$.
Then $f' = \vee H'$. So
$dnfRank(f) = dnfRank(\vee H)$
$= (AndOr(\vee H), N(\vee, \vee H), N(\wedge, \vee H))$
$= (\sum(AndOr(h) : h \in H), 1+N(\vee, H), N(\wedge, H))$
$= (\sum(AndOr(h) : h \in H), 1+\sum(N(\vee, h) : h \in H), \sum(N(\wedge, h) : h \in H))$
$= (\sum(AndOr(h) : h \in H - \{h_0\}) + AndOr(h_0),$
$\quad 1+\sum(N(\vee,h) : h \in H - \{h_0\}) + N(\vee, h_0), \sum(N(\wedge,h) : h \in H - \{h_0\}) + N(\wedge, h_0))$
$> (\sum(AndOr(h) : h \in H - \{h_0\}) + AndOr(h'_0),$
$\quad 1+\sum(N(\vee,h) : h \in H - \{h_0\}) + N(\vee, h'_0), \sum(N(\wedge,h) : h \in H - \{h_0\}) + N(\wedge, h'_0))$
$= (\sum(AndOr(h) : h \in H'), 1+\sum(N(\vee,h) : h \in H'), \sum(N(\wedge,h) : h \in H'))$
$= (\sum(AndOr(h) : h \in H'), 1+N(\vee, H'), N(\wedge, H'))$
$= (AndOr(\vee H'), N(\vee, \vee H'), N(\wedge, \vee H'))$
$= dnfRank(\vee H')$
$= dnfRank(f')$.

Thus by structural induction, for all nnf-formulas f, $dnfRank(f) > dnfRank(f')$.
(6)

Let $(f_1, f_2, ...)$ be a dnf-conversion of the nnf-formula f_1. If f_i and f_{i+1} are in $(f_1, f_2, ...)$ then there exists r in $\{dd, ncd, ndd\}$ such that $f_i \overset{r}{\longmapsto} f_{i+1}$. By part (5), $cnfRank(f_i) > cnfRank(f_{i+1})$.

But for any nnf-formula f, $dnfRank(f) \geq (0,0,0)$. Thus every dnf-conversion of an nnf-formula is finite.
EndProofLemma2.7.10

2.7.4 Correctness

Now that we know that conversions are finite, we must show that they are correct. That is, we must show that the last formula in a cnf-conversion is a cnf-formula, and that the last formula in a dnf-conversion is a dnf-formula.

Before we do that, we shall need a little terminology. An nnf-formula that cannot be rewritten is called a terminal formula, as defined below.

Definition 2.7.11.
1) A formula f is a *cnf-terminal formula* iff f is an nnf-formula and each subformula of f is not in $Ddc \cup Dndc \cup Dncc$.
   CnfTerminals denotes the set of all cnf-terminal formulas.
2) A formula f is a *dnf-terminal formula* iff f is an nnf-formula and each subformula of f is not in $Ddd \cup Dncd \cup Dndd$.
   DnfTerminals denotes the set of all dnf-terminal formulas.

Let f be an nnf-formula. Then $f \in CnfTerminals$ iff for each r in $\{dc, ndc, ncc\}$, r is not applicable to f. Also, $f \in DnfTerminals$ iff for each r in $\{dd, ncd, ndd\}$, r is not applicable to f.

2.7 Normal Forms

Clearly the last formula in a cnf-conversion is a cnf-terminal formula, and the last formula in a dnf-conversion is a dnf-terminal formula. Part (2) of the next lemma shows that the last formula in a cnf-conversion is a cnf-formula; and part (3) shows that the last formula in a dnf-conversion is a dnf-formula.

Lemma 2.7.12.
1) $CnfFml = \{c : c \text{ is a clause}\} \cup \{\wedge C : C \text{ is a finite set of clauses}\}$.
   $DnfFml = \{m : m \text{ is a meet}\} \cup \{\vee M : M \text{ is a finite set of meets}\}$.
2) $CnfTerminals = CnfFml$.
3) $DnfTerminals = DnfFml$.

Proof
(1) By Definition 2.7.1, a cnf-formula is either a clause or the conjunction, $\wedge C$, of a finite set, C, of clauses. By Definition 2.7.1, a dnf-formula is either a meet or the disjunction, $\vee M$, of a finite set, M, of meets.

(2) First we show $CnfFml \subseteq CnfTerminals$. Take any f in $CnfFml$. Using part (1) it is easy to see that each subformula of f is in $CnfFml$. So by Lemma 2.7.4(1), each subformula of f is not in $Ddc \cup Dndc \cup Dncc$. Hence $f \in CnfTerminals$. Therefore $CnfFml \subseteq CnfTerminals$.

Conversely, let $S(f)$ be the following statement. "If $f \in CnfTerminals$ then $f \in CnfFml$." We prove $S(f)$ by induction on the structure of f.

The base case is when f is a literal, say l. Then l is a cnf-formula. So $S(l)$ holds.

Take any formula f and suppose that $S(g)$ holds for all strict subformulas g of f. We show $S(f)$ holds. Suppose $f \in CnfTerminals$. Then f is an nnf-formula and each subformula of f is not in $Ddc \cup Dndc \cup Dncc$. We must show $f \in CnfFml$. There are two cases, $f = \wedge G$ and $f = \vee G$ where G is a finite set of nnf-formulas. Also if $g \in G$ then each subformula of g is not in $Ddc \cup Dndc \cup Dncc$, and so $g \in CnfTerminals$. Moreover if $g \in G$, then g is a strict subformula of f, hence $S(g)$ holds, and so $g \in CnfFml$. Therefore $G \subseteq CnfFml$.

Case 1: $f = \wedge G$.
If $g \in G$ then either g is a clause or $g = \wedge C$ where C is a finite set of clauses. If $\wedge C \in G$ then $f \in Dncc$, which is a contradiction. So G is a finite set of clauses. Hence $f = \wedge G \in CnfFml$.

Case 2: $f = \vee G$.
If $g \in G$ then either g is a clause or $g = \wedge C$ where C is a finite set of clauses.
If $g \in G$ and $g = \vee L$ where L is a finite set of literals then $f \in Dndc$, which is a contradiction.

So if $g \in G$ then either g is a literal or $g = \wedge C$ where C is a finite set of clauses. If $\wedge C \in G$ then $f \in Ddc$, which is a contradiction. So G is a finite set of literals. Hence $f = \vee G \in CnfFml$.

Therefore by induction $CnfTerminals \subseteq CnfFml$.
Thus $CnfTerminals = CnfFml$.

(3) First we show $DnfFml \subseteq DnfTerminals$. Take any f in $DnfFml$. Using part (1) it is easy to see that each subformula of f is in $DnfFml$. So by Lemma 2.7.4(2), each subformula of f is not in $Ddd \cup Dncd \cup Dndd$. Hence $f \in DnfTerminals$. Therefore $DnfFml \subseteq DnfTerminals$.

Conversely, let $S(f)$ be the following statement. "If $f \in DnfTerminals$ then $f \in DnfFml$." We prove $S(f)$ by induction on the structure of f.

The base case is when f is a literal, say l. Then l is a dnf-formula. So $S(l)$ holds.

Take any formula f and suppose that $S(g)$ holds for all strict subformulas g of f. We show $S(f)$ holds. Suppose $f \in DnfTerminals$. Then f is an nnf-formula and each subformula of f is not in $Ddd \cup Dncd \cup Dndd$. We must show $f \in DnfFml$. There are two cases, $f = \wedge G$ and $f = \vee G$ where G is a finite set of nnf-formulas. Also if $g \in G$ then each subformula of g is not in $Ddd \cup Dncd \cup Dndd$, and so $g \in DnfTerminals$. Moreover if $g \in G$, then g is a strict subformula of f, hence $S(g)$ holds, and so $g \in DnfFml$. Therefore $G \subseteq DnfFml$.

Case 1: $f = \vee G$.

If $g \in G$ then either g is a meet or $g = \vee M$ where M is a finite set of meets. If $\vee M \in G$ then $f \in Dndd$, which is a contradiction. So G is a finite set of meets. Hence $f = \vee G \in DnfFml$.

Case 2: $f = \wedge G$.

If $g \in G$ then either g is a meet or $g = \vee M$ where M is a finite set of meets.

If $g \in G$ and $g = \wedge L$ where L is a finite set of literals then $f \in Dncd$, which is a contradiction.

So if $g \in G$ then either g is a literal or $g = \vee M$ where M is a finite subset of meets. If $\vee M \in G$ then $f \in Ddd$, which is a contradiction. So G is a finite set of literals. Hence $f = \wedge G \in DnfFml$.

Therefore by induction $DnfTerminals \subseteq DnfFml$.

Thus $DnfTerminals = DnfFml$.

EndProofLemma2.7.12

2.8 Stable Normal Forms

Unfortunately a formula may have many different, but equivalent, conjunctive normal forms; and also many different, but equivalent, disjunctive normal forms. There are unique conjunctive normal forms and unique disjunctive normal forms but they contain more symbols than necessary. So we shall introduce two new normal forms, called stable conjunctive normal form and stable disjunctive normal form, that are relatively lean and have some desirable properties. Most importantly we shall show that a formula has exactly one stable conjunctive normal form and exactly one stable disjunctive normal form.

Recall from Definition 2.6.6 that if G is either a set of clauses or a set of meets then G is defined to be stable iff $CorRes(G) = G$. Also by Lemma 2.6.8(5), $CorRes(G)$ is stable.

Stability is important because it has the following uniqueness property. If two stable sets of clauses are equivalent then they are equal. This, and a similar uniqueness property for meets, is the subject of the following theorem. Moreover Theorem 2.8.1(3) shows that $CorRes(C)$ is a unique normal form of a set C of clauses.

2.8 Stable Normal Forms

Theorem 2.8.1. Let C and C' be two sets of clauses. Let M and M' be two sets of meets.
1) $\wedge C \equiv \wedge C'$ iff $C \equiv C'$ iff $C_\models(C) = C_\models(C')$.
2) If C and C' are both stable then $C \equiv C'$ iff $C = C'$. (uniqueness)
3) $C \equiv C'$ iff $CorRes(C) = CorRes(C')$.
4) $\vee M \equiv \vee M'$ iff $Imp(\vee M) = Imp(\vee M')$.
5) If M and M' are both stable then $\vee M \equiv \vee M'$ iff $M = M'$. (uniqueness)
6) $\vee M \equiv \vee M'$ iff $CorRes(M) = CorRes(M')$.

Proof

Let C and C' be two sets of clauses. Let M and M' be two sets of meets.

(1) By Lemma 2.2.18(7), $C \equiv C'$ iff $\wedge C \equiv \wedge C'$.

Suppose $C \equiv C'$. Then $C_\models(C) = \{f \in Fml : C \models f\} = \{f \in Fml : C' \models f\} = C_\models(C')$.

Conversely suppose $C_\models(C) = C_\models(C')$. Since $C \models C_\models(C)$, we have $C \models C_\models(C')$. But $C' \subseteq C_\models(C')$, so $C \models C'$. Similarly, since $C' \models C_\models(C')$, $C' \models C_\models(C)$. But $C \subseteq C_\models(C)$, so $C' \models C$. Thus $C \equiv C'$.

(2) If $C = C'$ then $C \equiv C'$.

Conversely suppose C and C' are both stable and $C \equiv C'$. Then $C = CorRes(C)$ and $C' = CorRes(C')$. By Theorem 2.8.1(1), $C_\models(C) = C_\models(C')$. Hence $ClsC_\models(C) = ClsC_\models(C')$ and so $CorClsC_\models(C) = CorClsC_\models(C')$. Therefore by using stability and Theorem 2.6.10(4)(core-completeness) we have, $C = CorRes(C) = CorClsC_\models(C) = CorClsC_\models(C') = CorRes(C') = C'$.

(3) Suppose $CorRes(C) = CorRes(C')$. Then by Lemma 2.6.8(1), $C \equiv CorRes(C) = CorRes(C') \equiv C'$.

Conversely suppose $C \equiv C'$. Then by Lemma 2.6.8(1), $CorRes(C) \equiv C \equiv C' \equiv CorRes(C')$. By Lemma 2.6.8(5), both $CorRes(C)$ and $CorRes(C')$ are stable. So by Theorem 2.8.1(2), $CorRes(C) = CorRes(C')$.

(4) Suppose that $\vee M \equiv \vee M'$. Then we have $Imp(\vee M) = \{f \in Fml : f \models \vee M\} = \{f \in Fml : f \models \vee M'\} = Imp(\vee M')$.

Conversely suppose $Imp(\vee M) = Imp(\vee M')$. Since $M \subseteq Imp(\vee M)$, $M \subseteq Imp(\vee M')$. Therefore $\vee M \models \vee M'$. Similarly, $M' \subseteq Imp(\vee M')$ and so $M' \subseteq Imp(\vee M)$. Therefore $\vee M' \models \vee M$. Thus $\vee M \equiv \vee M'$.

(5) If $M = M'$ then $\vee M \equiv \vee M'$.

Conversely suppose M and M' are both stable and that $\vee M \equiv \vee M'$. Then $M = CorRes(M)$ and $M' = CorRes(M')$. By Theorem 2.8.1(4), $Imp(\vee M) = Imp(\vee M')$. Hence $MtsImp(\vee M) = MtsImp(\vee M')$ and so $CorMtsImp(\vee M) = CorMtsImp(\vee M')$. Hence by stability and Theorem 2.6.10(5)(core-completeness), $M = CorRes(M) = CorMtsImp(\vee M) = CorMtsImp(\vee M') = CorRes(M') = M'$.

(6) Suppose that $CorRes(M) = CorRes(M')$. Then by Lemma 2.6.8(1), $\vee M \equiv \vee CorRes(M) = \vee CorRes(M') \equiv \vee M'$.

Conversely suppose $\vee M \equiv \vee M'$. By Lemma 2.6.8(1), $\vee CorRes(M) \equiv \vee M \equiv \vee M' \equiv \vee CorRes(M')$. By Lemma 2.6.8(5), both $CorRes(M)$ and $CorRes(M')$ are stable. So by Theorem 2.8.1(5), $CorRes(M) = CorRes(M')$.

EndProofTheorem2.8.1

We are now able to define the promised new normal forms. But first we note that the definitions are not ambiguous. By Theorem 2.8.1(3), Definition 2.8.2(1) below does not depend on the choice of C. Also by Theorem 2.8.1(6), Definition 2.8.2(5) below does not depend on the choice of M.

Definition 2.8.2. Let f be any formula and F a set of formulas.
Suppose C is any finite set of clauses such that $f \equiv \wedge C$.
Suppose M is any finite set of meets such that $f \equiv \vee M$.
1) $Cl(f) = CorRes(C)$.
2) $Cl(F) = \bigcup \{Cl(f) : f \in F\}$.
3) $scnf(f) = ns(\wedge Cl(f))$.
4) f is in *stable conjunctive normal form* iff f is an scnf-formula iff $scnf(f) = f$.
5) $Mt(f) = CorRes(M)$.
6) $Mt(F) = \bigcup \{Mt(f) : f \in F\}$.
7) $sdnf(f) = ns(\vee Mt(f))$.
8) f is in *stable disjunctive normal form* iff f is an sdnf-formula iff $sdnf(f) = f$.

Some examples of the functions in Definition 2.8.2 applied to simple formulas are in the following lemma.

Lemma 2.8.3. Let $\mathfrak{t}$ be any tautology, and $\mathfrak{c}$ be any contradiction, and l be a literal. Suppose L is a finite contingent set of literals such that $|L| \geq 2$.
1) $Cl(\mathfrak{t}) = \{\}$. $scnf(\mathfrak{t}) = \wedge\{\}$.
   $scnf(\wedge\{\}) = \wedge\{\}$. $\wedge\{\}$ is an scnf-formula.
2) $Mt(\mathfrak{t}) = \{\wedge\{\}\}$. $sdnf(\mathfrak{t}) = \wedge\{\}$.
   $sdnf(\wedge\{\}) = \wedge\{\}$. $\wedge\{\}$ is an sdnf-formula.
3) $Cl(\mathfrak{c}) = \{\vee\{\}\}$. $scnf(\mathfrak{c}) = \vee\{\}$.
   $scnf(\vee\{\}) = \vee\{\}$. $\vee\{\}$ is an scnf-formula.
4) $Mt(\mathfrak{c}) = \{\}$. $sdnf(\mathfrak{c}) = \vee\{\}$.
   $sdnf(\vee\{\}) = \vee\{\}$. $\vee\{\}$ is an sdnf-formula.
5) $Cl(l) = \{l\}$. $scnf(l) = l$. l is an scnf-formula.
   $Mt(l) = \{l\}$. $sdnf(l) = l$. l is an sdnf-formula.
6) $Cl(\vee\{l\}) = \{l\}$. $scnf(\vee\{l\}) = l$. $\vee\{l\}$ is not an scnf-formula.
   $Mt(\vee\{l\}) = \{l\}$. $sdnf(\vee\{l\}) = l$. $\vee\{l\}$ is not an sdnf-formula.
7) $Cl(\wedge\{l\}) = \{l\}$. $scnf(\wedge\{l\}) = l$. $\wedge\{l\}$ is not an scnf-formula.
   $Mt(\wedge\{l\}) = \{l\}$. $sdnf(\wedge\{l\}) = l$. $\wedge\{l\}$ is not an sdnf-formula.
8) $Cl(\vee L) = \{\vee L\}$. $scnf(\vee L) = \vee L$. $\vee L$ is an scnf-formula.
   $Mt(\vee L) = L$. $sdnf(\vee L) = \vee L$. $\vee L$ is an sdnf-formula.
9) $Cl(\wedge L) = L$. $scnf(\wedge L) = \wedge L$. $\wedge L$ is an scnf-formula.
   $Mt(\wedge L) = \{\wedge L\}$. $sdnf(\wedge L) = \wedge L$. $\wedge L$ is an sdnf-formula.

Proof

Let $\mathfrak{t}$ be any tautology, and $\mathfrak{c}$ be any contradiction, and l be a literal. Suppose L is a finite contingent set of literals such that $|L| \geq 2$.

2.8 Stable Normal Forms

(1) Since $\mathfrak{t} \equiv \wedge\{\}$, by Definition 2.8.2 and Lemma 2.6.7(1), $Cl(\mathfrak{t}) = CorRes(\{\}) = \{\}$. Hence $scnf(\mathfrak{t}) = ns(\wedge Cl(\mathfrak{t})) = \wedge\{\}$. Since $\wedge\{\}$ is a tautology, $scnf(\wedge\{\}) = \wedge\{\}$. Hence $\wedge\{\}$ is an scnf-formula.

(2) Since $\mathfrak{t} \equiv \vee\{\wedge\{\}\}$, by Definition 2.8.2 and Lemma 2.6.7(3) we get, $Mt(\mathfrak{t}) = CorRes(\{\wedge\{\}\}) = \{\wedge\{\}\}$. Hence $sdnf(\mathfrak{t}) = ns(\vee Mt(\mathfrak{t})) = ns(\vee\{\wedge\{\}\}) = \wedge\{\}$. Since $\wedge\{\}$ is a tautology, $sdnf(\wedge\{\}) = \wedge\{\}$. Hence $\wedge\{\}$ is an sdnf-formula.

(3) Since $\mathfrak{c} \equiv \wedge\{\vee\{\}\}$, by Definition 2.8.2 and Lemma 2.6.7(2) we have, $Cl(\mathfrak{c}) = CorRes(\{\vee\{\}\}) = \{\vee\{\}\}$. Hence $scnf(\mathfrak{c}) = ns(\wedge Cl(\mathfrak{c})) = ns(\wedge\{\vee\{\}\}) = \vee\{\}$. Since $\vee\{\}$ is a contradiction, $scnf(\vee\{\}) = \vee\{\}$. Hence $\vee\{\}$ is an scnf-formula.

(4) Since $\mathfrak{c} \equiv \vee\{\}$, by Definition 2.8.2 and Lemma 2.6.7(1), $Mt(\mathfrak{c}) = CorRes(\{\}) = \{\}$. Hence $sdnf(\mathfrak{c}) = ns(\vee Mt(\mathfrak{c})) = ns(\vee\{\}) = \vee\{\}$. Since $\vee\{\}$ is a contradiction, $sdnf(\vee\{\}) = \vee\{\}$. Hence $\vee\{\}$ is an sdnf-formula.

(5) Since $l \equiv \wedge\{l\}$, by Definition 2.8.2 and Lemma 2.6.7(4), $Cl(l) = CorRes(\{l\}) = \{l\}$. Hence $scnf(l) = ns(\wedge Cl(l)) = ns(\wedge\{l\}) = l$, and so l is an scnf-formula.

Since $l \equiv \vee\{l\}$, by Definition 2.8.2 and Lemma 2.6.7(4), $Mt(l) = CorRes(\{l\}) = \{l\}$. Hence $sdnf(l) = ns(\vee Mt(l)) = ns(\vee\{l\}) = l$, and so l is an sdnf-formula.

(6) Since $\vee\{l\} \equiv \wedge\{\vee\{l\}\}$, by Definition 2.8.2, $Cl(\vee\{l\}) = CorRes(\{\vee\{l\}\}) = Cor(\{\vee\{l\}\}) = nsMinCtge(\{\vee\{l\}\}) = nsMin(\{\vee\{l\}\}) = ns(\{\vee\{l\}\}) = \{ns(\vee\{l\})\} = \{l\}$. Hence $scnf(\vee\{l\}) = ns(\wedge Cl(\vee\{l\})) = ns(\wedge\{l\}) = l$, and so $\vee\{l\}$ is not an scnf-formula.

By Definition 2.8.2 and Lemma 2.6.7(4), $Mt(\vee\{l\}) = CorRes(\{l\}) = \{l\}$. Hence $sdnf(\vee\{l\}) = ns(\vee Mt(\vee\{l\})) = ns(\vee\{l\}) = l$, and so $\vee\{l\}$ is not an sdnf-formula.

(7) By Definition 2.8.2 and Lemma 2.6.7(4), $Cl(\wedge\{l\}) = CorRes(\{l\}) = \{l\}$. Hence $scnf(\wedge\{l\}) = ns(\wedge Cl(\wedge\{l\})) = ns(\wedge\{l\}) = l$, and so $\wedge\{l\}$ is not an scnf-formula.

Since $\wedge\{l\} \equiv \vee\{\wedge\{l\}\}$, by Definition 2.8.2, $Mt(\wedge\{l\}) = CorRes(\{\wedge\{l\}\}) = Cor(\{\wedge\{l\}\}) = nsMinCtge(\{\wedge\{l\}\}) = nsMin(\{\wedge\{l\}\}) = ns(\{\wedge\{l\}\}) = \{ns(\wedge\{l\})\} = \{l\}$. Hence $sdnf(\wedge\{l\}) = ns(\vee Mt(\wedge\{l\})) = ns(\vee\{l\}) = l$, and so $\wedge\{l\}$ is not an sdnf-formula.

(8) Since $\vee L \equiv \wedge\{\vee L\}$, by Definition 2.8.2 and Lemma 2.2.16(5), $Cl(\vee L) = CorRes(\{\vee L\}) = Cor(\{\vee L\}) = nsMinCtge(\{\vee L\}) = nsMin(\{\vee L\}) = ns(\{\vee L\}) = \{\vee L\}$. Hence $scnf(\vee L) = ns(\wedge Cl(\vee L)) = ns(\wedge\{\vee L\}) = \vee L$, and so $\vee L$ is an scnf-formula.

By Definition 2.8.2, $Mt(\vee L) = CorRes(L) = Cor(L) = nsMinCtge(L) = nsMin(L) = ns(L) = L$. Hence $sdnf(\vee L) = ns(\vee Mt(\vee L)) = ns(\vee L) = \vee L$, and so $\vee L$ is an sdnf-formula.

(9) By Definition 2.8.2, $Cl(\wedge L) = CorRes(L) = Cor(L) = nsMinCtge(L) = nsMin(L) = ns(L) = L$. Hence $scnf(\wedge L) = ns(\wedge Cl(\wedge L)) = ns(\wedge L) = \wedge L$, and so $\wedge L$ is an scnf-formula.

Since $\wedge L \equiv \vee\{\wedge L\}$, by Definition 2.8.2 and Lemma 2.2.16(5) we get, $Mt(\wedge L) = CorRes(\{\wedge L\}) = Cor(\{\wedge L\}) = nsMinCtge(\{\wedge L\}) = nsMin(\{\wedge L\}) = ns(\{\wedge L\}) = \{\wedge L\}$. Hence $sdnf(\wedge L) = ns(\vee Mt(\wedge L)) = ns(\vee\{\wedge L\}) = \wedge L$, and so $\wedge L$ is an sdnf-formula.

EndProofLemma2.8.3

Before we prove several results about the *Cl* function defined in Definition 2.8.2, we shall introduce the following convenient notation.

Definition 2.8.4. Let C be any set of clauses. Let X be any formula or set of formulas.
$ClCor(C) = Cl(Cor(C))$.
$CorCl(X) = Cor(Cl(X))$.

Lemma 2.8.5. Let f be any formula and C be any set of clauses. Suppose F is any set of formulas and G is any set of formulas.
1) $Cl(f)$ and $Mt(f)$ are both stable.
2) $Cl(\{f\}) = Cl(f)$ and $Mt(\{f\}) = Mt(f)$.
3) $Cl(f) \equiv f \equiv scnf(f)$ and $\vee Mt(f) \equiv f \equiv sdnf(f)$.
4) $F \equiv Cl(F) \equiv CorCl(F)$.
5) $F \equiv G$ iff $Cl(F) \equiv Cl(G)$.
6) If $F \subseteq G$ then $Cl(F) \subseteq Cl(G)$. (*Cl* is monotonic.)
7) $Cl(F \cup G) = Cl(F) \cup Cl(G)$. (*Cl* distributes over $\cup$.)
8) $Cl(f) = Cl(f) - Taut$. That is, $Cl(f)$ is tautology-free.
9) $Cl(F) = Cl(F) - Taut$. That is, $Cl(F)$ is tautology-free.
10) $Cl(F - Taut) = Cl(F)$.
11) If a clause c is not a tautology then $Cl(c) = \{ns(c)\}$.
12) $Cl(C) = ns(C - Taut) = ns(C) - Taut$.
13) $ClCor(C) = Cor(C) = CorCl(C)$.
14) $Cor(C) \subseteq Cl(C)$.
15) $CorCl(f) = Cl(f)$.
16) If $c \in Cl(f)$ then $ns(c) = c$.
17) $ns(Cl(f)) = Cl(f)$ and $ns(Cl(F)) = Cl(F)$.
18) $Cl(Fml) = ns(Cls) - Taut$.
19) If $C \subseteq Cl(F)$ then $Cl(C) = C$.
20) $Cl(Cl(F)) = Cl(F)$ and $Cl(Cl(f)) = Cl(f)$. (*Cl* is idempotent.)
21) $CorCl(CorCl(F)) = CorCl(F)$. (*CorCl* is idempotent.)
22) $CorCl(CorCl(f)) = CorCl(f) = Cl(f)$. (*CorCl* is idempotent.)

Proof

Let f be any formula and C be any set of clauses. Suppose F is any set of formulas and G is any set of formulas.

(1) This follows from Lemma 2.6.8(5).

(2) By Definition 2.8.2(2), $Cl(\{f\}) = \bigcup\{Cl(f)\} = Cl(f)$. By Definition 2.8.2(6), $Mt(\{f\}) = \bigcup\{Mt(f)\} = Mt(f)$.

(3) Suppose C is any finite set of clauses such that $f \equiv \wedge C$. By Lemma 2.6.8(1), $f \equiv \wedge C \equiv \wedge CorRes(C) = Cl(f)$. By Lemma 2.5.6(14), $scnf(f) = ns(\wedge Cl(f)) \equiv \wedge Cl(f) \equiv Cl(f) \equiv f$.

Suppose M is any finite set of meets such that $f \equiv \vee M$. By Lemma 2.6.8(1), $f \equiv \vee M \equiv \vee CorRes(M) = \vee Mt(f)$. By Lemma 2.5.6(14), $sdnf(f) = ns(\vee Mt(f)) \equiv \vee Mt(f) \equiv f$.

2.8 Stable Normal Forms

(4) Recall $Cl(F) = \bigcup\{Cl(f) : f \in F\}$. Let v be any valuation. By Lemma 2.8.5(3), $v \models Cl(F)$ iff for all f in F, $v \models Cl(f)$ iff for all f in F, $v \models f$ iff $v \models F$. So $F \equiv Cl(F)$.

By Lemma 2.6.5(3), $Cor(Cl(F)) \equiv Cl(F)$.

(5) By part 4, and the transitivity of $\equiv$, $F \equiv G$ iff $Cl(F) \equiv G$ iff $Cl(F) \equiv Cl(G)$.

(6) Suppose $F \subseteq G$. Then $Cl(F) = \bigcup\{Cl(f) : f \in F\} \subseteq \bigcup\{Cl(g) : g \in G\} = Cl(G)$.

(7) $Cl(F \cup G) = \bigcup\{Cl(h) : h \in F \cup G\} = \bigcup\{Cl(f) : f \in F\} \cup \bigcup\{Cl(g) : g \in G\} = Cl(F) \cup Cl(G)$.

(8) If C is any set of clauses then by Lemma 2.6.5(6), there are no tautologies in $Cor(C)$. For each f in F let C_f be a finite set of clauses such that $f \equiv \wedge C_f$. Then $Cl(f) = CorRes(C_f)$ and so there are no tautologies in $Cl(f)$.

(9) $Cl(F) = \bigcup\{Cl(f) : f \in F\}$. So by part (8), there are no tautologies in $Cl(F)$.

(10) By Lemma 2.8.3(1), $Cl(\mathfrak{t}) = \{\}$ and so $\bigcup\{Cl(f) : f \in F \cap Taut\} = \{\}$. Hence $Cl(F) = \bigcup\{Cl(f) : f \in F\} = \bigcup\{Cl(f) : f \in F - Taut\} \cup \bigcup\{Cl(f) : f \in F \cap Taut\} = \bigcup\{Cl(f) : f \in F - Taut\} = Cl(F - Taut)$.

(11) Suppose c is a clause that is not a tautology. Then either c is a literal or $c = \vee L$ where L is a finite set of literals.

Case 1: $c = l$ where l is a literal.
By Lemma 2.8.3(5) and Lemma 2.5.7(1), $Cl(c) = Cl(l) = \{l\} = \{ns(l)\} = \{ns(c)\}$.

Case 2: $c = \vee L$ where L is a finite set of literals.
Then $c \equiv \wedge\{\vee L\}$. Therefore $Cl(c) = Cl(\vee L) = CorRes(\{\vee L\}) = Cor(\{\vee L\}) = nsMinCtge(\{\vee L\}) = nsMin(\{\vee L\}) = ns(\{\vee L\}) = \{ns(\vee L)\} = \{ns(c)\}$.

So in both cases $Cl(c) = \{ns(c)\}$.

(12) $Cl(C) = Cl(C - Taut)$ by part (10),
$= \bigcup\{Cl(c) : c \in C - Taut\} = \bigcup\{\{ns(c)\} : c \in C - Taut\}$ by part (11),
$= \{ns(c) : c \in C - Taut\} = ns(C - Taut)$.
$= \{ns(c) : c \in C\} - Taut = ns(C) - Taut$.

(13) $ClCor(C) = ns(Cor(C)) - Taut$ by part (12),
$= Cor(C) - Taut$ by Lemma 2.5.6(8),
$= Cor(C)$ by Lemma 2.6.5(6).

$CorCl(C) = Cor(ns(C) - Taut)$ by part (12),
$= nsMinCtge(ns(C) - Taut)$
$= nsMinCtge(Ctge(ns(C)))$ by Lemma 2.6.2(2),
$= nsMin(Ctge(ns(C)))$ by Lemma 2.6.2(3),
$= nsMin(ns(C) - Taut)$ by Lemma 2.6.2(2),
$= nsMin(ns(C - Taut))$ by part (12.1),
$= ns(nsMin(C - Taut))$ by Lemma 2.6.3(8),
$= nsMin(C - Taut)$ by Lemma 2.5.6(8),
$= nsMinCtge(C)$ by Lemma 2.6.2(2),
$= Cor(C)$.

(14) $Cor(C) = nsMinCtge(C)$
$= nsMin(C - Taut)$ by Lemma 2.6.2(2),
$= Min(ns(C - Taut))$ by Lemma 2.6.3(8),

$\subseteq ns(C-Taut)$
$= Cl(C)$ by part (12).

(15) Let C_f be a finite set of clauses such that $f \equiv \wedge C_f$. Then
$CorCl(f) = Cor(CorRes(C_f))$
$= CorRes(C_f)$ by Lemma 2.6.5(8),
$= Cl(f)$.

(16) Suppose C is a finite set of clauses such that $f \equiv \wedge C$. Then $Cl(f) = CorRes(C) = nsMinCtgeRes(C)$.

Take any c in $Cl(f)$. Then there is a clause c' in $MinCtgeRes(C)$ such that $c = ns(c')$. By Lemma 2.5.6(8), $ns(c) = ns(ns(c')) = ns(c') = c$.

(17) Suppose C is a finite set of clauses such that $f \equiv \wedge C$. Then $Cl(f) = CorRes(C) = nsMinCtgeRes(C)$.

By Lemma 2.5.6(8), $ns(Cl(f)) = ns(nsMinCtgeRes(C)) = nsMinCtgeRes(C) = Cl(f)$.

By part (16), $ns(Cl(F)) = ns(\bigcup\{Cl(f) : f \in F\}) = ns(\{c \in Cl(f) : f \in F\}) = \{ns(c) : c \in Cl(f) \text{ and } f \in F\} = \{c \in Cl(f) : f \in F\} = \bigcup\{Cl(f) : f \in F\} = Cl(F)$.

(18) By Lemma 2.8.5(9), there are no tautologies in $Cl(Fml)$. By Lemma 2.8.5(17), $Cl(Fml) = ns(Cl(Fml)) = ns(\bigcup\{Cl(f) : f \in Fml\}) \subseteq ns(Cls)$. Hence $Cl(Fml) \subseteq ns(Cls) - Taut$.

Take any clause c in $ns(Cls) - Taut$. Then c is not a tautology. Also there is a clause c' in Cls such that $c = ns(c')$. By Lemma 2.5.6(8), $ns(c) = ns(ns(c')) = ns(c') = c$. By parts (2 and 11), $Cl(\{c\}) = Cl(c) = \{ns(c)\} = \{c\}$. But $\{c\} \subseteq Fml$, so by part (6), $\{c\} = Cl(\{c\}) \subseteq Cl(Fml)$. Hence $c \in Cl(Fml)$. Therefore $ns(Cls) - Taut \subseteq Cl(Fml)$.

Thus $Cl(Fml) = ns(Cls) - Taut$.

(19) Suppose $C \subseteq Cl(F)$. Then $C \subseteq Cl(F) = \bigcup\{Cl(f) : f \in F\}$. Suppose $c \in C$. Then there exists f in F such that $c \in Cl(f)$. So c is a clause and by Lemma 2.8.5(9), c is not a tautology. By Lemma 2.8.5(11), $Cl(c) = \{ns(c)\}$. By Lemma 2.8.5(16), $ns(c) = c$. So $Cl(c) = \{c\}$. Hence $Cl(C) = \bigcup\{Cl(c) : c \in C\} = \bigcup\{\{c\} : c \in C\} = C$.

(20) If $C = Cl(F)$ then by part (19), $Cl(Cl(F)) = Cl(F)$.

If $F = \{f\}$ then by Lemma 2.8.5(2), $Cl(F) = Cl(\{f\}) = Cl(f)$. So $Cl(Cl(f)) = Cl(f)$.

(21) $CorCl(CorCl(F)) = Cor(Cl(Cor(Cl(F))))$
$= Cor(Cor(Cl(F)))$ by part (13),
$= CorCl(F)$ by Lemma 2.6.5(8).

(22) If $F = \{f\}$ then by Lemma 2.8.5(2), $Cl(F) = Cl(\{f\}) = Cl(f)$. Therefore $CorCl(CorCl(f)) = CorCl(f)$ by part (21),
$= Cl(f)$ by part (15).

EndProofLemma2.8.5

Among other things the following theorem shows that stable conjunctive normal form is unique and that stable disjunctive normal form is unique. In particular, Theorem 2.8.6(2), shows that $scnf(f)$ is an scnf-formula and $sdnf(f)$ is an sdnf-formula. Hence $scnf(f)$ is in stable conjunctive normal form and $sdnf(f)$ is in stable disjunctive normal form. Therefore Theorem 2.8.6(1) shows that $scnf(f)$ is the unique *stable*

2.8 Stable Normal Forms

conjunctive normal form of f and that $sdnf(f)$ is the unique *stable disjunctive normal form* of f.

Theorem 2.8.6. Let f and g be two formulas.
1) The following seven statements are all equivalent. (uniqueness)
    1.1) $f \equiv g$
    1.2) $Cl(f) = Cl(g)$
    1.3) $Cl(f) \equiv Cl(g)$
    1.4) $scnf(f) = scnf(g)$
    1.5) $Mt(f) = Mt(g)$
    1.6) $sdnf(f) = sdnf(g)$
    1.7) $scnf(f) \equiv scnf(g)$
    1.8) $sdnf(f) \equiv sdnf(g)$
2) $scnf(scnf(f)) = scnf(f)$. ($scnf$ is idempotent.)
   $sdnf(sdnf(f)) = sdnf(f)$. ($sdnf$ is idempotent.)
3) $Cl(\wedge Cl(f)) = Cl(f)$ and $Mt(\vee Mt(f)) = Mt(f)$.

Proof

Let f and g be two formulas.

(1)

(1.1) implies (1.2). Suppose $f \equiv g$. By Lemma 2.8.5(3), $Cl(f) \equiv f \equiv g \equiv Cl(g)$. By Lemma 2.8.5(1), both $Cl(f)$ and $Cl(g)$ are stable. By Theorem 2.8.1(2), $Cl(f) = Cl(g)$.

(1.2) implies (1.1). Suppose $Cl(f) = Cl(g)$. Then $Cl(f) \equiv Cl(g)$. Therefore by Lemma 2.8.5(3) we get, $f \equiv Cl(f) \equiv Cl(g) \equiv g$.

(1.1) is equivalent to (1.3). By Lemma 2.8.5(3), $f \equiv Cl(f)$ and $g \equiv Cl(g)$. So by the transitivity of $\equiv$, $f \equiv g$ iff $Cl(f) \equiv g$ iff $Cl(f) \equiv Cl(g)$.

(1.1) implies (1.4). Suppose $f \equiv g$. By (1.1) implies (1.2) above, we have $Cl(f) = Cl(g)$. Hence $scnf(f) = ns(\wedge Cl(f)) = ns(\wedge Cl(g)) = scnf(g)$.

(1.4) implies (1.1). Suppose $scnf(f) = scnf(g)$. By Lemma 2.8.5(3), $f \equiv scnf(f) = scnf(g) \equiv g$.

(1.1) implies (1.5). Suppose $f \equiv g$. By Lemma 2.8.5(3), $\vee Mt(f) \equiv f \equiv g \equiv \vee Mt(g)$. By Lemma 2.8.5(1), both $Mt(f)$ and $Mt(g)$ are stable. By Theorem 2.8.1(5), $Mt(f) = Mt(g)$.

(1.5) implies (1.1). Suppose $Mt(f) = Mt(g)$. Then $\vee Mt(f) = \vee Mt(g)$ and so $\vee Mt(f) \equiv \vee Mt(g)$. By Lemma 2.8.5(3), $f \equiv \vee Mt(f) \equiv \vee Mt(g) \equiv g$.

(1.1) implies (1.6). Suppose $f \equiv g$. By (1.1) implies (1.5) above, we have $Mt(f) = Mt(g)$. Hence $sdnf(f) = ns(\vee Mt(f)) = ns(\vee Mt(g)) = sdnf(g)$.

(1.6) implies (1.1). Suppose $sdnf(f) = sdnf(g)$. By Lemma 2.8.5(3), $f \equiv sdnf(f) = sdnf(g) \equiv g$.

(1.1) implies (1.7). Suppose $f \equiv g$. By Lemma 2.8.5(3), $scnf(f) \equiv f \equiv g \equiv scnf(g)$.

(1.7) implies (1.1). Suppose $scnf(f) \equiv scnf(g)$. By Lemma 2.8.5(3), $f \equiv scnf(f) \equiv scnf(g) \equiv g$.

(1.1) implies (1.8). Suppose $f \equiv g$. By Lemma 2.8.5(3), $sdnf(f) \equiv f \equiv g \equiv sdnf(g)$.

(1.8) implies (1.1). Suppose $sdnf(f) \equiv sdnf(g)$. By Lemma 2.8.5(3), $f \equiv sdnf(f) \equiv sdnf(g) \equiv g$.

(2) By Lemma 2.8.5(3), $scnf(f) \equiv f$ and $sdnf(f) \equiv f$. By Theorem 2.8.6(1), $scnf(scnf(f)) = scnf(f)$ and $sdnf(sdnf(f)) = sdnf(f)$.

(3) By Lemma 2.8.5(3), $\wedge Cl(f) \equiv Cl(f) \equiv f \equiv \vee Mt(f)$. By Theorem 2.8.6(1), $Cl(\wedge Cl(f)) = Cl(f)$ and $Mt(\vee Mt(f)) = Mt(f)$.

EndProofTheorem2.8.6

2.9 Proof Using Resolution

A first attempt at proving a formula f from a set of formulas F might be to convert F to a set of clauses, say $Cl(F)$, and then to check if $Cl(f) \subseteq Res(Cl(F))$ holds. If it does then we could declare that f is proved from F. Moreover we could say that the collection of resolution derivations of each clause in $Cl(f)$ constituted a proof of f from F. Such a method of proof is sound but unfortunately not complete (Definition 2.6.9).

But by Theorem 2.6.10(4), $CorRes(Cl(F)) = CorClsC_{\vdash}(Cl(F))$. So a sound and complete method of proving f from F is to check if every clause in $Cl(f)$ has a subclause which is in $CorRes(Cl(F))$. If it does then we could declare that f is proved from F, and if it does not then we could declare that f cannot be proved from F. This is a direct method of proof. That is, we start with F and see if we can derive (all parts of) f.

The more common way is to use an indirect method of proof, called a refutation proof. We start with F and $\neg f$ and see if we can derive a contradiction, that is, we show that F refutes f. In particular we try to generate the contradiction $\vee \{\}$ from $Cl(F \cup \{\neg f\})$ using resolution. The details and the notation we use are given in the next definition.

Definition 2.9.1. Let F be any set of formulas and f be any formula. We say f is *resolution-provable* from F, denoted by $F \vdash f$, iff $\vee\{\} \in Res(Cl(\neg f) \cup Cl(F))$. As usual, the negation of $F \vdash f$ is written $F \nvdash f$.
Define $C_{\vdash}(F) = \{f \in Fml : F \vdash f\}$ to be the set of $\vdash$-consequences of F.
For simplicity define $C_{\vdash}(f) = C_{\vdash}(\{f\})$.
Also if G is a set of formulas then define $F \vdash G$ iff for all g in G, $F \vdash g$.

If no ambiguity results we often abbreviate 'resolution-provable' to 'provable'.

The following theorem shows that the proof method in Definition 2.9.1 is sound and complete. We also give several useful properties of the functions $C_{\vdash}(.)$ and $C_{\vDash}(.)$.

Theorem 2.9.2 (Soundness and Completeness). Let F and G be any sets of formulas, and f be any formula.
1) $F \vdash f$ iff $F \vDash f$; and so $C_{\vdash}(F) = C_{\vDash}(F)$. (soundness and completeness)
2) $F \equiv C_{\vDash}(F)$. So F is satisfiable iff $C_{\vDash}(F)$ is satisfiable.
   $F \equiv C_{\vdash}(F)$. So F is satisfiable iff $C_{\vdash}(F)$ is satisfiable.

2.9 Proof Using Resolution

3) If $F \equiv G$ then $C_\models(F) = C_\models(G)$. ($C_\models$ satisfies left equivalence.)
   If $F \equiv G$ then $C_\vdash(F) = C_\vdash(G)$. ($C_\vdash$ satisfies left equivalence.)
4) $F \subseteq C_\models(F)$. ($C_\models$ is inclusive.)
   $F \subseteq C_\vdash(F)$. ($C_\vdash$ is inclusive.)
5) If $F \subseteq G$ then $C_\models(F) \subseteq C_\models(G)$. ($C_\models$ is monotonic.)
   If $F \subseteq G$ then $C_\vdash(F) \subseteq C_\vdash(G)$. ($C_\vdash$ is monotonic.)
6) If $F \subseteq C_\models(G)$ then $C_\models(F \cup G) = C_\models(G)$. ($C_\models$ is cumulative.)
   If $F \subseteq C_\vdash(G)$ then $C_\vdash(F \cup G) = C_\vdash(G)$. ($C_\vdash$ is cumulative.)
7) $C_\models(C_\models(F)) = C_\models(F)$. ($C_\models$ is idempotent.)
   $C_\vdash(C_\vdash(F)) = C_\vdash(F)$. ($C_\vdash$ is idempotent.)
8) If $F \subseteq C_\models(G)$ then $C_\models(F) \subseteq C_\models(G)$. ($C_\models$ is classically closed.)
   If $F \subseteq C_\vdash(G)$ then $C_\vdash(F) \subseteq C_\vdash(G)$. ($C_\vdash$ is classically closed.)
9) $C_\models(G) = \bigcup\{C_\models(F) : F \text{ is a finite subset of } G\}$. ($C_\models$ is compact.)
   $C_\vdash(G) = \bigcup\{C_\vdash(F) : F \text{ is a finite subset of } G\}$. ($C_\vdash$ is compact.)

Proof

Let F be any set of formulas, and f be any formula.
(1) $F \vdash f$ iff $\vee\{\} \in Res(Cl(\neg f) \cup Cl(F))$
iff $\vee\{\} \in Res(Cl(\{\neg f\}) \cup Cl(F))$ by Lemma 2.8.5(2),
iff $\vee\{\} \in Res(Cl(\{\neg f\} \cup F))$ by Lemma 2.8.5(7),
iff $Cl(\{\neg f\} \cup F)$ is not satisfiable by Lemma 2.4.12(5),
iff $\{\neg f\} \cup F$ is not satisfiable by Lemma 2.8.5(4),
iff $F \models f$.

Hence $C_\vdash(F) = C_\models(F)$.
(2) This follows from Lemma 2.2.21(3) and Theorem 2.9.2(1).
(3) This follows from Lemma 2.2.21(5) and Theorem 2.9.2(1).
(4) This follows from Lemma 2.2.21(2) and Theorem 2.9.2(1).
(5) This follows from Lemma 2.2.21(6) and Theorem 2.9.2(1).
(6) This follows from Lemma 2.2.21(7) and Theorem 2.9.2(1).
(7) This follows from Lemma 2.2.21(8) and Theorem 2.9.2(1).
(8) Suppose $F \subseteq C_\models(G)$. By Theorem 2.9.2(5), $C_\models(F) \subseteq C_\models(C_\models(G))$. By using Theorem 2.9.2(7), $C_\models(C_\models(G)) = C_\models(G)$. Hence $C_\models(F) \subseteq C_\models(G)$.

Suppose $F \subseteq C_\vdash(G)$. By using Theorem 2.9.2(5), $C_\vdash(F) \subseteq C_\vdash(C_\vdash(G))$. By using Theorem 2.9.2(7), $C_\vdash(C_\vdash(G)) = C_\vdash(G)$. Hence $C_\vdash(F) \subseteq C_\vdash(G)$.

(9) By Lemma 2.2.21(6), $\bigcup\{C_\models(F) : F \text{ is a finite subset of } G\} \subseteq C_\models(G)$.

Conversely, take any g in $C_\models(G)$. Then $G \models g$. So $G \cup \{\neg g\}$ is unsatisfiable. By Theorem 2.3.11(1), there is a finite subset F' of $G \cup \{\neg g\}$ such that F' is unsatisfiable. If $\neg g \in F'$ then let $F = F' - \{\neg g\}$; otherwise let $F = F'$. Then F is a finite subset of G such that $F \cup \{\neg g\}$ is unsatisfiable. Hence $F \models g$ and so $g \in C_\models(F)$. Therefore $g \in \bigcup\{C_\models(F) : F \text{ is a finite subset of } G\}$. But g was arbitrary, and therefore we have $C_\models(G) \subseteq \bigcup\{C_\models(F) : F \text{ is a finite subset of } G\}$.

Thus $C_\models(G) = \bigcup\{C_\models(F) : F \text{ is a finite subset of } G\}$.

The second part follows from Theorem 2.9.2(1) and what we have just proved.

EndProofTheorem2.9.2

Finally we can formally define what a propositional resolution logic is.

Definition 2.9.3. A *propositional resolution logic* is a triple $(\mathcal{A}, Fml(\mathcal{A}), \vdash)$ where $\mathcal{A}$ is an alphabet as defined in Definition 2.2.1,
$Fml(\mathcal{A})$ is the set of formulas using $\mathcal{A}$ as defined in Definition 2.2.2, and $\vdash$ is the proof relation as defined in Definition 2.9.1.

We have now covered all the topics of classical propositional logic that we will need later. However there is one more idea that is useful, namely the concept of an adequate set of connectives. Since this concept is not essential for what follows, we shall only define it and not go into details or proofs.

Definition 2.9.4. If A is a set of atoms then define $Val(A)$, the set of valuations which are false outside A, by $Val(A) = \{v \in Val : \text{for all } a \text{ in } Atm - A,\ v(a) = \mathbf{F}\}$.

Let A be any finite set of atoms such that $|A| = n$. Then $|Val(A)| = 2^n$ and there are $2^{(2^n)}$ formulas such that any two are not equivalent. Not all sets of connectives are able to express all these formulas. A set of connectives that is (for any A) able to express all these formulas is said to be *adequate*. It can be shown that, if we limit our attention to just the common connectives of negation, conjunction, disjunction, and material implication, then a set of connectives is adequate iff it contains negation and at least one of conjunction, disjunction, or material implication. By a *propositionally adequate logic* we shall mean a propositional logic that has an adequate set of connectives.

Chapter 3
Rational Reasoning

Abstract The sections of Chapter 3 Rational Reasoning are: 3.1 Introduction, 3.2 Four Satisfiable Subsets of a Set of Clauses, 3.3 Rational Propositional Logics, and 3.4 Consequence Functions.

The second section defines four satisfiable subsets of an unsatisfiable set of clauses, two of which are new. It also shows that each of these four subsets have four very useful properties. The third section uses these properties and a consideration of tautologies to form two hierarchies of rational logics.

The last section considers consequence functions, a commonly used tool for studying logics. Although the concept of a consequence function is clear, the definitions in the literature fall well short of capturing this concept. A new definition is suggested and some of its properties are considered.

3.1 Introduction

There are many different proof theories for classical propositional logic; that is, there are many different ways of defining the proof relation $\vdash$. But they all satisfy soundness and completeness, Theorem 2.9.2. Unfortunately this means that propositional logic has two problems that we want to avoid. The most important problem is explosiveness.

A logic is *explosive* iff whenever a set F of formulas is unsatisfiable then every formula is provable from F, in symbols $C_\vdash(F) = Fml$. Classical propositional logic is explosive.

When faced with an unsatisfiable set of formulas, an explosive logic panics and does about the most stupid thing possible: it allows every formula to be proved. A rational logic, when faced with an unsatisfiable set F of formulas, should keep calm and behave rationally.

There are at least two ways of behaving rationally. We could accept that F is unsatisfiable and change the way a logic reasons about F; this is what paraconsistent logics do, see for example [55]. Or we could replace F with a satisfiable set and reason as usual. This is sometimes called 'consolidation' and it is the approach taken in Belief Revision. The problem considered in Belief Revision is how to restore satisfiability after a belief (formula) has been added to a set of beliefs, producing the set F. The researchers in Belief Revision seek to remove the smallest number of beliefs from F that will restore satisfiability. That is, they try to minimise the loss of information. So their result is a superset of the intersection of all the maximal satisfiable subsets of F.

We regard a set F of formulas as a set of facts, possibly describing part of the real world. So if F is unsatisfiable then something is wrong with F. Ideally the logic should inform its user that F is unsatisfiable and ask for F to be corrected. But the user may not be able to access the correct version of F. In this case the logic should replace F with a suitable satisfiable set S. Such an S could be a subset of F, or we could convert each formula f in F to a suitable equivalent formula f' and then let S be a suitable satisfiable subset of $\{f' : f \in F\}$. Our motivation is to minimise the risk of including faulty information. So our result is a subset of the intersection of all the maximal satisfiable subsets of F or $\{f' : f \in F\}$.

3.2 Four Satisfiable Subsets of a Set of Clauses

Imagine that we have a set F of formulas about cats and dogs. Suppose that a mistake was made when entering the formulas about cats, making the formulas about cats unsatisfiable. But the formulas about dogs were satisfiable. Then F is unsatisfiable. We would like a logic to give reasonable answers to questions about dogs, but refuse to answer questions about cats; perhaps even politely suggesting that the formulas about cats be corrected. Generally, if F is unsatisfiable then we would like to prove what is not disputed and only what is not disputed.

3.2.1 Maximal Satisfiable, Minimal Unsatisfiable

Let F be an unsatisfiable set of formulas, converted to a suitable form if required. In the introduction to this chapter we said that we needed to find a suitable satisfiable set S. An extreme option is to make S empty. Except for tautologies, this agrees with the statement "ex contradictione nihil sequitur" (from a contradiction nothing follows). (The word 'nil', used in Definition 3.2.1(1) is a contraction of the Latin word 'nihil'.) But then the logic cannot answer questions we would like it to. In the cats and dogs example it could not answer questions about dogs. However, this option is reasonable as it could be thought of as the logic saying "I do not know what F should be, ask someone who knows".

The other extreme is to make S a maximal satisfiable subset of F. (This is called maxichoice in Belief Revision.) But what if there are many maximal satisfiable sub-

3.2 Four Satisfiable Subsets of a Set of Clauses

sets? One could be chosen arbitrarily, or we could take the intersection of all the maximal satisfiable subsets of F. Both these cases can be covered by choosing some (possibly one, possibly all) maximal satisfiable subsets of F and intersecting them. (In Belief Revision this is called the partial meet contraction of F by a contradiction. If all the maximal satisfiable subsets of F are chosen then the intersection is called the full meet contraction of F by a contradiction.)

Another choice for S is to remove at least one element from each of the minimal unsatisfiable subsets of F. (This approach has received some attention from a Belief Revision perspective, see for example [52], [30], and its two page summary [31].)

For example, let $F_1 = \{a, \neg a, \vee\{a,b\}\}$. The intersection of all the maximal satisfiable subsets of F_1 is $\{\vee\{a,b\}\}$. The (only) minimal unsatisfiable subset of F_1 is $\{a, \neg a\}$. Removing this from F_1 leaves $\{\vee\{a,b\}\}$. That these two methods give the same answer is not a coincidence. Indeed we shall show that intersecting maximal satisfiable subsets of F and removing at least one element from each of the minimal unsatisfiable subsets of F give the same result. But first we need some definitions.

Definition 3.2.1. Let F be any set of formulas.
1) If F is satisfiable then $Nil(F) = F$.
   If F is unsatisfiable then $Nil(F) = \{\}$.
2) $Sat(F) = \{S \subseteq F : S \text{ is satisfiable}\}$.
   $Unsat(F) = \{U \subseteq F : U \text{ is unsatisfiable}\}$.
3) $MaxSat(F) = \{S \in Sat(F) : \text{if } X \in Sat(F) \text{ then } S \not\subset X\}$. So
   $MaxSat(F)$ is the set of all *maximal satisfiable* subsets of F.
4) $MinUnsat(F) = \{U \in Unsat(F) : \text{if } X \in Unsat(F) \text{ then } X \not\subset U\}$. So
   $MinUnsat(F)$ is the set of all *minimal unsatisfiable* subsets of F.

We note that $\{\} \in Sat(F)$ and so both $Sat(F)$ and $MaxSat(F)$ are not empty. Also F is satisfiable iff $Unsat(F)$ is empty iff $MinUnsat(F)$ is empty.

Theorem 3.2.2. Suppose each of F and G is a set of formulas. Let $SubMaxSat(F)$ be any non-empty subset of $MaxSat(F)$. Let v be any valuation.
1) $\bigcap MaxSat(F) \cup \bigcup MinUnsat(F) = F$ and
   $\bigcap MaxSat(F) \cap \bigcup MinUnsat(F) = \{\}$.
2) $\bigcup \{F - S : S \in MaxSat(F)\} = \bigcup MinUnsat(F)$.
3) $\bigcap SubMaxSat(F) \cup \bigcup \{F - S : S \in SubMaxSat(F)\} = F$ and
   $\bigcap SubMaxSat(F) \cap \bigcup \{F - S : S \in SubMaxSat(F)\} = \{\}$.
4) If $S \in MaxSat(F)$ and $U \in MinUnsat(F)$ then $(F - S) \cap U \neq \{\}$.
5) $\bigcup MinUnsat(Cl(Fml)) = Cl(Fml)$.
6) $\bigcap MaxSat(Cl(Fml)) = \{\}$.
7) If $F' \subseteq F$ then $Sat(F') \subseteq Sat(F)$.
8) If $S \in MaxSat(F)$ then there exists G' such that $G' \subseteq G - F$ and
   $S \cup G' \in MaxSat(F \cup G)$.
9) Let $M = \{A \cup G_A \in MaxSat(F \cup G) : A \in MaxSat(F) \text{ and } G_A \subseteq G - F\}$.
   If $v \models \bigcap MaxSat(F)$ and $v \models G$ then $v \models \bigcap M$ and so $v \models \bigcap MaxSat(F \cup G)$.

Proof

Suppose each of F and G is a set of formulas. Let $SubMaxSat(F)$ be any non-empty subset of $MaxSat(F)$. Let v be any valuation.

(1) The first equation.

From the definitions $\bigcap MaxSat(F) \cup \bigcup MinUnsat(F) \subseteq F$. Conversely take any f in F. Suppose $f \notin \bigcap MaxSat(F)$. Then there is a set S in $MaxSat(F)$ such that $f \notin S$. So $S \cup \{f\}$ is unsatisfiable. Hence there is a set U in $MinUnsat(F)$ such that $U \subseteq S \cup \{f\}$. Moreover $f \in U$; as if not then $U \subseteq S$ and so U would be satisfiable. Hence $f \in \bigcup MinUnsat(F)$. Therefore $F \subseteq \bigcap MaxSat(F) \cup \bigcup MinUnsat(F)$.

Thus $\bigcap MaxSat(F) \cup \bigcup MinUnsat(F) = F$.

The second equation.

Take any f in $\bigcup MinUnsat(F)$. Then there exists U in $MinUnsat(F)$ such that $f \in U$. So $U - \{f\}$ is satisfiable. Hence there is a set S in $MaxSat(F)$ such that $U - \{f\} \subseteq S$. Moreover $f \notin S$; as if not then $U \subseteq S$ and so U would be satisfiable. Hence $f \notin \bigcap MaxSat(F)$. Thus $\bigcap MaxSat(F) \cap \bigcup MinUnsat(F) = \{\}$.

(2) Take any f in $\bigcup \{F - S : S \in MaxSat(F)\}$. Then there exists S in $MaxSat(F)$ such that $f \in F - S$. So $S \cup \{f\}$ is unsatisfiable. Hence there is a U in $MinUnsat(F)$ such that $U \subseteq S \cup \{f\}$. Moreover $f \in U$; as if not then $U \subseteq S$ and so U would be satisfiable. Hence $f \in \bigcup MinUnsat(F)$. Therefore $\bigcup \{F - S : S \in MaxSat(F)\} \subseteq \bigcup MinUnsat(F)$.

Conversely, take any f in $\bigcup MinUnsat(F)$. Then there exists U in $MinUnsat(F)$ such that $f \in U$. So $U - \{f\}$ is satisfiable. Hence there is a set S in $MaxSat(F)$ such that $U - \{f\} \subseteq S$. Moreover $f \notin S$; as if not then $U \subseteq S$ and so U would be satisfiable. But $f \in F$. Hence $f \in F - S$ and so $f \in \bigcup \{F - S : S \in MaxSat(F)\}$. Therefore $\bigcup MinUnsat(F) \subseteq \bigcup \{F - S : S \in MaxSat(F)\}$.

Thus $\bigcup \{F - S : S \in MaxSat(F)\} = \bigcup MinUnsat(F)$.

(3) The first equation.

From the definitions $\bigcap SubMaxSat(F) \cup \bigcup \{F - S : S \in SubMaxSat(F)\} \subseteq F$. Conversely take any f in F. Suppose $f \notin \bigcap SubMaxSat(F)$. Then there is a set S in $SubMaxSat(F)$ such that $f \notin S$. Therefore $f \in F - S$ and hence $f \in \bigcup \{F - S : S \in SubMaxSat(F)\}$. Therefore $F \subseteq \bigcap SubMaxSat(F) \cup \bigcup \{F - S : S \in SubMaxSat(F)\}$.

Thus $\bigcap SubMaxSat(F) \cup \bigcup \{F - S : S \in SubMaxSat(F)\} = F$.

The second equation.

Take any f in $\bigcup \{F - S : S \in SubMaxSat(F)\}$. Then there exists S in $SubMaxSat(F)$ such that $f \in F - S$. So $f \notin S$. Hence $f \notin \bigcap SubMaxSat(F)$. Thus $\bigcap SubMaxSat(F) \cap \bigcup \{F - S : S \in SubMaxSat(F)\} = \{\}$.

(4) Take any S in $MaxSat(F)$ and any U in $MinUnsat(F)$. Hence $U \subseteq F$. Therefore if $(F - S) \cap U = \{\}$ then $U \subseteq S$; which contradicts the unsatisfiability of U and the satisfiability of S. Therefore $(F - S) \cap U \neq \{\}$.

(5) Clearly $\bigcup MinUnsat(Cl(Fml)) \subseteq Cl(Fml)$.

Conversely, take any clause c in $Cl(Fml)$. By Lemma 2.8.5(18), $Cl(Fml) = ns(Cls) - Taut$; so $c \in ns(Cls) - Taut$. Hence c is not a tautology. We now consider three cases.

Case 1: $c = \vee\{\}$.

Then c is unsatisfiable and $\{c\} \in MinUnsat(Cl(Fml))$. So $c \in \bigcup MinUnsat(Cl(Fml))$.

3.2 Four Satisfiable Subsets of a Set of Clauses

Case 2: c is a literal.
Then $\{c, \sim c\}$ is unsatisfiable and $\{c, \sim c\} \in MinUnsat(Cl(Fml))$. Therefore we have $c \in \bigcup MinUnsat(Cl(Fml))$.

Case 3: $c = \vee L$, where L is a finite set of two or more literals.
Let $L = \{l_1, l_2, ..., l_n\}$. Then n is a positive integer greater than or equal to 2, and each l_i is a literal. For simplicity we shall suppose that any subscript of l will be in $[1..n]$. If $\{i, j\} \subseteq [1..n]$ then $i \neq j$ iff $l_i \neq l_j$ since L is a set. If $\{i, j\} \subseteq [1..n]$ then $l_i \neq \sim l_j$ since c is not a tautology.

If f is a function from $[1..n]$ to $\{0, \sim\}$ then define $f(L) = \{l_i \in L : f(i) = 0\} \cup \{\sim l_i \in \sim L : f(i) = \sim\}$. Let $U = \{\vee f(L) : f \text{ is a function from } [1..n] \text{ to } \{0, \sim\}\}$. Then $c \in U$ and $U \subseteq Cl(Fml)$.

We show U is unsatisfiable. Assume U is satisfiable. Then there is a valuation v such that $v \models U$. Define the function f' from $[1..n]$ to $\{0, \sim\}$ as follows. If $v(l_i) = \mathbf{T}$ then $f'(i) = \sim$; and if $v(l_i) = \mathbf{F}$ then $f'(i) = 0$. Then $\vee f'(L) \in U$ and $v \not\models \vee f'(L)$. Hence $v \not\models U$. This contradiction shows that U is unsatisfiable and so $U \in Unsat(Cl(Fml))$.

We now show $U \in MinUnsat(Cl(Fml))$. Take any function f from $[1..n]$ to $\{0, \sim\}$. Define a valuation v_f as follows. If $f(i) = 0$ then $v_f(l_i) = \mathbf{F}$; and if $f(i) = \sim$ then $v_f(l_i) = \mathbf{T}$. Then $v_f \not\models \vee f(L)$ and $v_f \models U - \{\vee f(L)\}$. Hence $U \in MinUnsat(Cl(Fml))$ and so $c \in \bigcup MinUnsat(Cl(Fml))$.

So in all cases $c \in \bigcup MinUnsat(Cl(Fml))$. Thus $Cl(Fml) \subseteq \bigcup MinUnsat(Cl(Fml))$ and so $\bigcup MinUnsat(Cl(Fml)) = Cl(Fml)$.

(6) This follows from parts (5 and 1).

(7) Suppose $F' \subseteq F$. Then $Sat(F') = \{S \subseteq F' : S \text{ is satisfiable}\} \subseteq \{S \subseteq F : S \text{ is satisfiable}\} = Sat(F)$.

(8) Suppose $S \in MaxSat(F)$. Then S is a satisfiable subset of $F \cup G$. If $S \in MaxSat(F \cup G)$ then let $G' = \{\}$.

If $S \notin MaxSat(F \cup G)$ then there exists G' such that $G' \subseteq F \cup G$ and $S \cup G' \in MaxSat(F \cup G)$. If $G' \cap F \neq \{\}$ then for each $f \in G' \cap F$, $S \cup \{f\}$ is satisfiable. But $S \subset S \cup \{f\} \subseteq F$, which contradicts $S \in MaxSat(F)$. Hence $G' \cap F = \{\}$ and so $G' \subseteq G - F$.

(9) Let $M = \{A \cup G_A \in MaxSat(F \cup G) : A \in MaxSat(F) \text{ and } G_A \subseteq G - F\}$. Suppose $v \models \bigcap MaxSat(F)$ and $v \models G$.

We show $v \models \bigcap M$. If $S \subseteq MaxSat(F)$ let $I_S = \bigcap (S \cup \{G_A : A \in MaxSat(F) - S\})$. By Lemma 1.2.1, with $MaxSat(F)$ replacing $\mathcal{A}$, and G_A replacing B_A, we have $\bigcap M = \bigcup \{I_S : S \subseteq MaxSat(F)\}$. Since $v \models G$, for all A in $MaxSat(F)$, $v \models G_A$. Also if $S = MaxSat(F)$ then $v \models \bigcap S$. Hence if $S \subseteq MaxSat(F)$ then $v \models I_S$. Therefore $v \models \bigcup \{I_S : S \subseteq MaxSat(F)\}$. So $v \models \bigcap M$.

Now observe that $M \subseteq MaxSat(F \cup G)$. Hence $\bigcap MaxSat(F \cup G) \subseteq \bigcap M$. Therefore $v \models \bigcap MaxSat(F \cup G)$.

EndProofTheorem3.2.2

By Theorem 3.2.2(1) we have $\bigcap MaxSat(F) = F - \bigcup MinUnsat(F)$. That is, the intersection of all the sets that are maximally satisfiable in F is the same as removing from F all the elements in the sets that are minimally unsatisfiable in F. More generally, if *SubMaxSat*(F) is any non-empty subset of *MaxSat*(F) then by Theorem 3.2.2(3) we have $\bigcap SubMaxSat(F) = F - \bigcup\{F-S : S \in SubMaxSat(F)\}$. So by Theorem 3.2.2(4), the intersection of at least one of the sets that are maximally satisfiable in F is the same as removing from F at least one element from each set that is minimally unsatisfiable in F.

3.2.2 No Errors, Some Non-errors

If there is no other information to guide us then replacing F with either *Nil*(F) or $\bigcap MaxSat(F)$ seems to be the most reasonable choice. However if each formula f in F is first converted to its equivalent set of clauses, $Cl(f)$, then arguably there are more reasonable choices than either *Nil*(F) or $\bigcap MaxSat(F)$ as a replacement for F.

First we convert F to the set of clauses $Cl(F)$. By Lemma 2.8.5(4), F is unsatisfiable iff $Cl(F)$ is unsatisfiable. By Lemma 2.4.12(5), $Cl(F)$ is unsatisfiable iff $\vee\{\} \in Res(Cl(F))$. And clearly $\vee\{\} \in Res(Cl(F))$ iff either $\vee\{\} \in Cl(F)$ or there is a literal, say l, such that $\vee\{l\} \in Res(Cl(F))$ and $\vee\{\sim l\} \in Res(Cl(F))$. In the latter case, at least one of l and $\sim l$ is an error. Because we do not know which is the error, we shall be cautious and regard both l and $\sim l$ as errors. Errors make any clause containing them unreliable and so it is risky to include them in a set of facts. As we have seen, rejecting all clauses in $Cl(F)$ to get $Nil(Cl(F))$ is too restrictive.

However there are now two new options. The first option is to removed from $Cl(F)$ the empty clause and all the clauses that contain any errors. We shall say that a clause is 'pure' iff it is non-empty and not contaminated by errors. So *Pure*($Cl(F)$) is the set of all pure clauses in $Cl(F)$. This option agrees with our desire to prove what is not disputed and only what is not disputed.

The second option is to removed from $Cl(F)$ the empty clause and all the clauses that contain only errors, to give *Dub*($Cl(F)$). So *Dub*($Cl(F)$) consists of all the clauses in $Cl(F)$ that contain at least one non-error literal. *Dub* is a contraction of 'dubious' because only some of the clauses in *Dub*($Cl(F)$) are contaminated with errors and hence are dubious.

We shall now formally define these and related ideas.

Definition 3.2.3. Let C be any set of clauses.
1) The set *Errlit*(C) of error literals of C is defined by
   $Errlit(C) = \{l \in Lit : \vee\{l\} \in Res(C)$ and $\vee\{\sim l\} \in Res(C)\}$.
2) The set *Errcl*(C) of error clauses in C is defined by
   $Errcl(C) = \{c \in C : c = \vee\{\}$ or $Lit(c) \cap Errlit(C) \neq \{\}\}$.
   So error clauses in C are either empty or contain at least one error literal of C.
3) The set *Errfulcl*(C) of errorful clauses in C is defined by
   $Errfulcl(C) = \{c \in C : Lit(c) \subseteq Errlit(C)\}$.
   So every literal in every errorful clause in C is an error literal of C.

3.2 Four Satisfiable Subsets of a Set of Clauses

4) The set $Pure(C)$ of *pure clauses in C* is defined by
$Pure(C) = C - Errcl(C) = \{c \in C : c \neq \vee\{\}$ and $Lit(c) \cap Errlit(C) = \{\}\}$.
So pure clauses in C are non-empty and free from error literals.
5) $Dub(C) = C - Errfulcl(C) = \{c \in C : Lit(c) \not\subseteq Errlit(C)\}$.
$Dub^*(C) = \{\vee L \in Cls :$ there exists c in $Dub(C)$ such that $Lit(c) \subseteq L\}$.

Roughly, $Dub^*(C)$ is the set of all superclauses of the clauses in $Dub(C)$.

Before proving some useful results about these concepts, we shall apply them to an earlier example.

Let $F_1 = \{a, \neg a, \vee\{a,b\}\}$. By Lemma 2.8.3(5,8),
$Cl(F_1) = \bigcup\{Cl(a), Cl(\neg a), Cl(\vee\{a,b\})\} = \{a, \neg a, \vee\{a,b\}\}$.
So $Res(Cl(F_1)) = \{\vee\{a\}, \vee\{\neg a\}, \vee\{a,b\}, \vee\{\}, \vee\{b\}\}$. Hence
$Errlit(Cl(F_1)) = \{a, \neg a\}$,
$Errcl(Cl(F_1)) = \{a, \neg a, \vee\{a,b\}\}$,
$Errfulcl(Cl(F_1)) = \{a, \neg a\}$,
$Pure(Cl(F_1)) = \{\}$, and
$Dub(Cl(F_1)) = \{\vee\{a,b\}\}$.

Lemma 3.2.4. Let C be a set of clauses.
1) $l \in Errlit(C)$ iff $\sim l \in Errlit(C)$.
2) If $Errlit(C) \neq \{\}$ then $\vee\{\} \in Res(C)$.
3) If C is satisfiable then $Errlit(C) = \{\}$.
4) If $C' \subseteq C$ then $Errlit(C') \subseteq Errlit(C)$.
5) If $C' \subseteq C$ then $Errcl(C') \subseteq Errcl(C)$.
6) If $C' \subseteq C$ then $Errfulcl(C') \subseteq Errfulcl(C)$.
7) $Errlit(Res(C)) = Errlit(C)$.
8) If $U \in MinUnsat(C)$ and $c \in U$ then $Lit(c) \subseteq Errlit(U) \subseteq Errlit(C)$.
9) If $c \in \bigcup MinUnsat(C)$ then $Lit(c) \subseteq Errlit(C)$.
10) $\bigcup MinUnsat(C) \subseteq Errfulcl(C) \subseteq Errcl(C)$.

Proof

Let C be a set of clauses.

(1, 2) These parts follow immediately from Definition 3.2.3.

(3) Suppose C is satisfiable. By Lemma 2.4.12(5), $\vee\{\} \notin Res(C)$. So by part (2), $Errlit(C) = \{\}$.

(4) Suppose $C' \subseteq C$. Take any l in $Errlit(C')$. Then we have $\vee\{l\} \in Res(C')$ and $\vee\{\sim l\} \in Res(C')$. By Lemma 2.4.11(5), $\vee\{l\} \in Res(C)$ and $\vee\{\sim l\} \in Res(C)$. Hence $l \in Errlit(C)$. Thus $Errlit(C') \subseteq Errlit(C)$.

(5) Suppose $C' \subseteq C$. $Errcl(C') = \{c \in C' : c = \vee\{\}$ or $Lit(c) \cap Errlit(C') \neq \{\}\}$
$\subseteq \{c \in C : c = \vee\{\}$ or $Lit(c) \cap Errlit(C) \neq \{\}\}$ by Lemma 3.2.4(4),
$= Errcl(C)$.

(6) Suppose $C' \subseteq C$. $Errfulcl(C') = \{c \in C' : Lit(c) \subseteq Errlit(C')\}$
$\subseteq \{c \in C : Lit(c) \subseteq Errlit(C)\}$ by Lemma 3.2.4(4),
$= Errfulcl(C)$.

(7) $Errlit(Res(C))$
$= \{l \in Lit : \vee\{l\} \in Res(Res(C))$ and $\vee\{\sim l\} \in Res(Res(C))\}$ by Definition 3.2.3(1),

$= \{l \in Lit : \vee\{l\} \in Res(C)$ and $\vee\{\sim l\} \in Res(C)\}$ by Lemma 2.4.11(6),
$= Errlit(C)$ by Definition 3.2.3(1).

(8) Suppose $U \in MinUnsat(C)$. Take any c in U. Then $U - \{c\}$ is satisfiable, and so $U - \{c\} \models \neg c$. Therefore for all l in $Lit(c)$, $U - \{c\} \models \sim l$. Therefore for all l in $Lit(c)$, $\sim l \in C_\models (U - \{c\})$. Since $U - \{c\}$ is satisfiable and $\sim l$ is a literal, for all l in $Lit(c)$, $\sim l \in MinCtgeClsC_\models(U - \{c\})$. By Theorem 2.6.10(4), for all l in $Lit(c)$, $\sim l \in CorRes(U - \{c\})$. Therefore for all l in $Lit(c)$, $\vee\{\sim l\} \in Res(U - \{c\})$. By Lemma 2.4.11(5), for all l in $Lit(c)$, $\vee\{\sim l\} \in Res(U)$.

Since $c \in U$, by resolving away all but one of the literals in c we have that for all l in $Lit(c)$, $\vee\{l\} \in Res(U)$. Therefore for all l in $Lit(c)$, $\{\vee\{l\}, \vee\{\sim l\}\} \subseteq Res(U)$. So for all l in $Lit(c)$, $l \in Errlit(U)$. Thus by Lemma 3.2.4(4), $Lit(c) \subseteq Errlit(U) \subseteq Errlit(C)$.

(9) This follows from Lemma 3.2.4(8).

(10) The first subset relationship follows from Lemma 3.2.4(9). The second subset relationship follows from Definition 3.2.3.
EndProofLemma3.2.4

Several subsets of a set C of clauses have been identified. We shall now show that the satisfiable subsets form a linear hierarchy.

Lemma 3.2.5. Let C be any set of clauses and suppose $SubMaxSat(C)$ is a non-empty subset of $MaxSat(C)$.
1) The following nine statements are all equivalent.
    1.1) C is satisfiable.
    1.2) $\bigcup MinUnsat(C) = \{\}$.
    1.3) $Errfulcl(C) = \{\}$.
    1.4) $Errcl(C) = \{\}$.
    1.5) $C = \bigcap SubMaxSat(C)$.
    1.6) $C = \bigcap MaxSat(C)$.
    1.7) $C = Dub(C)$.
    1.8) $C = Pure(C)$.
    1.9) $C = Nil(C)$.
2) $Nil(C) \subseteq Pure(C) \subseteq Dub(C) \subseteq \bigcap MaxSat(C) \subseteq \bigcap SubMaxSat(C) \subseteq C$.
3) $Nil(C)$, $Pure(C)$, $Dub(C)$, $\bigcap MaxSat(C)$, and $\bigcap SubMaxSat(C)$ are all satisfiable.
Proof

Let C be any set of clauses and suppose $SubMaxSat(C)$ is a non-empty subset of $MaxSat(C)$.

(1)

(1.1) iff (1.2). From the definitions, C is satisfiable iff $\bigcup MinUnsat(C) = \{\}$.

(1.1) iff (1.4). C is satisfiable iff $\vee\{\} \notin Res(C)$, by Lemma 2.4.12(5) iff $\vee\{\} \notin C$ and $Errlit(C) = \{\}$
iff $Errcl(C) = \{\}$.

(1.4) implies (1.3). This follows from $Errfulcl(C) \subseteq Errcl(C)$.

(1.3) implies (1.2). By Lemma 3.2.4(10), $\bigcup MinUnsat(C) \subseteq Errfulcl(C)$. Hence the result.

3.2 Four Satisfiable Subsets of a Set of Clauses

(1.5) iff (1.1) iff (1.6). $C = \bigcap SubMaxSat(C)$ iff $SubMaxSat(C) = \{C\}$
iff C is satisfiable
iff $MaxSat(C) = \{C\}$
iff $C = \bigcap MaxSat(C)$.
  (1.3) iff (1.7). By Definition 3.2.3(5), $Dub(C) = C - Errfulcl(C)$. Hence the result.
  (1.4) iff (1.8). By Definition 3.2.3(4), $Pure(C) = C - Errcl(C)$. Hence the result.
  (1.1) iff (1.9). If C is satisfiable then $Nil(C) = C$. If C is empty then C is satisfiable. If C is not empty and $Nil(C) = C$ then C is not unsatisfiable; hence C is satisfiable.
  (2) By Lemma 3.2.4(10) and Theorem 3.2.2(1), $Pure(C) = C - Errcl(C) \subseteq C - Errfulcl(C) = Dub(C)$; and $Dub(C) = C - Errfulcl(C) \subseteq C - \bigcup MinUnsat(C) = \bigcap MaxSat(C)$. Also $\bigcap MaxSat(C) \subseteq \bigcap SubMaxSat(C) \subseteq C$. If C is unsatisfiable then $Nil(C) = \{\} \subseteq Pure(C)$. By Lemma 3.2.5(1), C is satisfiable iff $C = Pure(C)$. So if C is satisfiable then $Nil(C) = C \subseteq Pure(C)$.
  (3) From its definition, $\bigcap SubMaxSat(C)$ is satisfiable. So by Lemma 3.2.5(2), $Nil(C)$, $Pure(C)$, $Dub(C)$, and $\bigcap MaxSat(C)$ are all satisfiable.
EndProofLemma3.2.5

The linear hierarchy in Lemma 3.2.5(2) is also a trustworthiness hierarchy; as the smaller members are at least as reliable as the bigger members.

The following example shows that the linear hierarchy in Lemma 3.2.5(2) can be entirely strict.

Example 3.2.6 (A strict hierarchy).
Let a, b, c, d, and e be five different atoms, let $C_4 = \{\vee\{a,b\}, \vee\{a,\neg b\}, \vee\{\neg a,c\}, \vee\{\neg a,\neg c\}\}$, and let $C_7 = C_4 \cup \{\vee\{\neg a,b\}, \vee\{a,d\}, e\}$.
  Then $\{\vee\{a\}, \vee\{\neg a\}, \vee\{b\}, \vee\{\neg b\}, \vee\{c\}, \vee\{\neg c\}, \vee\{\}\} \subseteq Res(C_4) \subseteq Res(C_7)$. So C_4 and C_7 are unsatisfiable. Hence $Nil(C_7) = \{\}$. Moreover C_4 is the only subset of C_7 that is minimally unsatisfiable in C_7. So $\bigcup MinUnsat(C_7) = C_4$ and so by Theorem 3.2.2(1), $\bigcap MaxSat(C_7) = C_7 - C_4 = \{\vee\{\neg a,b\}, \vee\{a,d\}, e\}$. Each set in $MaxSat(C_7)$ has the form $C_7 - \{\text{an element of } C_4\}$. So $|MaxSat(C_7)| = 4$. Suppose $\{\} \subset SubMaxSat(C_7) \subset MaxSat(C_7)$. Also $Errlit(C_7) = \{a, \neg a, b, \neg b, c, \neg c\}$, so $Errcl(C_7) = C_7 - \{e\}$, and hence $Pure(C_7) = C_7 - Errcl(C_7) = \{e\}$. But $Errfulcl(C_7) = C_4 \cup \{\vee\{\neg a,b\}\}$. So $Dub(C_7) = C_7 - Errfulcl(C_7) = \{\vee\{a,d\}, e\}$.
  Thus $Nil(C_7) \subset Pure(C_7) \subset Dub(C_7) \subset \bigcap MaxSat(C_7) \subset \bigcap SubMaxSat(C_7) \subset C_7$.

3.2.3 Pre-cumulativity

The ability to use previously proved results to prove new results is fundamental in a reasoning system. This property is called cumulativity, or lemma addition, because it allows the accumulation, or addition, of previously proved results, or lemmas. The following property, which we shall call pre-cumulativity, ensures that the logics defined in Section 3.3 are cumulative.

Definition 3.2.7. Suppose C is a function that takes a set F of formulas and outputs the set $C(F)$ of formulas.
Pre-cumulativity: C is pre-cumulative iff for any sets of formulas, F and G, if $C(F) \models G$ then $C(F \cup G) \equiv C(F)$.

In the above definition, we can think of C as converting F to a suitable set of clauses $C(F)$. The set F can be thought of as premises, and so $C(F)$ are also premises. The set G contains the previously proved results or lemmas. So pre-cumulativity says that adding lemmas is equivalent to the original set of premises.

Let c be any clause and C be any set of clauses. Recall from Definition 2.4.5 that if l is a literal then $\check{}(l) = \check{l} = \vee\{l\}$; and if c is not a literal then $\check{}(c) = \check{c} = c$. Also $\check{}(C) = \check{C} = \{\check{c} : c \in C\}$.

We shall need the following two technical lemmas before we can prove the main result of this subsection. The first concerns $\check{}$.

Lemma 3.2.8. Let c be any clause. Let C and D be any two sets of clauses.
1) $\check{c} = \vee Lit(c)$ and $\check{C} = \vee.Lit(C)$.
2) $\check{}(C \cup D) = \check{C} \cup \check{D}$.
3) $Lit(\check{c}) = Lit(c)$ and $Lit(\check{C}) = Lit(C)$.
4) $\check{c} \equiv c$ and $\check{C} \equiv C$.
5) $Res(\check{C}) = Res(C)$.
6) $\check{C} \subseteq Res(\check{C})$.
7) $Errlit(\check{C}) = Errlit(C)$.

Proof

Let c be any clause. Let C and D be any two sets of clauses.
(1) This follows immediately from the definitions.
(2) $\check{}(C \cup D) = \{\check{c} : c \in C \cup D\} = \{\check{c} : c \in C\} \cup \{\check{c} : c \in D\} = \check{C} \cup \check{D}$.
(3) $Lit(\check{c}) = Lit(c)$ follows immediately from the definitions. Therefore $Lit(\check{C}) = Lit(\{\check{c} : c \in C\}) = \{Lit(\check{c}) : c \in C\} = \{Lit(c) : c \in C\} = Lit(C)$.
(4) $\check{c} \equiv c$ follows immediately from the definitions. Therefore $\check{C} = \{\check{c} : c \in C\} \equiv \{c : c \in C\} = C$.
(5) $Res(C) = \vee.ResLit(C)$
$= \vee.ResLit(\check{C})$ by part 3,
$= Res(\check{C})$.
(6) The elements of $\check{C}$ are not literals. So by Lemma 2.4.11(3), $\check{C} \subseteq Res(\check{C})$.
(7) $Errlit(\check{C}) = \{l \in Lit : \vee\{l\} \in Res(\check{C}) \text{ and } \vee\{\sim l\} \in Res(\check{C})\}$
$= \{l \in Lit : \vee\{l\} \in Res(C) \text{ and } \vee\{\sim l\} \in Res(C)\}$ by part 5,
$= Errlit(C)$.

EndProofLemma3.2.8

Lemma 3.2.9. Let C be a set of clauses and G be a set of formulas.
1) $ns(Dub(C)) \subseteq ns(Dub(Res(C)))$.
2) $Res(Dub(Res(C))) = Dub(Res(C))$.
3) $Res(Dub(C)) \subseteq Dub(Res(C))$.
4) $Res(Errfulcl(Res(C))) = Errfulcl(Res(C))$.

3.2 Four Satisfiable Subsets of a Set of Clauses

5) $Dub(Res(C)) \subseteq Dub^*(Res(C))$.
6) $Res(Dub^*(Res(C))) = Dub^*(Res(C))$.
7) $Res(Errfulcl(Res(C)) \cup Dub^*(Res(C))) = Errfulcl(Res(C)) \cup Dub^*(Res(C))$.
8) If $Dub(C) \models G$ then $`(Cl(G)) \subseteq Dub^*(Res(C))$.
9) $Errlit(Errfulcl(Res(C)) \cup Dub^*(Res(C))) \subseteq Errlit(C)$.
10) If $Dub(C) \models G$ then $Errlit(C \cup Cl(G)) = Errlit(C)$.
11) If $Pure(C) \models G$ then $Errlit(C \cup Cl(G)) = Errlit(C)$.

Proof

Let C be a set of clauses and G be a set of formulas.

(1) $ns(Dub(C))$
$= ns(\{c \in C : Lit(c) \not\subseteq Errlit(C)\})$ by Definition 3.2.3(5),
$= \{ns(c) \in ns(C) : Lit(c) \not\subseteq Errlit(C)\}$ by Definition 2.5.3,
$= \{ns(c) \in ns(C) : Lit(c) \not\subseteq Errlit(Res(C))\}$ by Lemma 3.2.4(7),
$\subseteq \{ns(c) \in ns(Res(C)) : Lit(c) \not\subseteq Errlit(Res(C))\}$ by Lemma 2.5.7(7),
$= ns(\{c \in Res(C) : Lit(c) \not\subseteq Errlit(Res(C))\})$ by Definition 2.5.3,
$= ns(Dub(Res(C)))$ by Definition 3.2.3(5).

(2) Since there are no literals in $Res(C)$, there are no literals in $Dub(Res(C))$, and so by Lemma 2.4.11(3), $Dub(Res(C)) \subseteq Res(Dub(Res(C)))$.

Conversely, we show that if two clauses in $Dub(Res(C))$ can be resolved then the resolvent of those two clauses is in $Dub(Res(C))$. Let c_1 and c_2 be any two clauses in $Dub(Res(C))$ such that the resolvent of c_1 and c_2 is c_3. Then $c_3 \in Res(C)$.

Assume $c_3 \notin Dub(Res(C))$. By Definition 3.2.3(5), $Lit(c_3) \subseteq Errlit(Res(C))$. For each i in $\{1,2\}$, $c_i \in Dub(Res(C))$, $c_i \in Res(C)$, and $Lit(c_i) \not\subseteq Errlit(Res(C))$. So for each i in $\{1,2\}$, there is a literal l_i in $Lit(c_i)$ such that $l_i \notin Errlit(Res(C))$. But $c_3 \notin Dub(Res(C))$, so $Lit(c_3) \subseteq Errlit(Res(C))$. Hence $l_1 = \sim l_2$, and for each i in $\{1,2\}$, $Lit(c_i) - \{l_i\} \subseteq Errlit(Res(C))$. By Lemma 2.4.11(6), $Res(Res(C)) = Res(C)$. So for each i in $\{1,2\}$, $c_i \in Res(C)$, and the complement of each literal in $Lit(c_i) - \{l_i\}$ is in $Res(C)$. Hence by resolving away each literal in $Lit(c_i) - \{l_i\}$ from c_i we have $\vee \{l_i\} \in Res(C)$. But $l_1 = \sim l_2$, so for each i in $\{1,2\}$, $l_i \in Errlit(Res(C))$. This contradiction shows that $c_3 \in Dub(Res(C))$.

Therefore $Dub(Res(C))$ is closed under resolution and so $Res(Dub(Res(C))) \subseteq Dub(Res(C))$.

Thus $Res(Dub(Res(C))) = Dub(Res(C))$.

(3) By part (1) and Lemma 2.4.11(5), $Res(ns(Dub(C))) \subseteq Res(ns(Dub(Res(C))))$. So by Lemma 2.5.7(6), $Res(Dub(C)) \subseteq Res(Dub(Res(C)))$. Hence by part (2), $Res(Dub(C)) \subseteq Dub(Res(C))$.

(4) By Definition 3.2.3(3), $Errfulcl(Res(C)) \subseteq Res(C)$. Therefore there are no literals in $Errfulcl(Res(C))$. So by Lemma 2.4.11(3), we have $Errfulcl(Res(C)) \subseteq Res(Errfulcl(Res(C)))$.

Take any two elements c_1 and c_2 of $Errfulcl(Res(C))$ such that c_3 is a resolvent of c_1 and c_2. Then $c_1 \in Res(C)$ and $c_2 \in Res(C)$ and so by Lemma 2.4.11(6), $c_3 \in Res(C)$. Since $Lit(c_3) \subseteq Lit(c_1) \cup Lit(c_2) \subseteq Errlit(Res(C))$, $c_3 \in Errfulcl(Res(C))$. Therefore $Res(Errfulcl(Res(C))) \subseteq Errfulcl(Res(C))$.

Thus $Res(Errfulcl(Res(C))) = Errfulcl(Res(C))$.

(5) This follows immediately from the definitions.

(6) From the definition, $Dub^*(Res(C)) = \{\vee L \in Cls$: there exists c in $Dub(Res(C))$ such that $Lit(c) \subseteq L\}$. Hence there are no literals in $Dub^*(Res(C))$. Therefore by Lemma 2.4.11(3), $Dub^*(Res(C)) \subseteq Res(Dub^*(Res(C)))$.

Take any two elements $\vee L_1$ and $\vee L_2$ of $Dub^*(Res(C))$ such that $\vee L_3$ is a resolvent of $\vee L_1$ and $\vee L_2$. Then there exists $\vee K_1$ in $Dub(Res(C))$ such that $L_1 = K_1 \cup (L_1 - K_1)$. Similarly, there exists $\vee K_2$ in $Dub(Res(C))$ such that $L_2 = K_2 \cup (L_2 - K_2)$. Suppose $L_3 = res(l; L_1, L_2)$. There are 3 cases to consider.

Case 1: $l \in K_1$ and $\sim l \in K_2$.

Then $\vee res(l; K_1, K_2) \in Res(Dub(Res(C)))$. By part 2, we have $Res(Dub(Res(C))) = Dub(Res(C))$. So $\vee res(l; K_1, K_2) \in Dub(Res(C))$. Since $res(l; K_1, K_2) \subseteq res(l; L_1, L_2) = L_3$, we see that $\vee L_3 \in Dub^*(Res(C))$.

Case 2: $l \in K_1$ and $\sim l \in (L_2 - K_2)$.

Then $K_2 \subseteq res(l; L_1, L_2) = L_3$. Hence $\vee L_3 \in Dub^*(Res(C))$.

Case 3: $l \in (L_1 - K_1)$ and $\sim l \in L_2$.

Then $K_1 \subseteq res(l; L_1, L_2) = L_3$. Hence $\vee L_3 \in Dub^*(Res(C))$.

So in all cases $\vee L_3 \in Dub^*(Res(C))$. Hence $Res(Dub^*(Res(C))) \subseteq Dub^*(Res(C))$. Thus $Res(Dub^*(Res(C))) = Dub^*(Res(C))$.

(7) There are no literals in $Errfulcl(Res(C)) \cup Dub^*(Res(C))$. By Lemma 2.4.11(3), $Errfulcl(Res(C)) \cup Dub^*(Res(C)) \subseteq Res(Errfulcl(Res(C)) \cup Dub^*(Res(C)))$.

Take any two elements $\vee L_1$ and $\vee L_2$ of $Errfulcl(Res(C)) \cup Dub^*(Res(C))$ such that $\vee L_3$ is a resolvent of $\vee L_1$ and $\vee L_2$. There are 3 cases to consider.

Case 1: $\{\vee L_1, \vee L_2\} \subseteq Errfulcl(Res(C))$.

Then $\vee L_3 \in Res(Errfulcl(Res(C)))$. By part 4 we have, $Res(Errfulcl(Res(C))) = Errfulcl(Res(C))$. So $\vee L_3 \in Errfulcl(Res(C)) \subseteq Errfulcl(Res(C)) \cup Dub^*(Res(C))$.

Case 2: $\{\vee L_1, \vee L_2\} \subseteq Dub^*(Res(C))$.

Then $\vee L_3 \in Res(Dub^*(Res(C)))$. By part 6, $Res(Dub^*(Res(C))) = Dub^*(Res(C))$. So $\vee L_3 \in Dub^*(Res(C)) \subseteq Errfulcl(Res(C)) \cup Dub^*(Res(C))$.

Case 3: $\vee L_1 \in Errfulcl(Res(C))$ and $\vee L_2 \in Dub^*(Res(C))$.

Then $\vee L_1 \in Res(C)$ and $L_1 \subseteq Errlit(Res(C))$. Also there exists $\vee K_2$ in $Dub(Res(C))$ such that $L_2 = K_2 \cup (L_2 - K_2)$. Hence $K_2 \not\subseteq Errlit(Res(C))$. Suppose $L_3 = res(l; L_1, L_2)$ where $l \in L_1$ and $\sim l \in L_2$. So $l \in Errlit(Res(C))$. By Lemma 2.4.11(6), $\vee \{l\} \in Res(C)$. By Lemma 3.2.4(1), $\sim l \in Errlit(Res(C))$. Hence $K_2 - \{\sim l\} \not\subseteq Errlit(Res(C))$. Since $\vee K_2 \in Res(C)$, by Lemma 2.4.11(6), $\vee(K_2 - \{\sim l\}) \in Res(C)$. So $\vee(K_2 - \{\sim l\}) \in Dub(Res(C))$. Hence $\vee(L_2 - \{\sim l\}) \in Dub^*(Res(C))$. So $\vee L_3 = \vee res(l; L_1, L_2) = \vee((L_1 - \{l\}) \cup (L_2 - \{\sim l\})) \in Dub^*(Res(C))$. Hence $\vee L_3 \in Dub^*(Res(C)) \subseteq Errfulcl(Res(C)) \cup Dub^*(Res(C))$.

So in all cases $\vee L_3 \in Errfulcl(Res(C)) \cup Dub^*(Res(C))$. Therefore we have $Res(Errfulcl(Res(C)) \cup Dub^*(Res(C))) \subseteq Errfulcl(Res(C)) \cup Dub^*(Res(C))$.

So $Res(Errfulcl(Res(C)) \cup Dub^*(Res(C))) = Errfulcl(Res(C)) \cup Dub^*(Res(C))$.

(8) Suppose $Dub(C) \models G$. By Lemma 2.8.5(4), $G \equiv Cl(G)$ and so $Dub(C) \models Cl(G)$. By Lemma 3.2.8(4), $\Upsilon(Cl(G)) \equiv Cl(G)$ and so $Dub(C) \models \Upsilon(Cl(G))$. So $\Upsilon(Cl(G)) \subseteq Cls(C_\models(Dub(C)))$. By Lemma 2.8.5(9), there are no tautologies in $Cl(G)$ and so there are no tautologies in $\Upsilon(Cl(G))$. So $\Upsilon(Cl(G)) \subseteq Cls(C_\models(Dub(C))) - Taut$. By

3.2 Four Satisfiable Subsets of a Set of Clauses 89

Lemma 2.6.5(7), if $c \in \Upsilon(Cl(G))$ then there exists c' in $CorCls(C_\models(Dub(C)))$ such that $Lit(c') \subseteq Lit(c)$. By Theorem 2.6.10(4), $CorRes(Dub(C)) = CorCls(C_\models(Dub(C)))$. So if $c \in \Upsilon(Cl(G))$ then there exists c' in $CorRes(Dub(C))$ such that $Lit(c') \subseteq Lit(c)$. By Lemma 2.6.5(6), $CorRes(Dub(C)) \subseteq ns(Res(Dub(C)))$. So $c' \in ns(Res(Dub(C)))$. Hence there exists c_1 in $Res(Dub(C))$ such that $c' = ns(c_1)$ and by Lemma 2.5.7(5), $Lit(c') = Lit(c_1)$. Therefore if $c \in \Upsilon(Cl(G))$ then there exists c_1 in $Res(Dub(C))$ such that $Lit(c_1) \subseteq Lit(c)$. By part (3), $Res(Dub(C)) \subseteq Dub(Res(C))$. Hence if $c \in \Upsilon(Cl(G))$ then there exists c_1 in $Dub(Res(C))$ such that $Lit(c_1) \subseteq Lit(c)$; and so $c \in Dub^*(Res(C))$.

Thus $\Upsilon(Cl(G)) \subseteq Dub^*(Res(C))$.

(9) Take any l in $Errlit(Errfulcl(Res(C)) \cup Dub^*(Res(C)))$. Then $\{\vee\{l\}, \vee\{\sim l\}\} \subseteq Res(Errfulcl(Res(C)) \cup Dub^*(Res(C)))$. So by part 7 we have, $\{\vee\{l\}, \vee\{\sim l\}\} \subseteq Errfulcl(Res(C)) \cup Dub^*(Res(C))$. If $c \in Errfulcl(Res(C))$ then $c \in Res(C)$. If $\vee\{l\} \in Dub^*(Res(C))$ then $\vee\{l\} \in Dub(Res(C)) \subseteq Res(C)$. Similarly, if $\vee\{\sim l\} \in Dub^*(Res(C))$ then $\vee\{\sim l\} \in Dub(Res(C)) \subseteq Res(C)$. So $\{\vee\{l\}, \vee\{\sim l\}\} \subseteq Res(C)$. Hence $l \in Errlit(C)$.

Thus $Errlit(Errfulcl(Res(C)) \cup Dub^*(Res(C))) \subseteq Errlit(C)$.

(10) By Lemma 3.2.4(4), $Errlit(C) \subseteq Errlit(C \cup Cl(G))$.

Suppose $Dub(C) \models G$. Then $Errlit(C \cup Cl(G))$
$= Errlit(\Upsilon(C \cup Cl(G)))$ by Lemma 3.2.8(7),
$= Errlit(\check{C} \cup \Upsilon(Cl(G)))$ by Lemma 3.2.8(2),
$\subseteq Errlit(Res(\check{C}) \cup \Upsilon(Cl(G)))$ by Lemma 3.2.8(6) and Lemma 3.2.4(4),
$= Errlit(Res(C) \cup \Upsilon(Cl(G)))$ by Lemma 3.2.8(5),
$= Errlit(Errfulcl(Res(C)) \cup Dub(Res(C)) \cup \Upsilon(Cl(G)))$ by Definition 3.2.3(5),
$\subseteq Errlit(Errfulcl(Res(C)) \cup Dub(Res(C)) \cup Dub^*(Res(C)))$ by part 8 and
    Lemma 3.2.4(4),
$= Errlit(Errfulcl(Res(C)) \cup Dub^*(Res(C)))$ by part 5,
$\subseteq Errlit(C)$ by part 9.

Therefore $Errlit(C \cup Cl(G)) \subseteq Errlit(C)$.

Thus $Errlit(C \cup Cl(G)) = Errlit(C)$.

(11) Suppose $Pure(C) \models G$. By Lemma 3.2.5(2), $Pure(C) \subseteq Dub(C)$. So $Dub(C) \models G$. By Lemma 3.2.9(10), $Errlit(C \cup Cl(G)) = Errlit(C)$.

EndProofLemma3.2.9

A set F of formulas will be converted to its set $Cl(F)$ of clauses; and so the following notation will be useful.

Definition 3.2.10. Let F be any set of formulas.
1) $NCl(F) = NilCl(F) = Nil(Cl(F))$.
2) $PCl(F) = PureCl(F) = Pure(Cl(F))$.
3) $DCl(F) = DubCl(F) = Dub(Cl(F))$.
4) $ICl(F) = \bigcap MaxSatCl(F) = \bigcap MaxSat(Cl(F))$.

The next lemma shows that for each X in $\{N, P, D, I\}$, XCl is pre-cumulative.

Lemma 3.2.11. Let F and G be any sets of formulas and suppose $X \in \{N,P,D,I\}$. If $XCl(F) \models G$ then $XCl(F \cup G) \equiv XCl(F)$.

Proof

Let F and G be any sets of formulas. Then either F is satisfiable or F is unsatisfiable. If F is unsatisfiable then we shall consider each value of X separately.

So suppose F is satisfiable. By Lemma 2.8.5(4), $F \equiv Cl(F)$, and so $Cl(F)$ is satisfiable. By Lemma 3.2.5(1), for each X in $\{N,P,D,I\}$, $XCl(F) = Cl(F)$. By Lemma 2.8.5(4), $G \equiv Cl(G)$ and by Definition 2.8.2(2), $Cl(F \cup G) = Cl(F) \cup Cl(G)$. Therefore $Cl(F \cup G) = Cl(F) \cup Cl(G) \models Cl(F)$. Also if $Cl(F) \models G$ then $Cl(F) \models Cl(G)$, and hence $Cl(F) \models Cl(F) \cup Cl(G) = Cl(F \cup G)$. Therefore, if $Cl(F) \models G$ then $Cl(F \cup G) \equiv Cl(F)$. Thus, if $XCl(F) \models G$ then $XCl(F \cup G) \equiv XCl(F)$.

Now suppose F is unsatisfiable and consider each value of X separately. By Lemma 2.8.5(4), $F \equiv Cl(F)$, and so $Cl(F)$ is unsatisfiable.

Case 1: $X = N$.

Suppose $NCl(F) \models G$. By Definition 3.2.1(1), $NCl(F) = Nil(Cl(F)) = \{\}$. Since $Cl(F)$ is unsatisfiable, $Cl(F) \cup Cl(G)$ is unsatisfiable too. Therefore $NCl(F \cup G) = Nil(Cl(F \cup G)) = Nil(Cl(F) \cup Cl(G)) = \{\} = NCl(F)$. Thus $NCl(F \cup G) \equiv NCl(F)$.

Case 2: $X = I$.

Suppose $ICl(F) \models G$. By Lemma 2.8.5(4), $G \equiv Cl(G)$. So $ICl(F) \models Cl(G)$.

We shall now show that $MaxSat(Cl(F) \cup Cl(G)) = \{S \cup Cl(G) : S \in MaxSatCl(F)\}$. Let $S_0 = ICl(F) = \bigcap MaxSatCl(F)$. Then $S_0 \models G$. For all S in $MaxSatCl(F)$, $S_0 \subseteq S$. Let v be any valuation and suppose $S \in MaxSatCl(F)$. If $v \models S$ then $v \models S_0$ and so $v \models G$ and hence $v \models Cl(G)$. Therefore $S \cup Cl(G) \equiv S$. Since S is maximal in $Sat(Cl(F))$, $S \cup Cl(G)$ is maximal in $Sat(Cl(F) \cup Cl(G))$. Therefore $\{S \cup Cl(G) : S \in MaxSatCl(F)\} \subseteq MaxSat(Cl(F) \cup Cl(G))$.

Conversely, take any S in $MaxSat(Cl(F) \cup Cl(G))$. Then S is satisfiable and $S = S_F \cup S_G$ where $S_F \subseteq Cl(F)$ and $S_G \subseteq Cl(G)$. Since S is satisfiable, S_F is satisfiable and so there exists S_M in $MaxSatCl(F)$ such that $S_F \subseteq S_M$. But in the previous paragraph we showed that $S_M \cup Cl(G) \equiv S_M$. Since S_M is satisfiable, $S_M \cup Cl(G) \in Sat(Cl(F) \cup Cl(G))$. But, $S = S_F \cup S_G \subseteq S_M \cup Cl(G)$. So $S = S_M \cup Cl(G) \in \{S \cup Cl(G) : S \in MaxSatCl(F)\}$. Therefore $MaxSat(Cl(F) \cup Cl(G)) \subseteq \{S \cup Cl(G) : S \in MaxSatCl(F)\}$.

Thus $MaxSat(Cl(F) \cup Cl(G)) = \{S \cup Cl(G) : S \in MaxSatCl(F)\}$.

So $ICl(F \cup G) = \bigcap MaxSatCl(F \cup G) = \bigcap MaxSat(Cl(F) \cup Cl(G)) = \bigcap \{S \cup Cl(G) : S \in MaxSatCl(F)\} = (\bigcap \{S : S \in MaxSatCl(F)\}) \cup Cl(G) = (\bigcap MaxSatCl(F)) \cup Cl(G) = ICl(F) \cup Cl(G)$.

Thus $ICl(F \cup G) \equiv ICl(F) \cup Cl(G) \equiv ICl(F)$.

Case 3: $X = P$.

Suppose $PCl(F) \models G$. We have $Cl(F \cup G) = Cl(F) \cup Cl(G)$. Hence $PCl(F \cup G) = Pure(Cl(F) \cup Cl(G))$. Let $Z = \{c \in Cl(G) : c \neq \vee\{\}$ and $Lit(c) \cap Errlit(Cl(F \cup G)) = \{\}\}$. Then $Z \subseteq Cl(G)$. From the definitions, $PCl(F \cup G)$
$= \{c \in Cl(F \cup G) : c \neq \vee\{\}$ and $Lit(c) \cap Errlit(Cl(F \cup G)) = \{\}\}$
$= \{c \in Cl(F) \cup Cl(G) : c \neq \vee\{\}$ and $Lit(c) \cap Errlit(Cl(F \cup G)) = \{\}\}$
$= \{c \in Cl(F) : c \neq \vee\{\}$ and $Lit(c) \cap Errlit(Cl(F \cup G)) = \{\}\} \cup$

3.2 Four Satisfiable Subsets of a Set of Clauses   91

$\{c \in Cl(G) : c \neq \vee\{\}$ and $Lit(c) \cap Errlit(Cl(F \cup G)) = \{\}\}$
$= \{c \in Cl(F) : c \neq \vee\{\}$ and $Lit(c) \cap Errlit(Cl(F \cup G)) = \{\}\} \cup Z$
$= \{c \in Cl(F) : c \neq \vee\{\}$ and $Lit(c) \cap Errlit(Cl(F)) = \{\}\} \cup Z$, by Lemma 3.2.9(11)
$= PCl(F) \cup Z$.

Therefore $PCl(F \cup G) \models PCl(F)$.

Conversely, by Lemma 2.8.5(4), $G \equiv Cl(G)$ and so $PCl(F) \models Cl(G)$. Hence $PCl(F) \models PCl(F) \cup Cl(G)$. But $Z \subseteq Cl(G)$, so $PCl(F) \models PCl(F) \cup Z$. Therefore $PCl(F) \models PCl(F \cup G)$.

Thus $PCl(F \cup G) \equiv PCl(F)$.

Case 4: $X = D$.
Suppose $DCl(F) \models G$. We have $Cl(F \cup G) = Cl(F) \cup Cl(G)$. Hence $DCl(F \cup G) = Dub(Cl(F) \cup Cl(G))$. Let $Z = \{c \in Cl(G) : Lit(c) \not\subseteq Errlit(Cl(F \cup G))\}$. Then $Z \subseteq Cl(G)$. From the definitions, $DCl(F \cup G)$
$= \{c \in Cl(F \cup G) : Lit(c) \not\subseteq Errlit(Cl(F \cup G))\}$
$= \{c \in Cl(F) \cup Cl(G) : Lit(c) \not\subseteq Errlit(Cl(F \cup G))\}$
$= \{c \in Cl(F) : Lit(c) \not\subseteq Errlit(Cl(F \cup G))\} \cup \{c \in Cl(G) : Lit(c) \not\subseteq Errlit(Cl(F \cup G))\}$
$= \{c \in Cl(F) : Lit(c) \not\subseteq Errlit(Cl(F \cup G))\} \cup Z$
$= \{c \in Cl(F) : Lit(c) \not\subseteq Errlit(Cl(F))\} \cup Z$, by Lemma 3.2.9(10)
$= DCl(F) \cup Z$.

Therefore $DCl(F \cup G) \models DCl(F)$.

Conversely, by Lemma 2.8.5(4), $G \equiv Cl(G)$ and so $DCl(F) \models Cl(G)$. Hence $DCl(F) \models DCl(F) \cup Cl(G)$. But $Z \subseteq Cl(G)$, so $DCl(F) \models DCl(F) \cup Z$. Therefore $DCl(F) \models DCl(F \cup G)$.

Thus $DCl(F \cup G) \equiv DCl(F)$.
EndProofLemma3.2.11

3.2.4 Disjunction

The defining property of disjunction is that a disjunction of a set of formulas is true (under some valuation) iff at least one of the formulas in the set is true (under the same valuation). Equivalently, a disjunction of a set of formulas is false (under some valuation) iff every formula in the set is false (under the same valuation). This leads to the following left disjunction property, where F is a set of formulas and f and g are formulas.

If $F \cup \{g\} \models f$ and $F \cup \{h\} \models f$ then $F \cup \{\vee\{g,h\}\} \models f$.

This can be rewritten as follows.

$C_\models(F \cup \{g\}) \cap C_\models(F \cup \{h\}) \subseteq C_\models(F \cup \{\vee\{g,h\}\})$.

This formulation is called "disjunction in the antecedent" in [64](page 47), and the "Or-rule" in [59](page 25).

When $h = \neg g$ the above property can be simplified to the following.

If $F \cup \{g\} \models f$ and $F \cup \{\neg g\} \models f$ then $F \models f$.

Or equivalently

$$C_{\models}(F\cup\{g\})\cap C_{\models}(F\cup\{\neg g\})\subseteq C_{\models}(F).$$

This formulation is called "proof by cases" in [64](page 47). It means that to prove f from F we may split the proof into two cases: one where g is true and the other where it is not. This is a common proof technique, often called "divide and conquer". We shall call this property "divisibility". These two properties are formally defined below.

Definition 3.2.12. Let C is a function from $\mathscr{P}(Fml)$ to $\mathscr{P}(Fml)$.
Left disjunction: C is *left disjunctive* iff for all $F\subseteq Fml$ and all $\{g,h\}\subseteq Fml$,
   $C(F\cup\{g\})\cap C(F\cup\{h\})\subseteq C(F\cup\{\vee\{g,h\}\})$.
Divisibility: C is *divisible* iff for all $F\subseteq Fml$ and all $g\in Fml$,
   $C(F\cup\{g\})\cap C(F\cup\{\neg g\})\subseteq C(F)$.

In propositional resolution logic we can form the disjunction of a set F of formulas, namely $\vee F$, but the disjunction of two sets F and G of formulas has not been defined. Although there are many reasonable definitions of the disjunction of F and G, we think that, whatever the operator is, it should be disjunctive, as defined in the next definition.

Definition 3.2.13. Let $\varovee$ be either a binary operation on $\mathscr{P}(Fml)$ or a binary operation on Fml. Then $\varovee$ is *disjunctive* iff for each valuation v, $v\models X\varovee Y$ iff $v\models X$ or $v\models Y$.

The symbol $\varovee$ is the disjunction symbol $\vee$ inside an O for operation or operator.

Before we use the concept of a disjunctive operator (or operation) we should show that at least one exists. A disjunctive operator on Fml is the usual disjunction operator $\vee$. A disjunctive operator $\curlyvee$ on $\mathscr{P}(Fml)$ is given in the next lemma. The symbol $\curlyvee$ is a curly version of $\vee$.

Lemma 3.2.14. Let each of F and G be a set of formulas. Let $F\curlyvee G=\{\vee\{f,g\}:f\in F$ and $g\in G\}$. If v is any valuation then $v\models F\curlyvee G$ iff $v\models F$ or $v\models G$. That is, $\curlyvee$ is disjunctive.
Proof

Let each of F and G be a set of formulas. Let v be any valuation.

If $v\models F$ then for each f in F, $v\models f$. Hence for each g in G, $v\models \vee\{f,g\}$. Therefore $v\models F\curlyvee G$.

If $v\models G$ then for each g in G, $v\models g$. Hence for each f in F, $v\models \vee\{f,g\}$. Therefore $v\models F\curlyvee G$.

Conversely, suppose not($v\models F$ or $v\models G$). Then $v\not\models F$ and $v\not\models G$. So there exists f in F and g in G such that $v\not\models f$ and $v\not\models g$. Hence $v\not\models \vee\{f,g\}$. Therefore $v\not\models F\curlyvee G$.

Thus $v\models F\curlyvee G$ iff $v\models F$ or $v\models G$.
EndProofLemma3.2.14

Some of the properties of a disjunctive operator are given in the next lemma. In particular, Lemma 3.2.15(10, 11, and 12) show that classical propositional logic satisfies stronger versions of left disjunction and divisibility. Moreover, considering the form of the left side of left disjunction, Lemma 3.2.15(9) shows the relevance of considering a disjunctive operation on $\mathscr{P}(Fml)$.

3.2 Four Satisfiable Subsets of a Set of Clauses

Lemma 3.2.15. Suppose each of F, F', G, G', and H is a set of formulas. Suppose each of f, f', g, g', and h is a formula. Let $\lor\!\!\!\!\lor$ be a disjunctive operation. (The domain of $\lor\!\!\!\!\lor$ will be clear from the context.)

1) If $T \subseteq \textit{Taut}$ then $T \lor\!\!\!\!\lor F \subseteq \textit{Taut}$ and $F \lor\!\!\!\!\lor T \subseteq \textit{Taut}$.
   If $\mathfrak{t} \in \textit{Taut}$ then $\mathfrak{t} \lor\!\!\!\!\lor f \in \textit{Taut}$ and $f \lor\!\!\!\!\lor \mathfrak{t} \in \textit{Taut}$.
2) $F \lor\!\!\!\!\lor F \equiv F$.
   $f \lor\!\!\!\!\lor f \equiv f$.
3) $F \lor\!\!\!\!\lor G \equiv G \lor\!\!\!\!\lor F$.
   $f \lor\!\!\!\!\lor g \equiv g \lor\!\!\!\!\lor f \equiv \lor\{f,g\} \equiv \{f\} \lor\!\!\!\!\lor \{g\}$.
4) If $F \models F'$ and $G \models G'$ then $F \lor\!\!\!\!\lor G \models F' \lor\!\!\!\!\lor G'$.
   If $f \models f'$ and $g \models g'$ then $f \lor\!\!\!\!\lor g \models f' \lor\!\!\!\!\lor g'$.
5) If $F \equiv F'$ and $G \equiv G'$ then $F \lor\!\!\!\!\lor G \equiv F' \lor\!\!\!\!\lor G'$.
   If $f \equiv f'$ and $g \equiv g'$ then $f \lor\!\!\!\!\lor g \equiv f' \lor\!\!\!\!\lor g'$.
6) $(F \cup G) \lor\!\!\!\!\lor (F \cup H) \equiv F \cup (G \lor\!\!\!\!\lor H)$.
7) If $F \models G$ or $F \models H$ then $F \models G \lor\!\!\!\!\lor H$.
8) $(F \models h$ and $G \models h)$ iff $F \lor\!\!\!\!\lor G \models h$.
9) $C_\models(F) \cap C_\models(G) = C_\models(F \lor\!\!\!\!\lor G)$.
10) $C_\models(F \cup G) \cap C_\models(F \cup H) = C_\models(F \cup (G \lor\!\!\!\!\lor H))$.
11) $C_\models(F \cup \{g\}) \cap C_\models(F \cup \{h\}) = C_\models(F \cup \{\lor\{g,h\}\})$.
12) $C_\models(F \cup \{g\}) \cap C_\models(F \cup \{\neg g\}) = C_\models(F)$.

Proof

Suppose each of F, F', G, G', and H is a set of formulas. Suppose each of f, f', g, g', and h is a formula. Let $\lor\!\!\!\!\lor$ be a disjunctive operation. (The domain of $\lor\!\!\!\!\lor$ will be clear from the context.) Let v be any valuation.

(1) Suppose $T \subseteq \textit{Taut}$. Then $v \models T \lor\!\!\!\!\lor F$ iff $v \models T$ or $v \models F$. But every valuation satisfies T, so every valuation satisfies $T \lor\!\!\!\!\lor F$. Hence $T \lor\!\!\!\!\lor F \subseteq \textit{Taut}$.

Similarly, $v \models F \lor\!\!\!\!\lor T$ iff $v \models F$ or $v \models T$. So every valuation satisfies $F \lor\!\!\!\!\lor T$. Hence $F \lor\!\!\!\!\lor T \subseteq \textit{Taut}$.

Suppose $\mathfrak{t} \in \textit{Taut}$. Then $v \models \mathfrak{t} \lor\!\!\!\!\lor f$ iff $v \models \mathfrak{t}$ or $v \models f$. But every valuation satisfies $\mathfrak{t}$, so every valuation satisfies $\mathfrak{t} \lor\!\!\!\!\lor f$. Hence $\mathfrak{t} \lor\!\!\!\!\lor f \in \textit{Taut}$.

Similarly, $v \models f \lor\!\!\!\!\lor \mathfrak{t}$ iff $v \models f$ or $v \models \mathfrak{t}$. So every valuation satisfies $f \lor\!\!\!\!\lor \mathfrak{t}$. Hence $f \lor\!\!\!\!\lor \mathfrak{t} \in \textit{Taut}$.

(2) Suppose either $X = F$ or $X = f$. By Definition 3.2.13, for each valuation v, we have $v \models X \lor\!\!\!\!\lor X$ iff $v \models X$. Hence $X \lor\!\!\!\!\lor X \equiv X$.

(3) Suppose either $X = F$ and $Y = G$, or $X = f$ and $Y = g$. By Definition 3.2.13, for each valuation v, $v \models X \lor\!\!\!\!\lor Y$ iff $v \models X$ or $v \models Y$ iff $v \models Y$ or $v \models X$ iff $v \models Y \lor\!\!\!\!\lor X$. Hence $X \lor\!\!\!\!\lor Y \equiv Y \lor\!\!\!\!\lor X$.

For each valuation v, $v \models f \lor\!\!\!\!\lor g$ iff $v \models f$ or $v \models g$ iff $v \models \lor\{f,g\}$. Hence $f \lor\!\!\!\!\lor g \equiv \lor\{f,g\}$. For each valuation v, $v \models \{f\} \lor\!\!\!\!\lor \{g\}$ iff $v \models \{f\}$ or $v \models \{g\}$ iff $v \models f$ or $v \models g$ iff $v \models \lor\{f,g\}$. Hence $\{f\} \lor\!\!\!\!\lor \{g\} \equiv \lor\{f,g\}$.

(4) Suppose either $X = F$, $X' = F'$, $Y = G$, and $Y' = G'$; or $X = f$, $X' = f'$, $Y = g$, and $Y' = g'$. Suppose $X \models X'$ and $Y \models Y'$. Then $v \models X \lor\!\!\!\!\lor Y$
iff $v \models X$ or $v \models Y$
implies $v \models X'$ or $v \models Y'$

iff $v \models X' \otimes Y'$.
Thus $X \otimes Y \models X' \otimes Y'$.

(5) Suppose either $X = F$, $X' = F'$, $Y = G$, and $Y' = G'$; or $X = f$, $X' = f'$, $Y = g$, and $Y' = g'$. Suppose $X \equiv X'$ and $Y \equiv Y'$. Then $v \models X \otimes Y$
iff $v \models X$ or $v \models Y$
iff $v \models X'$ or $v \models Y'$
iff $v \models X' \otimes Y'$.
Thus $X \otimes Y \equiv X' \otimes Y'$.

(6) $v \models (F \cup G) \otimes (F \cup H)$
iff $v \models (F \cup G)$ or $v \models (F \cup H)$
iff $v \models F$ and $v \models G$, or $v \models F$ and $v \models H$
iff $v \models F$, and $v \models G$ or $v \models H$
iff $v \models F$, and $v \models G \otimes H$
iff $v \models (F \cup (G \otimes H))$.
Thus $(F \cup G) \otimes (F \cup H) \equiv F \cup (G \otimes H)$.

(7) Suppose $F \models G$ or $F \models H$. If $v \models F$ then $v \models G$ or $v \models H$; and so $v \models G \otimes H$. Thus $F \models G \otimes H$.

(8) Suppose $F \models h$ and $G \models h$. Suppose $v \models F \otimes G$. Then $v \models F$ or $v \models G$. So in either case, $v \models h$. Hence $F \otimes G \models h$.

Conversely, suppose $F \otimes G \models h$. If $v \models F$, then $v \models F \otimes G$, and so $v \models h$. Therefore $F \models h$. Similarly, if $v \models G$, then $v \models F \otimes G$, and so $v \models h$. Therefore $G \models h$.

Thus $(F \models h$ and $G \models h)$ iff $F \otimes G \models h$.

(9) Take any h in $C_\models(F) \cap C_\models(G)$. Then $F \models h$ and $G \models h$. By part (8), $F \otimes G \models h$, and so $h \in C_\models(F \otimes G)$. Therefore $C_\models(F) \cap C_\models(G) \subseteq C_\models(F \otimes G)$.

Conversely, take any h in $C_\models(F \otimes G)$. Then $F \otimes G \models h$. By part (8), $F \models h$ and $G \models h$. Hence $h \in C_\models(F)$ and $h \in C_\models(G)$. Therefore $h \in C_\models(F) \cap C_\models(G)$ and so $C_\models(F \otimes G) \subseteq C_\models(F) \cap C_\models(G)$.

Thus $C_\models(F) \cap C_\models(G) = C_\models(F \otimes G)$.

(10) By Lemma 3.2.15(6), $F \cup (G \otimes H) \equiv (F \cup G) \otimes (F \cup H)$. So $C_\models(F \cup (G \otimes H))$
$= C_\models((F \cup G) \otimes (F \cup H))$ by Lemma 2.2.21(5),
$= C_\models(F \cup G) \cap C_\models(F \cup H)$ by Lemma 3.2.15(9).

(11) $f \in C_\models(F \cup \{g\}) \cap C_\models(F \cup \{h\})$
iff $F \cup \{g\} \models f$ and $F \cup \{h\} \models f$
iff $F \cup \{\vee\{g,h\}\} \models f$
iff $f \in C_\models(F \cup \{\vee\{g,h\}\})$.
Thus $C_\models(F \cup \{g\}) \cap C_\models(F \cup \{h\}) = C_\models(F \cup \{\vee\{g,h\}\})$.

(12) $C_\models(F \cup \{g\}) \cap C_\models(F \cup \{\neg g\})$
$= C_\models(F \cup \{\vee\{g, \neg g\}\})$ by part (11),
$= C_\models(F)$ by Lemma 2.2.21(5).

EndProofLemma3.2.15

We would like to see the extent to which the logics defined in Section 3.3 satisfy a version of left disjunction. Consider the very strong version of left disjunction in Lemma 3.2.15(10), $C_\models(F \cup G) \cap C_\models(F \cup H) = C_\models(F \cup (G \otimes H))$. For the logics in Section 3.3 to satisfy this we would need $XCl(F \cup G) \otimes XCl(F \cup H) \equiv XCl(F \cup (G \otimes H))$,

3.2 Four Satisfiable Subsets of a Set of Clauses

where $X \in \{N,P,D,I\}$. But the counter-example below in Lemma 3.2.16(5) shows that, in general, we do not have $XCl(F \cup G) \varovee XCl(F \cup H) \models XCl(F \cup (G \varovee H))$. So the best we can hope for is $XCl(F \cup (G \varovee H)) \models XCl(F \cup G) \varovee XCl(F \cup H)$, which still yields a version of left disjunction.

If $X = N$ then, by Lemma 3.2.16(2, 6, and 8) below, we have, $NCl(F \cup (G \varovee H)) \models NCl(F \cup G) \varovee NCl(F \cup H)$.

If at least one of $F \cup G$ and $F \cup H$ is satisfiable, then by Lemma 3.2.16(2 and 6) below, $XCl(F \cup (G \varovee H)) \models XCl(F \cup G) \varovee XCl(F \cup H)$.

But if both $F \cup G$ and $F \cup H$ are unsatisfiable, and $X \in \{P,D,I\}$, then, in general, we do not have $XCl(F \cup (G \varovee H)) \models XCl(F \cup G) \varovee XCl(F \cup H)$; as shown by the counter-example in Lemma 3.2.16(9) below.

Lemma 3.2.16. Let each of F, G, and H be a set of formulas. Let each of a and b be atoms. Let $\varovee$ be a disjunctive operation. Suppose $X \in \{N,P,D,I\}$.
1) If both G and H are satisfiable then $XCl(G \varovee H) \equiv G \varovee H \equiv XCl(G) \varovee XCl(H)$.
2) If both $F \cup G$ and $F \cup H$ are satisfiable then
$XCl(F \cup (G \varovee H)) \equiv F \cup (G \varovee H)$
$\equiv (F \cup G) \varovee (F \cup H)$
$\equiv XCl(F \cup G) \varovee XCl(F \cup H)$.
3) If $G = \{a\}$ and $H = \{a, \neg a\}$ then $XCl(G) \varovee XCl(H) \not\models XCl(G \varovee H)$.
4) Let one of G and H be satisfiable and the other be unsatisfiable. Then $XCl(G \varovee H) \models XCl(G) \varovee XCl(H)$.
5) If $F = \{a\} = G$ and $H = \{\neg a\}$ then we have $XCl(F \cup G) \varovee XCl(F \cup H) \subseteq$ Taut and $XCl(F \cup (G \varovee H)) \not\subseteq$ Taut. Hence $XCl(F \cup G) \varovee XCl(F \cup H) \not\models XCl(F \cup (G \varovee H))$.
6) Let one of $F \cup G$ and $F \cup H$ be satisfiable and the other be unsatisfiable. Then $F \cup (G \varovee H)$ is satisfiable; and $XCl(F \cup (G \varovee H)) \models XCl(F \cup G) \varovee XCl(F \cup H)$.
7) If both G and H are unsatisfiable then $NCl(G) = \{\} = NCl(H)$, $NCl(G \varovee H) = \{\}$, and $NCl(G) \varovee NCl(H) = \{\} \varovee \{\} \subseteq$ Taut. So $NCl(G \varovee H) \equiv NCl(G) \varovee NCl(H)$.
8) Let both $F \cup G$ and $F \cup H$ be unsatisfiable. Then $F \cup (G \varovee H)$ is unsatisfiable, $NCl(F \cup (G \varovee H)) = \{\}$, $NCl(F \cup G) = \{\}$, and $NCl(F \cup H) = \{\}$. So $NCl(F \cup (G \varovee H)) \equiv NCl(F \cup G) \varovee NCl(F \cup H)$.
9) If $X \in \{P,D,I\}$, $F = \{a,b\}$, $G = \{\neg a\}$, and $H = \{\neg b\}$ then
$XCl(F \cup G) \varovee XCl(F \cup H) \equiv \vee \{b,a\}$, and $\vee \{\neg a, \neg b\} \models XCl(F \cup (G \varovee H))$. Hence $XCl(F \cup (G \varovee H)) \not\models XCl(F \cup G) \varovee XCl(F \cup H)$.

Proof

Let each of F, G, and H be a set of formulas. Let each of a and b be atoms. Let $\varovee$ be a disjunctive operation. Suppose $X \in \{N,P,D,I\}$.

(1) Suppose both G and H are satisfiable. By Lemma 2.8.5(4), both $Cl(G)$ and $Cl(H)$ are satisfiable. By Lemma 3.2.5(1), $XCl(G) = Cl(G)$ and $XCl(H) = Cl(H)$. Also by Definition 3.2.13, $G \varovee H$ is satisfiable. By Lemma 2.8.5(4), $Cl(G \varovee H)$ is satisfiable. Therefore $XCl(G \varovee H)$
$= Cl(G \varovee H)$ by Lemma 3.2.5(1),
$\equiv G \varovee H$ by Lemma 2.8.5(4),
$\equiv Cl(G) \varovee Cl(H)$ by Lemma 2.8.5(4) and Lemma 3.2.15(5),
$= XCl(G) \varovee XCl(H)$.

(2) Let both $F \cup G$ and $F \cup H$ be satisfiable. By Definition 3.2.13, $(F \cup G) \varotimes (F \cup H)$ is satisfiable. So by Lemma 3.2.15(6), $F \cup (G \varotimes H)$ is satisfiable. By Lemma 2.8.5(4), $Cl(F \cup (G \varotimes H))$ is satisfiable. Therefore $XCl(F \cup G) \varotimes XCl(F \cup H)$
$\equiv (F \cup G) \varotimes (F \cup H)$ by part (1),
$\equiv F \cup (G \varotimes H)$ by Lemma 3.2.15(6),
$\equiv Cl(F \cup (G \varotimes H))$ by Lemma 2.8.5(4),
$= XCl(F \cup (G \varotimes H))$ by Lemma 3.2.5(1).

(3) Suppose $G = \{a\}$ and $H = \{a, \neg a\}$. Then G is satisfiable, and therefore by Lemma 2.8.5(4), $Cl(G)$ is satisfiable. Therefore $XCl(G)$
$= Cl(G)$ by Lemma 3.2.5(1),
$= \bigcup \{Cl(a)\}$
$= Cl(a)$
$= \{a\}$ by Lemma 2.8.3(5),
$= G$.
And $Cl(H)$
$= \bigcup \{Cl(a), Cl(\neg a)\}$
$= Cl(a) \cup Cl(\neg a)$
$= \{a, \neg a\}$ by Lemma 2.8.3(5),
$= H$.
So $MaxSat(Cl(H)) = MaxSat(\{a, \neg a\}) = \{\{a\}, \{\neg a\}\}$. So $\bigcap MaxSat(Cl(H)) = \{\}$. By Lemma 3.2.5(2), $XCl(H) = \{\}$. Therefore $XCl(G) \varotimes XCl(H) = \{a\} \varotimes \{\} \subseteq \textit{Taut}$ by Lemma 3.2.15(1).

But $G \varotimes H = \{a\} \varotimes \{a, \neg a\}$. By Definition 3.2.13, $G \varotimes H$ is satisfiable. Hence by Lemma 2.8.5(4), $Cl(G \varotimes H)$ is satisfiable. Therefore $XCl(G \varotimes H)$
$= Cl(G \varotimes H)$ by Lemma 3.2.5(1),
$\equiv G \varotimes H$ by Lemma 2.8.5(4),
$= \{a\} \varotimes \{a, \neg a\}$.

So any valuation that makes a false will satisfy $XCl(G) \varotimes XCl(H)$, but will not satisfy $\{a\} \varotimes \{a, \neg a\}$ and so will not satisfy $XCl(G \varotimes H)$. Thus $XCl(G) \varotimes XCl(H) \not\models XCl(G \varotimes H)$.

(4) Suppose one of G and H is satisfiable and the other is unsatisfiable. By Lemma 3.2.15(3), $G \varotimes H \equiv H \varotimes G$. So without loss of generality we may suppose that G is satisfiable and H is unsatisfiable.

By Lemma 2.8.5(4) we have, $Cl(G)$ is satisfiable and $Cl(H)$ is unsatisfiable. From Lemma 3.2.5(1), $XCl(G) = Cl(G)$. By Lemma 3.2.5(2), $XCl(H) \subseteq Cl(H)$. Hence $Cl(H) \models XCl(H)$. So by Lemma 3.2.15(4), $Cl(G) \varotimes Cl(H) \models XCl(G) \varotimes XCl(H)$.

By Definition 3.2.13, $G \varotimes H$ is satisfiable, and so by Lemma 2.8.5(4), $Cl(G \varotimes H)$ is satisfiable. Therefore $XCl(G \varotimes H)$
$= Cl(G \varotimes H)$ by Lemma 3.2.5(1),
$\equiv G \varotimes H$ by Lemma 2.8.5(4),
$\equiv Cl(G) \varotimes Cl(H)$ by Lemma 2.8.5(4) and Lemma 3.2.15(5),
$\models XCl(G) \varotimes XCl(H)$ by previous paragraph.

(5) Suppose $F = \{a\} = G$ and $H = \{\neg a\}$. Then $XCl(F \cup G)$
$= XCl(\{a\})$

3.2 Four Satisfiable Subsets of a Set of Clauses

$= XCl(a)$ by Lemma 2.8.5(2),
$= X(\{a\})$ by Lemma 2.8.3(5),
$= \{a\}$ by Lemma 3.2.5(1).
And $XCl(F \cup H)$
$= XCl(\{a, \neg a\}) = X(\bigcup\{Cl(a), Cl(\neg a)\})$
$= X(\bigcup\{\{a\}, \{\neg a\}\})$ by Lemma 2.8.3(5),
$= X(\{a, \neg a\})$
$= \{\}$ because by Lemma 3.2.5(2),
$$\{\} \subseteq X(\{a, \neg a\}) \subseteq \bigcap MaxSat(\{a, \neg a\}) = \{a\} \cap \{\neg a\} = \{\}.$$
Therefore $XCl(F \cup G) \otimes XCl(F \cup H) = \{a\} \otimes \{\} \subseteq Taut$ by Lemma 3.2.15(1).

By Lemma 3.2.15(6), $F \cup (G \otimes H) \equiv (F \cup G) \otimes (F \cup H) = \{a\} \otimes \{a, \neg a\}$. By Definition 3.2.13, $\{a\} \otimes \{a, \neg a\}$ is satisfiable, and so by Lemma 2.8.5(4), $Cl(\{a\} \otimes \{a, \neg a\})$ is satisfiable. Therefore $XCl(F \cup (G \otimes H))$
$= Cl(F \cup (G \otimes H))$ by Lemma 3.2.5(1),
$\equiv F \cup (G \otimes H)$ by Lemma 2.8.5(4),
$= \{a\} \otimes \{a, \neg a\}$.
So any valuation that makes a false will not satisfy $\{a\} \otimes \{a, \neg a\}$ and so will not satisfy $XCl(F \cup (G \otimes H))$. Hence $XCl(F \cup (G \otimes H)) \not\subseteq Taut$.

Thus $XCl(F \cup G) \otimes XCl(F \cup H) \not\models XCl(F \cup (G \otimes H))$.

(6) Let one of $F \cup G$ and $F \cup H$ be satisfiable and the other be unsatisfiable.

By Lemma 3.2.15(6), $(F \cup G) \otimes (F \cup H) \equiv F \cup (G \otimes H)$. So by Definition 3.2.13, $F \cup (G \otimes H)$ is satisfiable.

$XCl(F \cup (G \otimes H))$
$= Cl(F \cup (G \otimes H))$ by Lemma 3.2.5(1),
$\equiv F \cup (G \otimes H)$ by Lemma 2.8.5(4),
$\equiv (F \cup G) \otimes (F \cup H)$ by Lemma 3.2.15(6),
$\equiv Cl((F \cup G) \otimes (F \cup H))$ by Lemma 2.8.5(4),
$= XCl((F \cup G) \otimes (F \cup H))$ by Lemma 3.2.5(1),
$\models XCl(F \cup G) \otimes XCl(F \cup H)$ by part (4).

(7) Let both G and H be unsatisfiable. By Definition 3.2.13, $G \otimes H$ is unsatisfiable. By Lemma 2.8.5(4), $G \equiv Cl(G)$, $H \equiv Cl(H)$, and $G \otimes H \equiv Cl(G \otimes H)$. By Definition 3.2.10 and Definition 3.2.1(1), $NCl(G) = \{\} = NCl(H)$, $NCl(G \otimes H) = \{\}$. By Lemma 3.2.15(1), $NCl(G) \otimes NCl(H) = \{\} \otimes \{\} \subseteq Taut$. So $NCl(G \otimes H) = \{\} \equiv \{\} \otimes \{\}$
$= NCl(G) \otimes NCl(H)$.

(8) Let both $F \cup G$ and $F \cup H$ be unsatisfiable. By Definition 3.2.13, $(F \cup G) \otimes (F \cup H)$ is unsatisfiable. By Lemma 3.2.15(6), $(F \cup G) \otimes (F \cup H) \equiv F \cup (G \otimes H)$. So $F \cup (G \otimes H)$ is unsatisfiable. By Lemma 2.8.5(4), $F \cup G \equiv Cl(F \cup G)$, $F \cup H \equiv Cl(F \cup H)$, and also $F \cup (G \otimes H) \equiv Cl(F \cup (G \otimes H))$. By Definition 3.2.10 and Definition 3.2.1(1) we have, $NCl(F \cup (G \otimes H)) = \{\}$, $NCl(F \cup G) = \{\}$, and $NCl(F \cup H) = \{\}$. By Lemma 3.2.15(1), $NCl(F \cup G) \otimes NCl(F \cup H) = \{\} \otimes \{\} \subseteq Taut$. So $NCl(F \cup (G \otimes H)) = \{\} \equiv \{\} \otimes \{\}$
$= NCl(F \cup G) \otimes NCl(F \cup H)$.

(9) Suppose $X \in \{P, D, I\}$, $F = \{a, b\}$, $G = \{\neg a\}$, and $H = \{\neg b\}$.

Then $F \cup G = \{a, b, \neg a\}$ and so $Cl(F \cup G)$
$= \bigcup\{Cl(a), Cl(b), Cl(\neg a)\}$

$= \{a\} \cup \{b\} \cup \{\neg a\}$ by Lemma 2.8.3(5),
$= \{a, b, \neg a\}$.
Hence $XCl(F \cup G) = \{b\}$.

Similarly, $F \cup H = \{a, b, \neg b\}$ and so $Cl(F \cup H)$
$= \bigcup \{Cl(a), Cl(b), Cl(\neg b)\}$
$= \{a\} \cup \{b\} \cup \{\neg b\}$ by Lemma 2.8.3(5),
$= \{a, b, \neg b\}$.
Hence $XCl(F \cup H) = \{a\}$.

Therefore $XCl(F \cup G) \otimes XCl(F \cup H) = \{b\} \otimes \{a\} \equiv \vee \{b, a\}$ by Lemma 3.2.15(3).

Also $F \cup (G \otimes H) = \{a, b\} \cup (\{\neg a\} \otimes \{\neg b\})$ and so $Cl(F \cup (G \otimes H)) = Cl(a) \cup Cl(b) \cup Cl(\{\neg a\} \otimes \{\neg b\}) = \{a, b\} \cup Cl(\{\neg a\} \otimes \{\neg b\})$ by Lemma 2.8.3(5).

Now $Cl(\{\neg a\} \otimes \{\neg b\})$
$\equiv \{\neg a\} \otimes \{\neg b\}$ by Lemma 2.8.5(4),
$\equiv \vee \{\neg a, \neg b\}$ by Lemma 3.2.15(3).

So $\{a, b\} \cup Cl(\{\neg a\} \otimes \{\neg b\})$ is unsatisfiable. Since $Cl(\{\neg a\} \otimes \{\neg b\}) \equiv \vee \{\neg a, \neg b\}$, we see that both $\{a\} \cup Cl(\{\neg a\} \otimes \{\neg b\})$ and $\{b\} \cup Cl(\{\neg a\} \otimes \{\neg b\})$ are maximal satisfiable subsets of $\{a, b\} \cup Cl(\{\neg a\} \otimes \{\neg b\})$. So $ICl(F \cup (G \otimes H)) \subseteq Cl(\{\neg a\} \otimes \{\neg b\})$. Hence $XCl(F \cup (G \otimes H))$
$\subseteq ICl(F \cup (G \otimes H))$ by Lemma 3.2.5(2),
$\subseteq Cl(\{\neg a\} \otimes \{\neg b\})$
$\equiv \vee \{\neg a, \neg b\}$.
Therefore $\vee \{\neg a, \neg b\} \models XCl(F \cup (G \otimes H))$.

So any valuation that makes a false and b false satisfies $XCl(F \cup (G \otimes H))$. But it does not satisfy $\vee \{b, a\}$ and so does not satisfy $XCl(F \cup G) \otimes XCl(F \cup H)$.

Thus $XCl(F \cup (G \otimes H)) \not\models XCl(F \cup G) \otimes XCl(F \cup H)$.
EndProofLemma3.2.16

Lemma 3.2.16(9) implies that when $X \in \{P, D, I\}$, in general, left disjunction fails for the logics defined in Section 3.3.

So let us consider divisibility. The following property, which we shall call pre-divisibility, ensures that the logics defined in Section 3.3 are divisible.

Definition 3.2.17. Suppose C is a function that takes a set F of formulas and outputs the set $C(F)$ of formulas.
Pre-divisibility: C is pre-divisible iff whenever $\otimes$ is any disjunctive operation, and F is any set of formulas, and g is any formula, then
$$C(F) \models C(F \cup \{g\}) \otimes C(F \cup \{\neg g\}).$$

In the above definition, we can think of C as converting F to a suitable set of clauses $C(F)$. So pre-divisibility says that if a valuation satisfies $C(F)$ then either it satisfies $C(F \cup \{g\})$ or it satisfies $C(F \cup \{\neg g\})$.

The next lemma shows that for each X in $\{N, P, D, I\}$, XCl is pre-divisible.

Lemma 3.2.18. Suppose $X \in \{N, P, D, I\}$. Let $\otimes$ be any disjunctive operation, F be any set of formulas, and g be any formula. Then
$XCl(F) \models XCl(F \cup \{g\}) \otimes XCl(F \cup \{\neg g\})$.

3.2 Four Satisfiable Subsets of a Set of Clauses

Proof

Suppose $X \in \{N,P,D,I\}$. Let F be any set of formulas, g be any formula, and $\otimes$ be any disjunctive operation. If $h \in \{g, \neg g\}$ and H is any set of formulas then by Lemma 2.8.5(3 and 4), $Cl(h) \equiv h$, and $Cl(H) \equiv H$; and by Definition 2.8.2(2), $Cl(F \cup H) = Cl(F) \cup Cl(H)$, and $Cl(F \cup \{h\}) = Cl(F) \cup Cl(h)$.

Suppose $X = N$.

If F is satisfiable then by Definition 3.2.1(1), $NCl(F) = Cl(F) \equiv F$. Suppose v is a valuation such that $v \models NCl(F)$. Then $v \models F$ and so there exists h in $\{g, \neg g\}$ such that $v \models F \cup \{h\}$. Hence $NCl(F \cup \{h\}) = Cl(F \cup \{h\}) \equiv F \cup \{h\}$. So $v \models NCl(F \cup \{h\})$. Hence $v \models NCl(F \cup \{g\}) \otimes NCl(F \cup \{\neg g\})$. Thus $NCl(F) \models NCl(F \cup \{g\}) \otimes NCl(F \cup \{\neg g\})$.

If F is unsatisfiable then by Definition 3.2.1(1), $NCl(F) = \{\}$. Also $F \cup \{g\}$ is unsatisfiable and $F \cup \{\neg g\}$ is unsatisfiable. Therefore $NCl(F \cup \{g\}) = \{\}$ and $NCl(F \cup \{\neg g\}) = \{\}$. Thus $NCl(F) \models NCl(F \cup \{g\}) \otimes NCl(F \cup \{\neg g\})$.

Suppose $X = P$. Suppose $h \in \{g, \neg g\}$. We show $PCl(F \cup \{h\}) \subseteq PCl(F) \cup Cl(h)$.
$PCl(F \cup \{h\})$
$= Pure(Cl(F \cup \{h\}))$ by Definition 3.2.10,
$= Pure(Cl(F) \cup Cl(h))$
$= (Cl(F) \cup Cl(h)) - Errcl(Cl(F) \cup Cl(h))$ by Definition 3.2.3,
$\subseteq (Cl(F) \cup Cl(h)) - Errcl(Cl(F))$ by Lemma 3.2.4(5),
$= (Cl(F) - Errcl(Cl(F))) \cup (Cl(h) - Errcl(Cl(F)))$
$= Pure(Cl(F)) \cup (Cl(h) - Errcl(Cl(F)))$ by Definition 3.2.3,
$\subseteq Pure(Cl(F)) \cup Cl(h)$
$= PCl(F) \cup Cl(h)$ by Definition 3.2.10.

Take any valuation v. Then either $v \models g$ or $v \models \neg g$. So we have either $v \models Cl(g)$ or $v \models Cl(\neg g)$. Hence there exists h in $\{g, \neg g\}$ such that $v \models Cl(h)$.

Suppose $v \models PCl(F)$. Then there exists h in $\{g, \neg g\}$ such that $v \models PCl(F) \cup Cl(h)$. So $v \models PCl(F \cup \{h\})$. Hence $v \models PCl(F \cup \{g\}) \otimes PCl(F \cup \{\neg g\})$.

Thus $PCl(F) \models PCl(F \cup \{g\}) \otimes PCl(F \cup \{\neg g\})$.

Suppose $X = D$. Suppose $h \in \{g, \neg g\}$. We show $DCl(F \cup \{h\}) \subseteq DCl(F) \cup Cl(h)$.
$DCl(F \cup \{h\})$
$= Dub(Cl(F \cup \{h\}))$ by Definition 3.2.10,
$= Dub(Cl(F) \cup Cl(h))$
$= (Cl(F) \cup Cl(h)) - Errfulcl(Cl(F) \cup Cl(h))$ by Definition 3.2.3,
$\subseteq (Cl(F) \cup Cl(h)) - Errfulcl(Cl(F))$ by Lemma 3.2.4(6),
$= (Cl(F) - Errfulcl(Cl(F))) \cup (Cl(h) - Errfulcl(Cl(F)))$
$= Dub(Cl(F)) \cup (Cl(h) - Errfulcl(Cl(F)))$ by Definition 3.2.3,
$\subseteq Dub(Cl(F)) \cup Cl(h)$
$= DCl(F) \cup Cl(h)$ by Definition 3.2.10.

Take any valuation v. Then either $v \models g$ or $v \models \neg g$. So we have either $v \models Cl(g)$ or $v \models Cl(\neg g)$. Hence there exists h in $\{g, \neg g\}$ such that $v \models Cl(h)$.

Suppose $v \models DCl(F)$. Then there exists h in $\{g, \neg g\}$ such that $v \models DCl(F) \cup Cl(h)$. So $v \models DCl(F \cup \{h\})$. Hence $v \models DCl(F \cup \{g\}) \otimes DCl(F \cup \{\neg g\})$.

Thus $DCl(F) \models DCl(F \cup \{g\}) \otimes DCl(F \cup \{\neg g\})$.

Suppose $X = I$. If $h \in \{g, \neg g\}$ then by Definition 3.2.10 we have, $ICl(F) = \bigcap MaxSat(Cl(F))$, and $ICl(F \cup \{h\}) = \bigcap MaxSat(Cl(F \cup \{h\}))$.

Take any valuation v. Then either $v \models g$ or $v \models \neg g$. So we have either $v \models Cl(g)$ or $v \models Cl(\neg g)$. Hence there exists h in $\{g, \neg g\}$ such that $v \models Cl(h)$.

Suppose $v \models ICl(F)$. Then $v \models \bigcap MaxSat(Cl(F))$. Since $v \models Cl(h)$, by Theorem 3.2.2(9), $v \models \bigcap MaxSat(Cl(F) \cup Cl(h))$. Hence $v \models \bigcap MaxSat(Cl(F \cup \{h\}))$, and so $v \models ICl(F \cup \{h\})$. Therefore $v \models ICl(F \cup \{g\}) \lor ICl(F \cup \{\neg g\})$.

Thus $ICl(F) \models ICl(F \cup \{g\}) \lor ICl(F \cup \{\neg g\})$.
EndProofLemma3.2.18

3.3 Rational Propositional Logics

If C is any set of clauses then we have the following four properties.
1) $Nil(C)$, $Pure(C)$, $Dub(C)$, and $\bigcap MaxSat(C)$ are all satisfiable subsets of C, by Lemma 3.2.5.
2) If C is satisfiable then by Lemma 3.2.5, $Nil(C) = C$, $Pure(C) = C$, $Dub(C) = C$, and $\bigcap MaxSat(C) = C$.
3) If $X \in \{N, P, D, I\}$ then XCl is pre-cumulative, by Lemma 3.2.11.
4) If $X \in \{N, P, D, I\}$ then XCl is pre-divisible, by Lemma 3.2.18.

These four properties are highly desirable for making rational deductions from a set of clauses. The example in Lemma 3.2.16(9) shows that, in general, left disjunction is not desirable for this purpose.

So if F is any set of formulas then $NCl(F)$, $PCl(F)$, $DCl(F)$, and $ICl(F)$, are suitable subsets of $Cl(F)$ for making rational deductions. But there may be others. So rather than develop logics for just these four examples, we shall generalise them.

3.3.1 Rational Clauses Functions

For any set F of formulas we want to select a suitable subset of $Cl(F)$ from which we can prove formulas. That is we want a function whose input is a set of formulas, whose output is a set of clauses, and which is the composition of Cl followed by a selection function. This property is captured by RC1 of Definition 3.3.1 below.

The second property that we want is for the selected subset of $Cl(F)$ to be satisfiable. This is done by RC2 below.

Of course if $Cl(F)$ is satisfiable then we want to select all of $Cl(F)$. RC3 below does this.

We also want this function to be pre-cumulative, which is done by RC4 below, and pre-divisible, which is done by RC5 below.

We shall call such a desired function a rational clauses function. The notation used to denote rational clauses functions is borrowed from Definition 3.2.10 and reminds

3.3 Rational Propositional Logics

us that they are the composition of *Cl* followed by a selection function. Examples of the notation are *RCl* and *SCl*.

Definition 3.3.1. A function *RCl* from $\mathscr{P}(Fml)$ to $\mathscr{P}(Cls)$ is called a *rational clauses function* iff RC1, RC2, RC3, RC4, and RC5 all hold.
RC1) If $F \subseteq Fml$ and $G \subseteq Fml$ and $Cl(F) = Cl(G)$ then $RCl(F) = RCl(G)$.
RC2) If $F \subseteq Fml$ then $RCl(F)$ is a satisfiable subset of $Cl(F)$.
RC3) If $F \subseteq Fml$ and F is satisfiable then $RCl(F) = Cl(F)$.
RC4) If $F \subseteq Fml$ and $G \subseteq Fml$ and $RCl(F) \models G$ then $RCl(F \cup G) \equiv RCl(F)$.
RC5) If $F \subseteq Fml$ and $g \in Fml$ and $\varovee$ is a disjunctive operation then
   $RCl(F) \models RCl(F \cup \{g\}) \varovee RCl(F \cup \{\neg g\})$.

Lemma 3.2.5(1,3), Lemma 3.2.11, and Lemma 3.2.18, show that *NCl*, *PCl*, *DCl*, and *ICl* are all rational clauses functions.

Some initial properties of rational clauses functions are proved in the following lemma.

Lemma 3.3.2. Let *RCl* be a rational clauses function. Let F be any set of formulas.
1) $RCl(F) - Taut = RCl(F)$.
2) $RCl(F - Taut) = RCl(F)$.
3) $RCl(RCl(F)) = RCl(F)$. (*RCl* is idempotent.)
4) RC4 is equivalent to the following condition.
   If $F \subseteq Fml$ and $G \subseteq Fml$ and $C_\models G \subseteq C_\models RCl(F)$ then $C_\models RCl(F \cup G) = C_\models RCl(F)$.
5) RC5 is equivalent to the following condition.
   If $F \subseteq Fml$ and $g \in Fml$ and $\varovee$ is a disjunctive operation then
   $C_\models(RCl(F \cup \{g\}) \varovee RCl(F \cup \{\neg g\})) \subseteq C_\models RCl(F)$.
Proof
Let *RCl* be a rational clauses function. Let each of F and G be any set of formulas.

(1) By Lemma 2.8.5(9), $Cl(F) - Taut = Cl(F)$. So there are no tautologies in $Cl(F)$. By Definition 3.3.1(RC2), $RCl(F) \subseteq Cl(F)$. Hence there are no tautologies in $RCl(F)$. Thus $RCl(F) - Taut = RCl(F)$.

(2) By Lemma 2.8.5(10), $Cl(F - Taut) = Cl(F)$. So by RC1, $RCl(F - Taut) = RCl(F)$.

(3) By Definition 3.3.1(RC2), $RCl(F)$ is satisfiable. So by RC3, $RCl(RCl(F)) = Cl(RCl(F))$. By Definition 3.3.1(RC2), $RCl(F) \subseteq Cl(F)$. So by Lemma 2.8.5(19), $Cl(RCl(F)) = RCl(F)$. Thus $RCl(RCl(F)) = RCl(F)$.

(4 and 5) By Lemma 2.2.21(4), $F \models G$ iff $C_\models(F) \supseteq C_\models(G)$. By Lemma 2.2.21(5), $F \equiv G$ iff $C_\models(F) = C_\models(G)$. Both parts (4) and (5) follow immediately from these equivalences.
EndProofLemma3.3.2

3.3.2 Follows From and Tautologies

Explosiveness is a symptom of the property that, from a set F of formulas, classical logic proves formulas that, arguably, do not follow from F. For example from $\wedge\{a, \neg a\}$ we can prove b. Because b has nothing to do with a or $\neg a$, our intuition is that b does not follow from $\wedge\{a, \neg a\}$. This section formally defines a concept which is closer to capturing the intuitive concept of 'follows from' than that obtained from an explosive logic. To refine our intuition let us consider tautologies.

Let F be a set of formulas, f and g be formulas, and $\mathfrak{t}$ be a tautology.

A) Syntax independence.

We would like 'follows from' to be syntax independent. That is, if f follows from F and f is equivalent to g then g follows from F. So if a tautology follows from F then all tautologies should follow from F.

B) Tautologies are independent.

A tautology stands on its own, it does not depend on any other formula or set of formulas, and so it does not follow from any other formulas or set of formulas.

C) Tautologies have no deductive power.

Adding a tautology to F should not change what can be deduced from F. That is, $F \cup \{\mathfrak{t}\}$ and F should deduce the same set of formulas.

In Chapter 4 we do not want to force tautologies to be provable. So we shall declare that tautologies do not 'follow from' any set of formulas. This is consistent with the intuition expressed in (B).

3.3.3 Two Hierarchies of Rational Logics

To make rational deductions from a set F of formulas, whether F is satisfiable or not, we start by applying a rational clauses function to F. We then proceed as we did in Definition 2.9.1. The details are given in Definition 3.3.3.

Definition 3.3.3. Let F and G be any sets of formulas and f be any formula. Let RCl and SCl be any two rational clauses functions. Suppose $X \in \{N, P, D, I, R, S\}$.

1) $F \vdash^X f$ iff $\vee\{\} \in Res(Cl(\neg f) \cup XCl(F))$.

   $F \vdash^X G$ iff for all g in G, $F \vdash^X g$.

   We read $\vdash^X$ as X-proves.

2) Define the set of $\vdash^X$-consequences of F, $C_{\vdash^X}(F)$, by $C_{\vdash^X}(F) = \{f \in Fml : F \vdash^X f\}$.
   For simplicity define $C_{\vdash^X}(f) = C_{\vdash^X}(\{f\})$.

3) $F \models^X f$ iff $XCl(F) \models f$.

   $F \models^X G$ iff $XCl(F) \models G$.

   We read $\models^X$ as X-implies.

4) Define the set of $\models^X$-consequences of F, $C_{\models^X}(F)$, by $C_{\models^X}(F) = \{f \in Fml : F \models^X f\}$.
   For simplicity define $C_{\models^X}(f) = C_{\models^X}(\{f\})$.

3.3 Rational Propositional Logics

5) Define the set $XFrom(F)$ by $XFrom(F) = C_{\not\models_X}(F) - Taut$.
   For simplicity define $XFrom(f) = XFrom(\{f\})$.

Depending on one's intuition or purpose, it is reasonable to regard any of the sets in Definition 3.3.3(5) as the set of all formulas that 'follow from' F.

Because a pure clause is not contaminated by any errors, the logic determined by $\models^P$ and $\not\models^P$ is our preferred logic. It accords with our desire to prove what is not disputed and only what is not disputed.

Some of the properties of the logics determined by Definition 3.3.3 are proved in the following result. In particular these logics are sound and complete.

Theorem 3.3.4. Let F be any set of formulas and each of f and g be any formula. Let RCl and SCl be any two rational clauses functions such that $RCl(F) \subseteq SCl(F)$.
1) $F \models^R f$ iff $F \not\models^R f$; and so $C_{\models^R}(F) = C_{\not\models^R}(F)$. (soundness and completeness)
   $RFrom(F) = C_{\models^R}(F) - Taut = C_{\not\models^R}(F) - Taut$.
2) $Taut \subseteq C_{\not\models^R}(F) \subseteq C_{\not\models^S}(F) \subseteq C_{\vdash}(F)$.
   $RFrom(F) \subseteq SFrom(F) \subseteq C_{\vdash}(F) - Taut$.
3) If F is satisfiable then $C_{\not\models^R}(F) = C_{\vdash}(F) = C_{\models}(F)$.
   If F is satisfiable then $RFrom(F) = C_{\vdash}(F) - Taut = C_{\models}(F) - Taut$.
4) If F is unsatisfiable then $C_{\not\models^N}(F) = Taut$ and $NFrom(F) = \{\}$.
5) $C_{\models^R}(F)$ and $C_{\not\models^R}(F)$ and $RFrom(F)$ are all satisfiable.
6) $F \models^R f$ iff $RCl(F) \vdash f$.
   $C_{\models^R}(F) = C_{\vdash}(RCl(F)) = C_{\models}(RCl(F)) = C_{\not\models^R}(F)$.
   $RFrom(F) = C_{\vdash}(RCl(F)) - Taut$.
7) $C_{\not\models^R}(C_{\not\models^R}(F)) = C_{\not\models^R}(F)$. ($C_{\not\models^R}$ is idempotent.)
   $RFrom(RFrom(F)) = RFrom(F)$. ($RFrom$ is idempotent.)
8) If $G \subseteq C_{\not\models^R}(F)$ then $C_{\not\models^R}(F \cup G) = C_{\not\models^R}(F)$. ($C_{\not\models^R}$ is cumulative.)
   If $G \subseteq RFrom(F)$ then $RFrom(F \cup G) = RFrom(F)$. ($RFrom$ is cumulative.)
9) $C_{\not\models^R}(F \cup \{g\}) \cap C_{\not\models^R}(F \cup \{\neg g\}) \subseteq C_{\not\models^R}(F)$. ($C_{\not\models^R}$ is divisible.)
   $RFrom(F \cup \{g\}) \cap RFrom(F \cup \{\neg g\}) \subseteq RFrom(F)$. ($RFrom$ is divisible.)

Proof

Let F be any set of formulas and f any formula. Let RCl and SCl be any two rational clauses functions such that $RCl(F) \subseteq SCl(F)$.

(1) By Definition 3.3.3, Lemma 2.4.12(5), and Lemma 2.8.5(3), $F \models^R f$
iff $\vee\{\} \in Res(Cl(\neg f) \cup RCl(F))$
iff $Cl(\neg f) \cup RCl(F)$ is not satisfiable
iff $\{\neg f\} \cup RCl(F)$ is not satisfiable
iff $RCl(F) \models f$
iff $F \not\models^R f$.
Hence $C_{\models^R}(F) = C_{\not\models^R}(F)$.

(2) Let $\mathfrak{t}$ be any tautology. Then $\neg \mathfrak{t}$ is a contradiction. By Lemma 2.8.5(3), $\neg \mathfrak{t} \equiv Cl(\neg \mathfrak{t})$. So $Cl(\neg \mathfrak{t})$ is unsatisfiable. By Lemma 2.4.12(5), $\vee\{\} \in Res(Cl(\neg \mathfrak{t}))$. So $\vee\{\} \in Res(Cl(\neg \mathfrak{t}) \cup RCl(F))$. Hence $F \models^R \mathfrak{t}$ and so $\mathfrak{t} \in C_{\not\models^R}(F)$. Since $\mathfrak{t}$ was arbitrary, $Taut \subseteq C_{\not\models^R}(F)$.

By the definitions we have the following two statements.
$C_{\vdash R}(F) = \{f \in Fml : F \vdash^R f\} = \{f \in Fml : \vee\{\} \in Res(Cl(\neg f) \cup RCl(F))\}$.
$C_{\vdash}(F) = \{f \in Fml : F \vdash f\} = \{f \in Fml : \vee\{\} \in Res(Cl(\neg f) \cup Cl(F))\}$.
By Lemma 2.4.11(5), $C_{\vdash R}(F)$
$= \{f \in Fml : \vee\{\} \in Res(Cl(\neg f) \cup RCl(F))\}$
$\subseteq \{f \in Fml : \vee\{\} \in Res(Cl(\neg f) \cup SCl(F))\} = C_{\vdash S}(F)$
$\subseteq \{f \in Fml : \vee\{\} \in Res(Cl(\neg f) \cup Cl(F))\} = C_{\vdash}(F)$.
Thus $C_{\vdash R}(F) \subseteq C_{\vdash S}(F) \subseteq C_{\vdash}(F)$.

So by Theorem 3.3.4(1), $RFrom(F) \subseteq SFrom(F) \subseteq C_{\vdash}(F) - Taut$.

(3) Suppose F is satisfiable.

By Lemma 2.8.5(4), $Cl(F)$ is satisfiable. By RC3, $RCl(F) = Cl(F)$. Therefore $C_{\vdash R}(F)$
$= \{f \in Fml : F \vdash^R f\}$
$= \{f \in Fml : \vee\{\} \in Res(Cl(\neg f) \cup RCl(F))\}$
$= \{f \in Fml : \vee\{\} \in Res(Cl(\neg f) \cup Cl(F))\}$
$= \{f \in Fml : F \vdash f\}$
$= C_{\vdash}(F)$.

So by Theorem 3.3.4(1), $RFrom(F) = C_{\vdash R}(F) - Taut = C_{\vdash}(F) - Taut$. Hence by Theorem 2.9.2(1), $C_{\vdash}(F) - Taut = C_{\vdash}(F) - Taut$.

(4) Let F be unsatisfiable. By Lemma 2.8.5(4), $F \equiv Cl(F)$, and so $Cl(F)$ is unsatisfiable. Therefore $C_{\models N}(F)$
$= C_{\vdash N}(F)$ by Theorem 3.3.4(1),
$= \{f \in Fml : F \models^N f\}$ by Definition 3.3.3(4),
$= \{f \in Fml : NCl(F) \models f\}$ by Definition 3.3.3(3),
$= \{f \in Fml : Nil(Cl(F)) \models f\}$ by Definition 3.2.10(1),
$= \{f \in Fml : \{\} \models f\}$ by Definition 3.2.1(1),
$= Taut$.

So by Theorem 3.3.4(1), $NFrom(F) = C_{\vdash N}(F) - Taut = Taut - Taut = \{\}$.

(5) By RC2, $RCl(F)$ is satisfiable. By Definition 3.3.3, $C_{\models R}(F) = \{f \in Fml : F \models^R f\}$
$= \{f \in Fml : RCl(F) \models f\}$. Thus $C_{\models R}(F)$ is satisfiable. By Theorem 3.3.4(1), $C_{\vdash R}(F)$ is satisfiable.

By Definition 3.3.3(5), $RFrom(F) = C_{\vdash R}(F) - Taut$. Hence $RFrom(F)$ is satisfiable.

(6) $F \models^R f$
iff $F \vdash^R f$ by Theorem 3.3.4(1),
iff $RCl(F) \models f$ by Definition 3.3.3(3),
iff $RCl(F) \vdash f$ by Theorem 2.9.2(1).
Hence $C_{\vdash R}(F) = C_{\vdash}(RCl(F)) = C_{\vdash}(RCl(F)) = C_{\models R}(F)$.

So by Theorem 3.3.4(1), $RFrom(F) = C_{\vdash R}(F) - Taut = C_{\vdash}(RCl(F)) - Taut$.

(7) $C_{\vdash R}(C_{\vdash R}(F))$
$= C_{\vdash}(C_{\vdash R}(F))$ by Theorem 3.3.4(5,3),
$= C_{\vdash}(C_{\vdash}(RCl(F)))$ by Theorem 3.3.4(6),
$= C_{\vdash}(RCl(F))$ by Theorem 2.9.2(7),
$= C_{\vdash R}(F)$ by Theorem 3.3.4(6).

3.3 Rational Propositional Logics

If G is satisfiable then by Theorem 3.3.4(3) and Theorem 2.9.2(1), $C_{\vdash^R}(G) = C_{\vdash}(G)$
$= C_{\vdash}(G) = C_{\vdash}(G-Taut) = C_{\vdash}(G-Taut) = C_{\vdash^R}(G-Taut)$. Let (*) denote the result:
(*) If G is satisfiable then $C_{\vdash^R}(G-Taut) = C_{\vdash^R}(G)$.
Then $RFrom(RFrom(F))$
$= C_{\vdash^R}(C_{\vdash^R}(F) - Taut) - Taut$ by Theorem 3.3.4(1),
$= C_{\vdash^R}(G - Taut) - Taut$ where $G = C_{\vdash^R}(F)$,
$= C_{\vdash^R}(C_{\vdash^R}(F)) - Taut$ by Theorem 3.3.4(5) and (*),
$= C_{\vdash^R}(F) - Taut$ by the previous paragraph,
$= RFrom(F)$ by Theorem 3.3.4(1).

(8) Suppose $G \subseteq C_{\vdash^R}(F)$. By Theorem 3.3.4(1), $G \subseteq C_{\vDash^R}(F) = \{f \in Fml : F \vDash^R f\}$
$= \{f \in Fml : RCl(F) \vDash f\}$. Hence $RCl(F) \vDash G$ and so by RC4, $RCl(F \cup G) \equiv RCl(F)$.
Therefore $C_{\vdash^R}(F \cup G)$
$= C_{\vDash^R}(F \cup G)$, by Theorem 3.3.4(1)
$= \{f \in Fml : F \cup G \vDash^R f\}$
$= \{f \in Fml : RCl(F \cup G) \vDash f\}$
$= \{f \in Fml : RCl(F) \vDash f\}$, as $RCl(F \cup G) \equiv RCl(F)$
$= \{f \in Fml : F \vDash^R f\}$
$= C_{\vDash^R}(F) = C_{\vdash^R}(F)$, by Theorem 3.3.4(1).

Suppose $G \subseteq RFrom(F)$. By Theorem 3.3.4(1), $G \subseteq C_{\vdash^R}(F) - Taut$. By the previous two paragraphs, $C_{\vdash^R}(F \cup G) = C_{\vdash^R}(F)$. Hence $C_{\vdash^R}(F \cup G) - Taut = C_{\vdash^R}(F) - Taut$.
So by Theorem 3.3.4(1), $RFrom(F \cup G) = RFrom(F)$.

(9) $C_{\vdash^R}(F \cup \{g\}) \cap C_{\vdash^R}(F \cup \{\neg g\})$
$= C_{\vdash} RCl(F \cup \{g\}) \cap C_{\vdash} RCl(F \cup \{\neg g\})$ by Theorem 3.3.4(6),
$= C_{\vdash}(RCl(F \cup \{g\}) \mathbin{\text{\textcircled{$\vee$}}} RCl(F \cup \{\neg g\}))$ where $\text{\textcircled{$\vee$}}$ is any disjunctive operation,
$\phantom{= C_{\vdash}(RCl(F \cup \{g\}) \mathbin{\text{\textcircled{$\vee$}}} RCl(F \cup \{\neg g\}))}$ by Lemma 3.2.15(9),
$\subseteq C_{\vdash} RCl(F)$ by Lemma 3.3.2(5),
$= C_{\vdash^R}(F)$ by Theorem 3.3.4(6).

$RFrom(F \cup \{g\}) \cap RFrom(F \cup \{\neg g\})$
$= (C_{\vdash^R}(F \cup \{g\}) - Taut) \cap (C_{\vdash^R}(F \cup \{\neg g\}) - Taut)$ by Theorem 3.3.4(1),
$= (C_{\vdash^R}(F \cup \{g\}) \cap C_{\vdash^R}(F \cup \{\neg g\})) - Taut$
$\subseteq C_{\vdash^R}(F) - Taut$ by the previous paragraph,
$= RFrom(F)$ by Theorem 3.3.4(1).

EndProofTheorem3.3.4

Theorem 3.3.4(2) shows that there are two hierarchies of rational propositional logics, namely, the $C_{\vdash^R}$ hierarchy that includes all tautologies, and the $RFrom$ hierarchy that excludes all tautologies. Both these hierarchies are cumulative, by Theorem 3.3.4(8), and divisible, by Theorem 3.3.4(9).

Let us conclude this section by considering the extent to which the logics defined in Definition 3.3.3(4 and 5), where $X \in \{N, P, D, I\}$, satisfy various versions of left disjunction.

Lemma 3.3.5. Let each of F, G, and H be a set of formulas. Let each of a and b be atoms. Let $\text{\textcircled{$\vee$}}$ be a disjunctive operation. Suppose $X \in \{N, P, D, I\}$.

1) If both $F \cup G$ and $F \cup H$ are satisfiable then
   $C_{\models}^{\underline{x}}(F \cup G) \cap C_{\models}^{\underline{x}}(F \cup H) = C_{\models}^{\underline{x}}(F \cup (G \otimes H))$, and
   $XFrom(F \cup G) \cap XFrom(F \cup H) = XFrom(F \cup (G \otimes H))$.
2) If $F = \{a\} = G$ and $H = \{\neg a\}$ then
   $C_{\models}^{\underline{x}}(F \cup G) \cap C_{\models}^{\underline{x}}(F \cup H) \not\supseteq C_{\models}^{\underline{x}}(F \cup (G \otimes H))$, and
   $XFrom(F \cup G) \cap XFrom(F \cup H) \not\supseteq XFrom(F \cup (G \otimes H))$.
3) Let one of $F \cup G$ and $F \cup H$ be satisfiable and the other be unsatisfiable. Then
   $C_{\models}^{\underline{x}}(F \cup G) \cap C_{\models}^{\underline{x}}(F \cup H) \subseteq C_{\models}^{\underline{x}}(F \cup (G \otimes H))$, and
   $XFrom(F \cup G) \cap XFrom(F \cup H) \subseteq XFrom(F \cup (G \otimes H))$.
4) If both $F \cup G$ and $F \cup H$ are unsatisfiable then
   $C_{\models}^{\underline{N}}(F \cup G) \cap C_{\models}^{\underline{N}}(F \cup H) = C_{\models}^{\underline{N}}(F \cup (G \otimes H))$, and
   $NFrom(F \cup G) \cap NFrom(F \cup H) = NFrom(F \cup (G \otimes H))$.
5) If $X \in \{P, D, I\}$, $F = \{a, b\}$, $G = \{\neg a\}$, and $H = \{\neg b\}$ then
   $C_{\models}^{\underline{x}}(F \cup G) \cap C_{\models}^{\underline{x}}(F \cup H) \not\subseteq C_{\models}^{\underline{x}}(F \cup (G \otimes H))$, and
   $XFrom(F \cup G) \cap XFrom(F \cup H) \not\subseteq XFrom(F \cup (G \otimes H))$.

Proof

Let each of F, G, and H be a set of formulas. Let each of a and b be atoms. Let $\otimes$ be a disjunctive operation. Suppose $X \in \{N, P, D, I\}$.

After Definition 3.3.1 we observed that each XCl is a rational clauses function. So by Theorem 3.3.4(6), $C_{\models}^{\underline{x}}(F) = C_{\models}(XCl(F))$. Hence
$C_{\models}^{\underline{x}}(F \cup (G \otimes H)) = C_{\models}(XCl(F \cup (G \otimes H)))$; and also
$C_{\models}^{\underline{x}}(F \cup G) \cap C_{\models}^{\underline{x}}(F \cup H) = C_{\models}(XCl(F \cup G)) \cap C_{\models}(XCl(F \cup H))$
$= C_{\models}(XCl(F \cup G) \otimes XCl(F \cup H))$ by Lemma 3.2.15(9).
By Lemma 2.2.21(5), $F \equiv G$ iff $C_{\models}(F) = C_{\models}(G)$.
By Lemma 2.2.21(4), $F \models G$ iff $C_{\models}(F) \supseteq C_{\models}(G)$.

We shall use these results, without referring to them, in what follows.

(1) Suppose both $F \cup G$ and $F \cup H$ are satisfiable. By Lemma 3.2.16(2) we have, $XCl(F \cup G) \otimes XCl(F \cup H) \equiv XCl(F \cup (G \otimes H))$. Hence
$C_{\models}(XCl(F \cup G) \otimes XCl(F \cup H)) = C_{\models}(XCl(F \cup (G \otimes H)))$. Therefore
$C_{\models}^{\underline{x}}(F \cup G) \cap C_{\models}^{\underline{x}}(F \cup H) = C_{\models}^{\underline{x}}(F \cup (G \otimes H))$.

Also $XFrom(F \cup G) \cap XFrom(F \cup H)$
$= (C_{\models}^{\underline{x}}(F \cup G) - Taut) \cap (C_{\models}^{\underline{x}}(F \cup H) - Taut)$
$= (C_{\models}^{\underline{x}}(F \cup G) \cap C_{\models}^{\underline{x}}(F \cup H)) - Taut$
$= C_{\models}^{\underline{x}}(F \cup (G \otimes H)) - Taut$
$= XFrom(F \cup (G \otimes H))$.

(2) Suppose $F = \{a\}$, $G = \{a\}$, and $H = \{\neg a\}$. By Lemma 3.2.16(5) we have, $XCl(F \cup G) \otimes XCl(F \cup H) \subseteq Taut$, $XCl(F \cup (G \otimes H)) \not\subseteq Taut$, and $XCl(F \cup G) \otimes XCl(F \cup H) \not\models XCl(F \cup (G \otimes H))$. Hence $C_{\models}(XCl(F \cup G) \otimes XCl(F \cup H)) \not\supseteq C_{\models}(XCl(F \cup (G \otimes H)))$. Therefore $C_{\models}^{\underline{x}}(F \cup G) \cap C_{\models}^{\underline{x}}(F \cup H) \not\supseteq C_{\models}^{\underline{x}}(F \cup (G \otimes H))$.

Also $XFrom(F \cup G) \cap XFrom(F \cup H)$
$= (C_{\models}^{\underline{x}}(F \cup G) - Taut) \cap (C_{\models}^{\underline{x}}(F \cup H) - Taut)$
$= (C_{\models}^{\underline{x}}(F \cup G) \cap C_{\models}^{\underline{x}}(F \cup H)) - Taut$
$= (C_{\models}(XCl(F \cup G) \otimes XCl(F \cup H))) - Taut$
$= Taut - Taut$

3.3 Rational Propositional Logics

$= \{\}$
$\not\supseteq C_{\not\models}^{\underline{X}}(F \cup (G \otimes H)) - Taut$
$= XFrom(F \cup (G \otimes H))$.

(3) Let one of $F \cup G$ and $F \cup H$ be satisfiable and the other be unsatisfiable. By Lemma 3.2.16(6), $F \cup (G \otimes H)$ is satisfiable; and
$XCl(F \cup (G \otimes H)) \models XCl(F \cup G) \otimes XCl(F \cup H)$. Hence
$C_{\models}(XCl(F \cup G) \otimes XCl(F \cup H)) \subseteq C_{\models}(XCl(F \cup (G \otimes H)))$. Therefore
$C_{\models}^{\underline{X}}(F \cup G) \cap C_{\models}^{\underline{X}}(F \cup H) \subseteq C_{\models}^{\underline{X}}(F \cup (G \otimes H))$.

Also $XFrom(F \cup G) \cap XFrom(F \cup H)$
$= (C_{\models}^{\underline{X}}(F \cup G) - Taut) \cap (C_{\models}^{\underline{X}}(F \cup H) - Taut)$
$= (C_{\models}^{\underline{X}}(F \cup G) \cap C_{\models}^{\underline{X}}(F \cup H)) - Taut$
$\subseteq C_{\models}^{\underline{X}}(F \cup (G \otimes H)) - Taut$
$= XFrom(F \cup (G \otimes H))$.

(4) Suppose both $F \cup G$ and $F \cup H$ are unsatisfiable. By Lemma 3.2.16(8) we have, $F \cup (G \otimes H)$ is unsatisfiable, $NCl(F \cup (G \otimes H)) = \{\}$, $NCl(F \cup G) = \{\}$, and $NCl(F \cup H) = \{\}$. So $NCl(F \cup (G \otimes H)) \equiv NCl(F \cup G) \otimes NCl(F \cup H)$. Hence
$C_{\models}(NCl(F \cup G) \otimes NCl(F \cup H)) = C_{\models}(NCl(F \cup (G \otimes H)))$. Therefore
$C_{\not\models}^{\underline{N}}(F \cup G) \cap C_{\not\models}^{\underline{N}}(F \cup H) = C_{\not\models}^{\underline{N}}(F \cup (G \otimes H))$.

Also $NFrom(F \cup G) \cap NFrom(F \cup H)$
$= (C_{\not\models}^{\underline{N}}(F \cup G) - Taut) \cap (C_{\not\models}^{\underline{N}}(F \cup H) - Taut)$
$= (C_{\not\models}^{\underline{N}}(F \cup G) \cap C_{\not\models}^{\underline{N}}(F \cup H)) - Taut$
$= C_{\not\models}^{\underline{N}}(F \cup (G \otimes H)) - Taut$
$= NFrom(F \cup (G \otimes H))$.

(5) Suppose that $X \in \{P, D, I\}$, $F = \{a, b\}$, $G = \{\neg a\}$, and $H = \{\neg b\}$. By Lemma 3.2.16(9), $XCl(F \cup G) \otimes XCl(F \cup H) \equiv \vee\{b, a\}$, $\vee\{\neg a, \neg b\} \models XCl(F \cup (G \otimes H))$, and so $XCl(F \cup (G \otimes H)) \not\models XCl(F \cup G) \otimes XCl(F \cup H)$. Hence
$C_{\models}(XCl(F \cup G) \otimes XCl(F \cup H)) \not\subseteq C_{\models}(XCl(F \cup (G \otimes H)))$. Therefore
$C_{\models}^{\underline{X}}(F \cup G) \cap C_{\models}^{\underline{X}}(F \cup H) \not\subseteq C_{\models}^{\underline{X}}(F \cup (G \otimes H))$.

For the second statement we have
$XFrom(F \cup G) \cap XFrom(F \cup H)$
$= (C_{\models}^{\underline{X}}(F \cup G) - Taut) \cap (C_{\models}^{\underline{X}}(F \cup H) - Taut)$
$= (C_{\models}^{\underline{X}}(F \cup G) \cap C_{\models}^{\underline{X}}(F \cup H)) - Taut$
$= C_{\models}(XCl(F \cup G) \otimes XCl(F \cup H)) - Taut$
$= C_{\models}(\vee\{b, a\}) - Taut$.

And $XFrom(F \cup (G \otimes H))$
$= C_{\models}^{\underline{X}}(F \cup (G \otimes H)) - Taut$
$= C_{\models}(XCl(F \cup (G \otimes H))) - Taut$
$\subseteq C_{\models}(\vee\{\neg a, \neg b\}) - Taut$.

But $\vee\{a, b\} \in C_{\models}(\vee\{b, a\}) - Taut$ and $\vee\{a, b\} \notin C_{\models}(\vee\{\neg a, \neg b\}) - Taut$. Therefore $\vee\{a, b\} \in XFrom(F \cup G) \cap XFrom(F \cup H)$ and $\vee\{a, b\} \notin XFrom(F \cup (G \otimes H))$. Thus $XFrom(F \cup G) \cap XFrom(F \cup H) \not\subseteq XFrom(F \cup (G \otimes H))$.

EndProofLemma3.3.5

By Lemma 3.3.5(1,3,4), we see that $C_{\not\models}^{\underline{N}}(F \cup G) \cap C_{\not\models}^{\underline{N}}(F \cup H) \subseteq C_{\not\models}^{\underline{N}}(F \cup (G \otimes H))$, and that $NFrom(F \cup G) \cap NFrom(F \cup H) \subseteq NFrom(F \cup (G \otimes H))$. So $C_{\not\models}^{\underline{N}}$ and $NFrom$

satisfy this version of left disjunction. However, Lemma 3.3.5(2) shows that equality does not hold.

Now suppose $X \in \{P,D,I\}$. Lemma 3.3.5(1 and 3), give conditions under which $C_{\not\models x}$ and *XFrom* satisfy the following version of left disjunction.
$C_{\not\models x}(F \cup G) \cap C_{\not\models x}(F \cup H) \subseteq C_{\not\models x}(F \cup (G \otimes H))$, and
$XFrom(F \cup G) \cap XFrom(F \cup H) \subseteq XFrom(F \cup (G \otimes H))$.
However, as Lemma 3.3.5(5) shows there are cases where
$C_{\not\models x}(F \cup G) \cap C_{\not\models x}(F \cup H) \not\subseteq C_{\not\models x}(F \cup (G \otimes H))$, and
$XFrom(F \cup G) \cap XFrom(F \cup H) \not\subseteq XFrom(F \cup (G \otimes H))$.
So in general, if $X \in \{P,D,I\}$ then $C_{\not\models x}$ and *XFrom* do not satisfy left disjunction.

To help clarify exactly why left disjunction fails, consider the following special case of the counter-example in Lemma 3.3.5(5). Suppose $X \in \{P,D,I\}$, $F = \{a,b\}$, $G = \{\neg a\}$, $H = \{\neg b\}$, and $\otimes = \curlyvee$, where $\curlyvee$ is defined in Lemma 3.2.14 by the equation $F \curlyvee G = \{\vee\{f,g\} : f \in F \text{ and } g \in G\}$.

By the proof of Lemma 3.2.16(9) we get $XCl(F \cup G) = \{b\}$ and $XCl(F \cup H) = \{a\}$. Also $F \cup (G \curlyvee H) = \{a,b,\vee\{\neg a,\neg b\}\}$. By Lemma 2.8.3(5 and 8), $Cl(F \cup (G \curlyvee H)) = \{a,b,\vee\{\neg a,\neg b\}\}$. So $XCl(F \cup (G \curlyvee H)) = \{\}$. Using the second paragraph of the proof of Lemma 3.3.5 we have
$C_{\not\models x}(F \cup (G \curlyvee H)) = C_{\models}(XCl(F \cup (G \curlyvee H))) = C_{\models}(\{\}) = \textit{Taut}$; and
$C_{\not\models x}(F \cup G) \cap C_{\not\models x}(F \cup H) = C_{\models}(XCl(F \cup G) \curlyvee XCl(F \cup H)) = C_{\models}(\vee\{b,a\})$.

In essence we can salvage $\{b\}$ from $F \cup G = \{a,b,\neg a\}$, and we can salvage $\{a\}$ from $F \cup H = \{a,b,\neg b\}$; but we cannot salvage anything from $F \cup (G \curlyvee H)) = \{a,b,\vee\{\neg a,\neg b\}\}$.

3.4 Consequence Functions

A consequence function takes a set F of formulas and returns all the consequences of F. A consequence relation relates a set F of formulas and a single formula g, such that g is regarded as a consequence of F. The following general notation will be used throughout this section.

Definition 3.4.1. Let $\models_{\overline{C}}$ be a relation from $\mathscr{P}(\textit{Fml})$ to *Fml*, and C be a function from $\mathscr{P}(\textit{Fml})$ to $\mathscr{P}(\textit{Fml})$ such that for all $\{F,G\} \subseteq \mathscr{P}(\textit{Fml})$, and for all $\{f,g\} \subseteq \textit{Fml}$, the following statements all hold.
1) $F \models_{\overline{C}} G$ iff for all g in G, $F \models_{\overline{C}} g$.
2) $C(F) = \{f \in \textit{Fml} : F \models_{\overline{C}} f\}$.
3) $F \models_{\overline{C}} f$ iff $f \in C(F)$.
4) $F \models_{\overline{C}} G$ iff $G \subseteq C(F)$.

In the above definition, 3.4.1(1) extends the definition of $\models_{\overline{C}}$ to also be a relation from $\mathscr{P}(\textit{Fml})$ to $\mathscr{P}(\textit{Fml})$. Given this extension, then 3.4.1(2), 3.4.1(3), and 3.4.1(4) are all equivalent, and can be used to convert between statements about C and state-

3.4 Consequence Functions

ments about $\models_C$. Given C we say $\models_C$ is its corresponding relation, and given $\models_C$ we say C is its corresponding function.

By Theorem 2.9.2(1), for every set F of formulas, $C_\vdash(F) = C_\models(F)$. The use of $C_\vdash(F)$ emphasises the proof-theoretic nature of the classical consequence function; whereas $C_\models(F)$ emphasises the truth-theoretic nature of the classical consequence function. In this section we want to refer to the classical consequence function without emphasising either nature. The following standard notation does this.

Definition 3.4.2. Let F be any set of formulas. The *classical consequence function*, Cn, is defined by $C_\vdash(F) = Cn(F) = C_\models(F)$.

3.4.1 A Review

Clearly not every function from $\mathscr{P}(Fml)$ to $\mathscr{P}(Fml)$ should be regarded as a consequence function. In order to find a suitable definition of a consequence function, we shall start by considering some definitions that are in the literature. These definitions refer to the following properties. We shall also define two other relevant properties, namely cumulativity and cautious monotonicity, and recall the definition of divisibility in Definition 3.2.12.

Definition 3.4.3. Let C be a function from $\mathscr{P}(Fml)$ to $\mathscr{P}(Fml)$, and $\models_C$ be its corresponding relation from $\mathscr{P}(Fml)$ to Fml.
Inclusion: C is *inclusive* iff for all $F \subseteq Fml$, $F \subseteq C(F)$.
Reflexivity: $\models_C$ is *reflexive* iff for all $F \subseteq Fml$, $F \models_C F$.
Idempotence: C is *idempotent* iff for all $F \subseteq Fml$, $C(C(F)) = C(F)$.
Compactness: C is *compact* iff for all $G \subseteq Fml$,
  $C(G) = \bigcup \{C(F) : F \text{ is a finite subset of } G\}$.
Monotonicity: C is *monotonic* iff for all $\{F,G\} \subseteq \mathscr{P}(Fml)$,
  if $F \subseteq G$ then $C(F) \subseteq C(G)$.
Cut: C satisfies *cut* iff for all $\{F,G\} \subseteq \mathscr{P}(Fml)$, if $F \subseteq C(G)$ then $C(F \cup G) \subseteq C(G)$.
Cumulativity: C is *cumulative* iff for all $\{F,G\} \subseteq \mathscr{P}(Fml)$,
  if $F \subseteq C(G)$ then $C(F \cup G) = C(G)$.
Cautious Monotonicity: C is *cautiously monotonic* iff for all $\{F,G\} \subseteq \mathscr{P}(Fml)$,
  if $F \subseteq C(G)$ then $C(F \cup G) \supseteq C(G)$.
Left Equivalence: C satisfies *left equivalence* iff for all $\{F,G\} \subseteq \mathscr{P}(Fml)$,
  if $F \equiv G$ then $C(F) = C(G)$.
Right Absorption: C is *right absorptive* iff for all $F \subseteq Fml$, $C(Cn(F)) = C(F)$.
Left Absorption: C is *left absorptive* iff for all $F \subseteq Fml$, $Cn(C(F)) = C(F)$.
Classical Closure: C is *classically closed* iff for all $\{F,G\} \subseteq \mathscr{P}(Fml)$,
  if $F \subseteq C(G)$ then $Cn(F) \subseteq C(G)$.
Divisibility: C is *divisible* iff for all $F \subseteq Fml$ and all $g \in Fml$,
  $C(F \cup \{g\}) \cap C(F \cup \{\neg g\}) \subseteq C(F)$.

The following relationships between these properties are worth noting.

Lemma 3.4.4. Let C be a function from $\mathscr{P}(Fml)$ to $\mathscr{P}(Fml)$. Let G be any set of formulas.
1) C is inclusive iff $\models_C$ is reflexive.
2) C satisfies inclusion, monotonicity, and cut iff
   C satisfies inclusion, monotonicity, and idempotence.
3) If C is monotonic then C is cautiously monotonic.
4) C is cumulative iff C satisfies cut and cautious monotonicity.
5) If C is compact then C is monotonic.
6) If C is monotonic then $\bigcup\{C(F) : F$ is a finite subset of $G\} \subseteq C(G)$.
7) C satisfies left equivalence iff C is right absorptive.
8) C is classically closed iff C is left absorptive.
9) If C satisfies inclusion, monotonicity, idempotence, and left equivalence then C satisfies classical closure.

Proof

Let C be a function from $\mathscr{P}(Fml)$ to $\mathscr{P}(Fml)$. Let G be any set of formulas.

(1) This follows immediately from the definitions.

(2) Suppose C satisfies inclusion, monotonicity, and cut. By inclusion, $G \subseteq C(G)$, and so $C(G) \cup G = C(G)$. By cut, $C(C(G) \cup G) \subseteq C(G)$ and so $C(C(G)) \subseteq C(G)$. By inclusion, $G \subseteq C(G)$, and so by monotonicity, $C(G) \subseteq C(C(G))$. Therefore $C(C(G)) = C(G)$. Thus C is idempotent.

Conversely, suppose C satisfies inclusion, monotonicity, and idempotence. Suppose $F \subseteq C(G)$. Then $F \cup G \subseteq C(G) \cup G$. By inclusion, $G \subseteq C(G)$, and so $C(G) \cup G = C(G)$. Hence $F \cup G \subseteq C(G)$. By monotonicity and idempotence, $C(F \cup G) \subseteq C(C(G)) = C(G)$. Thus C satisfies cut.

(3) Suppose C is monotonic. If F is a set of formulas then $G \subseteq F \cup G$. By monotonicity, $C(G) \subseteq C(F \cup G)$. Thus C is cautiously monotonic.

(4) This follows immediately from the definitions.

(5) Suppose C is compact. Also suppose $F \subseteq G$. We show $C(F) \subseteq C(G)$. Take any f in $C(F)$. Then there is a finite subset F' of F such that $f \in C(F')$. But F' is a finite subset of G. Hence $C(F') \subseteq C(G)$ and so $f \in C(G)$. Therefore $C(F) \subseteq C(G)$. Thus C is monotonic.

(6) Suppose C is monotonic. If F is a finite subset of G then $F \subseteq G$ and so $C(F) \subseteq C(G)$. Thus $\bigcup\{C(F) : F$ is a finite subset of $G\} \subseteq C(G)$.

(7) Suppose C satisfies left equivalence. That is, for all $\{F, G\} \subseteq \mathscr{P}(Fml)$, if $F \equiv G$ then $C(F) = C(G)$. We must show that for all $F \subseteq Fml$, $C(Cn(F)) = C(F)$. Take any set F of formulas. By Lemma 2.2.21(3), $F \equiv Cn(F)$. Hence $C(F) = C(Cn(F))$. Therefore C satisfies right absorption.

Conversely, suppose C satisfies right absorption. That is, for all $F \subseteq Fml$, $C(Cn(F)) = C(F)$. We must show that for all $\{F, G\} \subseteq \mathscr{P}(Fml)$, if $F \equiv G$ then $C(F) = C(G)$. Take any sets, F and G, of formulas. Suppose $F \equiv G$. By Lemma 2.2.21(5), $Cn(F) = Cn(G)$. So $C(F) = C(Cn(F)) = C(Cn(G)) = C(G)$. Therefore C satisfies left equivalence.

3.4 Consequence Functions   111

(8) Suppose C is classically closed. That is, for all $\{F,G\} \subseteq \mathscr{P}(Fml)$, if $F \subseteq C(G)$ then $Cn(F) \subseteq C(G)$. We must show that for all $G \subseteq Fml$, $Cn(C(G)) = C(G)$. Take any set G of formulas. By letting $F = C(G)$ we have $Cn(C(G)) \subseteq C(G)$. By Lemma 2.2.21(2), $C(G) \subseteq Cn(C(G))$. Hence $Cn(C(G)) = C(G)$. Therefore C satisfies left absorption.

Conversely, suppose C satisfies left absorption. That is, for all $F \subseteq Fml$, $Cn(C(F)) = C(F)$. We must show that for all $\{F,G\} \subseteq \mathscr{P}(Fml)$, if $F \subseteq C(G)$ then $Cn(F) \subseteq C(G)$. Take any sets, F and G, of formulas. Suppose $F \subseteq C(G)$. By Lemma 2.2.21(6), $Cn(F) \subseteq Cn(C(G))$. So by left absorption, $Cn(F) \subseteq Cn(C(G)) = C(G)$. Therefore C is classically closed.

(9) Suppose C satisfies inclusion, monotonicity, idempotence, and left equivalence. Suppose $F \subseteq C(G)$. We must show $Cn(F) \subseteq C(G)$. $Cn(F)$
$\subseteq C(Cn(F))$ by inclusion,
$= C(F)$ by left equivalence and $F \equiv Cn(F)$ by Lemma 2.2.21(3),
$\subseteq C(C(G))$ by monotonicity and $F \subseteq C(G)$,
$= C(G)$ by idempotence.

Thus C satisfies classical closure.
EndProofLemma3.4.4

It seems that the earliest account of a consequence function, C, was given in a lecture by Alfred Tarski in 1928, see Chapter III in [80]. In this chapter, page 30, Tarski says that the consequences of a set of sentences are obtained by certain 'rules of inference'. He then gives 10 axioms, pages 31-32, that define the notions of 'sentence' and 'set of consequences'. Axioms 6 to 10 concern the connectives 'negation' and 'implication'. In our notation Axiom 1 says that Fml is countable. Axiom 2 says C is inclusive. Axiom 3 says C is idempotent. Axiom 4 says C is compact. We note that if C is compact then it is monotonic. Axiom 5 says C is explosive; that is, there is a formula f such that $C(\{f\}) = Fml$. We have already expressed our thoughts about the absurdity of explosiveness. All the above axioms, with the exception of this explosiveness axiom, seem reasonable, at first sight.

In [78](page 414), Scott considers a relation $\vdash$ where both sides of the turnstile are finite sets of formulas. The intuition being that $F \vdash G$ means that if every formula in F is accepted then at least one formula in G is accepted. He notes that $\vdash$ satisfies corresponding versions of reflexivity, monotonicity, and cut. Gabbay in [38](pages 6-7) calls the same relation $\vdash$ a consequence relation iff it satisfies ten conditions. The first three are the same as in [78](page 414), the rest concern connectives and quantifiers.

In [37](pages 441-442), Gabbay says that a relation from $\mathscr{P}(Fml)$ to Fml is a provability relation of some monotonic logical system iff its corresponding function from $\mathscr{P}(Fml)$ to $\mathscr{P}(Fml)$ satisfies Inclusion, Monotonicity, and Cut. In [63](page 1) and [64](page 42-43), Makinson says that a function from $\mathscr{P}(Fml)$ to $\mathscr{P}(Fml)$ is a consequence operation iff it satisfies Inclusion, Idempotence, and Monotonicity. In [36](page 164, definition 3), Freund and Lehmann state that a function from $\mathscr{P}(Fml)$ to $\mathscr{P}(Fml)$ is an inference operation iff it satisfies Inclusion, Classical Closure (called

Left Absorption), and Left Equivalence (called Right Absorption). Also in [64](page 45), Makinson suggests that a non-monotonic inference operation deserves the name 'logical' only if it satisfies Classical Closure (called Left Absorption) and Left Equivalence (called Right Absorption).

The classical consequence function Cn satisfies
Classical Closure by Theorem 2.9.2(8).
Compactness by Theorem 2.9.2(9),
Cumulativity by Theorem 2.9.2(6),
Cut by Theorem 2.9.2(6),
Idempotence by Theorem 2.9.2(7),
Inclusion by Theorem 2.9.2(4),
Left Equivalence by Theorem 2.9.2(3), and
Monotonicity by Theorem 2.9.2(5),
So Cn satisfies all of the above definitions of a consequence function.

However there are functions that should not be regarded as consequence functions but which satisfy Compactness, Cumulativity, Cut, Idempotence, Inclusion, and Monotonicity; as the next example shows.

Example 3.4.5 (Non-consequence functions).
If S is a set of positive integers then define $\min(S)$ to be the least element of S; and define $m(S) = \{i \in \mathbb{Z}^+ : \min(S) \leq i\}$. Let b be any bijection from Fml to $\mathbb{Z}^+$. Then define the function C_b from $\mathscr{P}(Fml)$ to $\mathscr{P}(Fml)$ as follows.
$$\text{If } F \subseteq Fml \text{ then } C_b(F) = b^{-1}(m(b(F))).$$
It is easy to see that C_b satisfies Compactness, Cumulativity, Cut, Idempotence, Inclusion, and Monotonicity.

But consider the formulas based on the set of atoms $\{a_1, a_2, a_3\}$. Define b to be any bijection such that $b(\vee\{a_1, a_2\}) = 1$, $b(a_1) = 2$, $b(a_2) = 3$, and $b(a_3) = 4$. Then
(i) $\vee\{a_1, a_2\} \notin C_b(\{a_1, a_2\})$,
(ii) $a_3 \in C_b(\{a_1, a_2\})$, and
(iii) $\neg a_1 \in C_b(\{a_1, a_2\})$.
But if C_b was a reasonable consequence function then we would expect the negation of each of (i), (ii), and (iii) to hold.

This example shows that even together Compactness, Cumulativity, Cut, Idempotence, Inclusion, and Monotonicity are not sufficient conditions for a reasonable consequence function.

In the previous section we introduced eight functions from $\mathscr{P}(Fml)$ to $\mathscr{P}(Fml)$, namely $C_{\vdash X}$ and $XFrom$ where $X \in \{N, P, D, I\}$. These are special cases of the two more general functions $C_{\vdash R}$ and $RFrom$ that are also defined in the previous two sections. Given a set F of formulas, all these functions produce a set of formulas that can be regarded as the set of all consequences of F. That we have many different functions each legitimately claiming to produce the set of all consequences of F is not surprising, because the set of all consequences of F depends on one's intuition and purpose.

As the following example shows $C_{\vdash P}$ and $PFrom$ do not satisfy any of the above definitions of a consequence function.

3.4 Consequence Functions

Example 3.4.6.
1) $C_{\vdash P}(\{a\}) = C_\vdash(\{a\})$, and $C_{\vdash P}(\{a, \neg a\}) = \textit{Taut}$.
   $C_{\vdash P}$ does not satisfy Inclusion, because $\{a, \neg a\} \not\subseteq \textit{Taut} = C_{\vdash P}(\{a, \neg a\})$.
   $C_{\vdash P}$ does not satisfy Monotonicity, because $C_{\vdash P}(\{a\}) \not\subseteq C_{\vdash P}(\{a, \neg a\})$.
2) $C_{\vdash P}(\{a, \neg a\}) = C_\vdash(\{\}) = \textit{Taut}$, and $C_{\vdash P}(\{a, \neg a, b\}) = C_\vdash(\{b\})$.
   Hence $C_{\vdash P}$ does not satisfy Left Equivalence.
3) $PFrom(\{a\}) = C_\vdash(\{a\}) - \textit{Taut}$, and $PFrom(\{a, \neg a\}) = \{\}$.
   PFrom does not satisfy Inclusion, because $\{a, \neg a\} \not\subseteq \{\} = PFrom(\{a, \neg a\})$.
   PFrom does not satisfy Monotonicity, because $PFrom(\{a\}) \not\subseteq PFrom(\{a, \neg a\})$.
4) $PFrom(\{\}) = \{\}$, and $Cn(\{\}) = \textit{Taut}$.
   Hence PFrom does not satisfy Classical Closure.
5) $PFrom(\{a, \neg a\}) = \textit{Taut} - \textit{Taut} = \{\}$, and $PFrom(\{a, \neg a, b\}) = C_\vdash(\{b\}) - \textit{Taut}$.
   Hence PFrom does not satisfy Left Equivalence.

This example shows that not one of Classical Closure, Inclusion, Left Equivalence, or Monotonicity is a necessary condition for a reasonable consequence function.

3.4.2 Rational Consequence Functions

Since it is reasonable to call $C_{\vdash P}$ and *PFrom* consequence functions, there is a need to create a new definition of what a consequence function should be. We could specify a set of properties and assert that these are the defining characteristics of a consequence function. But there would always be the chance that there could be a function with these properties that should not be regarded as a consequence function. Also there would always be the chance that a reasonable consequence function did not satisfy one of the characteristic properties. So we shall use a different approach, one based on the familiar classical consequence function *Cn*.

As we saw earlier, *Cn* behaves irrationally on unsatisfiable input sets. Also as shown in Subsection 3.3.2(A), syntax independence requires that the set of consequences of any set of formulas should contain either all tautologies or no tautologies. But apart from unsatisfiable input sets and tautologies, does *Cn* have any other flaws?

If g is a consequence of F then it seems unintuitive that a valuation can make F true and g false. That is, if $g \in C(F)$ then $F \models g$. Conversely, if g is not a tautology and F is satisfiable and $F \models g$ then whenever a valuation makes F true it is forced to make g true too. So intuitively g is a consequence of F. That is, provided g is not a tautology and F is satisfiable, we have if $F \models g$ then $g \in C(F)$. So apart from the two flaws mentioned earlier, $g \in C(F)$ iff $F \models g$; that is, *Cn* is just what a consequence function C should be.

Therefore, if F is satisfiable then we would like to have either $C(F) = Cn(F)$ or $C(F) = Cn(F) - \textit{Taut}$. But this is not enough; because if F is satisfiable and $F \equiv G$ then we want $C(F) = C(G)$. As we saw in Subsection 3.3.2(B), whether tautologies are consequences or not, does not depend on the input set F, but on the consequence function C itself. We shall call such a C essentially classical and formally define this idea in Definition 3.4.7.

As noted before Definition 3.3.1, we should first apply a rational clauses function to F and then apply C to the result. We shall call such a C rational and also formally define this idea in Definition 3.4.7.

Definition 3.4.7. Let C be a function from $\mathscr{P}(Fml)$ to $\mathscr{P}(Fml)$.
Essential Classicality: C is *essentially classical* iff either (1) or (2).
    (1) for each satisfiable set F of formulas, $C(F) = Cn(F)$.
    (2) for each satisfiable set F of formulas, $C(F) = Cn(F) - Taut$.
Rationality: C is *rational* iff there is a rational clauses function, RCl, such that for each set F of formulas, $C(F) = C(RCl(F))$.
Rational Classicality: C is *rationally classical* iff there is a rational clauses function, RCl, such that
    either for each set F of formulas, $C(F) = Cn(RCl(F))$;
    or for each set F of formulas, $C(F) = Cn(RCl(F)) - Taut$.

As may have been expected, rational classicality is equivalent to the conjunction of rationality and essential classicality, as we now show.

Lemma 3.4.8. Let C be a function from $\mathscr{P}(Fml)$ to $\mathscr{P}(Fml)$. Then C is rationally classical iff C is both essentially classical and rational.
Proof
    Let C be a function from $\mathscr{P}(Fml)$ to $\mathscr{P}(Fml)$.
    Suppose C is rationally classical. Then there is a rational clauses function, RCl, such that one of two cases holds.
    Case 1: For each set G of formulas, $C(G) = Cn(RCl(G))$.
By Definition 3.3.1(RC3), if F is any satisfiable set of formulas then $RCl(F) = Cl(F)$. By Lemma 2.8.5(4), if F is any set of formulas then $F \equiv Cl(F)$. So for each satisfiable set F of formulas, $C(F) = Cn(RCl(F)) = Cn(Cl(F)) = Cn(F)$. Hence C is essentially classical.
    Let F be any set of formulas. By Definition 3.3.1(RC2), $RCl(F)$ is satisfiable. So by the preceding paragraph, $C(RCl(F)) = Cn(RCl(F)) = C(F)$. Hence C is rational.
    Case 2: For each set G of formulas, $C(G) = Cn(RCl(G)) - Taut$.
By Definition 3.3.1(RC3), if F is any satisfiable set of formulas then $RCl(F) = Cl(F)$. By Lemma 2.8.5(4), if F is any set of formulas then $F \equiv Cl(F)$. So for each satisfiable set F of formulas, $C(F) = Cn(RCl(F)) - Taut = Cn(Cl(F)) - Taut = Cn(F) - Taut$. Hence C is essentially classical.
    Let F be any set of formulas. By Definition 3.3.1(RC2), $RCl(F)$ is satisfiable. So by the preceding paragraph, $C(RCl(F)) = Cn(RCl(F)) - Taut = C(F)$. Hence C is rational.
    Conversely, suppose C is both essentially classical and rational. By rationality there is a rational clauses function, RCl, such that for each set G of formulas, we have, $C(G) = C(RCl(G))$. By essential classicality there are two cases.

3.4 Consequence Functions

Case 1: For each satisfiable set G of formulas, $C(G) = Cn(G)$.
Take any set F of formulas.

Subcase 1.1: F is satisfiable. By essential classicality, $C(F) = Cn(F)$. By Definition 3.3.1(RC3), $RCl(F) = Cl(F)$. By Lemma 2.8.5(4), $F \equiv Cl(F)$. Therefore $C(F) = Cn(F) = Cn(Cl(F)) = Cn(RCl(F))$.

Subcase 1.2: F is unsatisfiable. By rationality, $C(F) = C(RCl(F))$. By Definition 3.3.1(RC2), $RCl(F)$ is satisfiable. So by essential classicality, $C(F) = C(RCl(F)) = Cn(RCl(F))$.

Therefore for each set F of formulas, $C(F) = Cn(RCl(F))$.

Case 2: For each satisfiable set G of formulas, $C(G) = Cn(G) - Taut$.
Take any set F of formulas.

Subcase 2.1: F is satisfiable. By essential classicality, $C(F) = Cn(F) - Taut$. By Definition 3.3.1(RC3), $RCl(F) = Cl(F)$. By Lemma 2.8.5(4), $F \equiv Cl(F)$. Therefore $C(F) = Cn(F) - Taut = Cn(Cl(F)) - Taut = Cn(RCl(F)) - Taut$.

Subcase 2.2: F is unsatisfiable. By rationality, $C(F) = C(RCl(F))$. By Definition 3.3.1(RC2), $RCl(F)$ is satisfiable. So by essential classicality, $C(F) = C(RCl(F)) = Cn(RCl(F)) - Taut$.

Therefore for each set F of formulas, $C(F) = Cn(RCl(F)) - Taut$.

Thus C is rationally classical.

EndProofLemma3.4.8

We think that rational classicality gives a reasonable characterisation of what a consequence function should be. But rather than give another definition of a consequence function we shall use the more descriptive term defined below.

Definition 3.4.9. A *rational consequence function* is a function from $\mathscr{P}(Fml)$ to $\mathscr{P}(Fml)$ that is rationally classical.

The functions $C_{\vdash^R}$ and $RFrom$ are rational consequence functions, as the next result shows.

Theorem 3.4.10. Suppose RCl is any rational clauses function. Let F be any set of formulas.
1) $C_{\vdash^R}(F) = Cn(RCl(F))$.
2) $RFrom(F) = Cn(RCl(F)) - Taut$.
3) $C_{\vdash^R}$ and $RFrom$ are both rationally classical and so both are rational consequence functions.

Proof

Let RCl be any rational clauses function. Let F be any set of formulas.
(1) $C_{\vdash^R}(F) = C_{\vdash}(RCl(F))$ by Theorem 3.3.4(6),
    $= Cn(RCl(F))$ by Definition 3.4.2.
(2) $RFrom(F) = C_{\vdash^R}(F) - Taut$ by Theorem 3.3.4(1),
    $= Cn(RCl(F)) - Taut$ by part (1).
(3) This follows directly from part (1) and part (2).

EndProofTheorem3.4.10

Since the definition of rational consequence functions was based on Cn and rational clauses functions, it is interesting to see what properties they have. This is the subject of the next subsection.

3.4.3 Properties of Rational Consequence Functions

Some initial properties of rational consequence functions are given in the following lemma.

Lemma 3.4.11. Suppose C is a rational consequence function. Let F be any set of formulas.
1) $C(F)$ is satisfiable.
2) $C(F) = C(F - Taut)$.
3) Either $Taut \subseteq C(F)$ or $Taut \cap C(F) = \{\}$.

Proof

Let C be a rational consequence function. Then there is a rational clauses function, RCl, such that either for each set F of formulas, $C(F) = Cn(RCl(F))$; or for each set F of formulas, $C(F) = Cn(RCl(F)) - Taut$.

(1) In both cases $C(F) \subseteq Cn(RCl(F))$. By Definition 3.3.1(RC2), $RCl(F)$ is satisfiable. Hence $Cn(RCl(F))$ is satisfiable, and so $C(F)$ is satisfiable.

(2) By Lemma 3.3.2(2), $RCl(F - Taut) = RCl(F)$.

Case 2.1: For each set F of formulas, $C(F) = Cn(RCl(F))$.
Let F be any set of formulas. Then $C(F) = Cn(RCl(F)) = Cn(RCl(F - Taut)) = C(F - Taut)$.

Case 2.2: For each set F of formulas, $C(F) = Cn(RCl(F)) - Taut$.
Suppose F is any set of formulas. Then we have $C(F) = Cn(RCl(F)) - Taut = Cn(RCl(F - Taut)) - Taut = C(F - Taut)$.

Thus in both cases $C(F) = C(F - Taut)$.

(3) Suppose for each set F of formulas, $C(F) = Cn(RCl(F))$. Then $Taut \subseteq Cn(RCl(F)) = C(F)$. Suppose for each set F of formulas, $C(F) = Cn(RCl(F)) - Taut$. Then $\{\} = Taut \cap (Cn(RCl(F)) - Taut) = Taut \cap C(F)$.

EndProofLemma3.4.11

We shall now investigate the properties listed in Definition 3.4.3 to see if they hold for rational consequence functions. As may be expected, several properties are not desirable when tautologies and unsatisfiable sets are concerned. And so only a restricted version of the property holds for rational consequence functions. We shall define the restricted property and then prove that rational consequence functions satisfy it.

Properties, or a set of related properties, are considered in separate subsubsections. Each subsubsection starts with the definition of the relevant properties.

3.4 Consequence Functions

3.4.3.1 Inclusion

C is inclusive iff for all $F \subseteq Fml$, $F \subseteq C(F)$.

Inclusion is not what we want if F is unsatisfiable or some tautologies are in F; as Example 3.4.6(1,3) shows. Hence the following definition.

Definition 3.4.12.
Satisfiable Tautology-free Inclusion: C satisfies *Satisfiable Tautology-free Inclusion* iff for all $F \subseteq Fml$, if F is satisfiable and does not contain any tautologies then $F \subseteq C(F)$.

Lemma 3.4.13. Let C be a rational consequence function. Then C satisfies satisfiable tautology-free inclusion.
Proof
Let C be a rational consequence function. Suppose F is a satisfiable set of formulas that does not contain any tautologies. We show $F \subseteq C(F)$.
$F \subseteq Cn(F)$
$= Cn(Cl(F))$ by Lemma 2.8.5(4),
$= Cn(RCl(F))$ by the satisfiability of F and Definition 3.3.1(RC3). So
$F \subseteq Cn(RCl(F)) - Taut$ as there are no tautologies in F,
$\subseteq C(F)$ as C is rationally classical.
EndProofLemma3.4.13

3.4.3.2 Idempotence

C is idempotent iff for all $F \subseteq Fml$, $C(C(F)) = C(F)$.

Lemma 3.4.14. Let C be a rational consequence function. Then C is idempotent.
Proof
Let C be a rational consequence function. Then there is a rational clauses function, RCl, such that either for each set F of formulas, $C(F) = Cn(RCl(F))$; or for each set F of formulas, $C(F) = Cn(RCl(F)) - Taut$.

Case 1: For each set F of formulas, $C(F) = Cn(RCl(F))$.
By Theorem 3.4.10(1), $C(F) = Cn(RCl(F)) = C_{\vdash^R}(F)$. But $C_{\vdash^R}$ is idempotent by Theorem 3.3.4(7) and so C is idempotent.

Case 2: For each set F of formulas, $C(F) = Cn(RCl(F)) - Taut$.
By Theorem 3.4.10(2), $C(F) = Cn(RCl(F)) - Taut = RFrom(F)$. But $RFrom$ is idempotent by Theorem 3.3.4(7) and so C is idempotent.

So in both cases C is idempotent.
EndProofLemma3.4.14

3.4.3.3 Monotonicity

C is monotonic iff for all $\{F, G\} \subseteq \mathscr{P}(Fml)$, if $F \subseteq G$ then $C(F) \subseteq C(G)$.

Example 3.4.6(1,3) shows that we do not want Monotonicity. However, if f follows from F then we expect that f will follow from $F \cup F'$ provided F' does not contradict any formula in F; which is guaranteed if $F \cup F'$ is satisfiable. This leads to the following weaker form of monotonicity.

Definition 3.4.15.
Satisfiable Monotonicity: C is *satisfiably monotonic* iff for all $\{F, G\} \subseteq \mathscr{P}(Fml)$, if $F \subseteq G$ and G is satisfiable then $C(F) \subseteq C(G)$.

Lemma 3.4.16. Let C be a rational consequence function. Then C is satisfiably monotonic.

Proof

Let C be a rational consequence function. Then there is a rational clauses function, RCl, such that either for each set F of formulas, $C(F) = Cn(RCl(F))$; or for each set F of formulas, $C(F) = Cn(RCl(F)) - Taut$.

Suppose $F \subseteq G$ and G is a satisfiable set of formulas.

Case 1: For each set F of formulas, $C(F) = Cn(RCl(F))$.
Then $C(F) = Cn(RCl(F))$
$= C_{\vdash\!\!R}(F)$ by Theorem 3.4.10(1),
$= C_{\vdash}(F)$ by Theorem 3.3.4(3),
$\subseteq C_{\vdash}(G)$ by Theorem 2.9.2(5),
$= C_{\vdash\!\!R}(G)$ by Theorem 3.3.4(3),
$= Cn(RCl(G))$ Theorem 3.4.10(1),
$= C(G)$ in this case.

Case 2: For each set F of formulas, $C(F) = Cn(RCl(F)) - Taut$.
Then $C(F) = Cn(RCl(F)) - Taut$
$\subseteq Cn(RCl(G)) - Taut$ by the reasoning in Case 1,
$= C(G)$ in this case.

So in both cases $C(F) \subseteq C(G)$. Hence C is satisfiably monotonic.
EndProofLemma3.4.16

3.4.3.4 Compactness

C is compact iff for all $G \subseteq Fml$, $C(G) = \bigcup \{C(F) : F \text{ is a finite subset of } G\}$.

By Lemma 3.4.4(5) compactness implies monotonicity, which we do not want, by Subsubsection 3.4.3.3. This leads to the following weaker form of compactness.

Definition 3.4.17.
Satisfiable Compactness: C is *satisfiably compact* iff whenever G is a satisfiable set of formulas, $C(G) = \bigcup \{C(F) : F \text{ is a finite subset of } G\}$.

3.4 Consequence Functions

Lemma 3.4.18. Let C be a rational consequence function. Then C is satisfiably compact.

Proof

Let C be a rational consequence function. Then there is a rational clauses function, RCl, such that either for each set F of formulas, $C(F) = Cn(RCl(F))$; or for each set F of formulas, $C(F) = Cn(RCl(F)) - Taut$.

Suppose G is a satisfiable set of formulas.

Case 1: For each set F of formulas, $C(F) = Cn(RCl(F))$.
Then $C(G) = Cn(RCl(G))$
$= C_{\vdash^R}(G)$ by Theorem 3.4.10(1),
$= C_{\vdash}(G)$ by Theorem 3.3.4(3),
$= \bigcup\{C_{\vdash}(F) : F \text{ is a finite subset of } G\}$ by Theorem 2.9.2(9),
$= \bigcup\{C_{\vdash^R}(F) : F \text{ is a finite subset of } G\}$ by Theorem 3.3.4(3),
$= \bigcup\{Cn(RCl(F)) : F \text{ is a finite subset of } G\}$ by Theorem 3.4.10(1),
$= \bigcup\{C(F) : F \text{ is a finite subset of } G\}$ in this case.

Case 2: For each set F of formulas, $C(F) = Cn(RCl(F)) - Taut$.
Then $C(G) = Cn(RCl(G)) - Taut$
$= \bigcup\{Cn(RCl(F)) : F \text{ is a finite subset of } G\} - Taut$ by the reasoning in Case 1,
$= \bigcup\{Cn(RCl(F)) - Taut : F \text{ is a finite subset of } G\}$
$= \bigcup\{C(F) : F \text{ is a finite subset of } G\}$ in this case.

So in both cases $C(G) = \bigcup\{C(F) : F \text{ is a finite subset of } G\}$. Hence C is satisfiably compact.

EndProofLemma3.4.18

3.4.3.5 Cumulativity, Cut, Cautious Monotonicity

C is cumulative iff for all $\{F, G\} \subseteq \mathscr{P}(Fml)$, if $F \subseteq C(G)$ then $C(F \cup G) = C(G)$.
C satisfies cut iff for all $\{F, G\} \subseteq \mathscr{P}(Fml)$, if $F \subseteq C(G)$ then $C(F \cup G) \subseteq C(G)$.
C is cautiously monotonic iff for all $\{F, G\} \subseteq \mathscr{P}(Fml)$, if $F \subseteq C(G)$ then $C(F \cup G) \supseteq C(G)$.

By Lemma 3.4.4(4), C is cumulative iff C satisfies cut and cautious monotonicity.

For convenience it is probably worthwhile rephrasing the part of the first paragraph in Subsection 3.2.3 concerning cumulativity.

Cumulativity is sometimes called 'Lemma Addition' because it is the property that any set of proved formulas (F) can be added to the knowledge set (G) without changing what can be proved. This is useful because it means that lemmas (proved formulas) can be used without reproving them. It is also a property that seems to be central to a 'logical' system.

Lemma 3.4.19. Let C be a rational consequence function. Then C is cumulative, cautiously monotonic, and satisfies cut.

Proof

Let C be a rational consequence function. Then there is a rational clauses function, RCl, such that either for each set F of formulas, $C(F) = Cn(RCl(F))$; or for each set F of formulas, $C(F) = Cn(RCl(F)) - Taut$.

Case 1: For each set F of formulas, $C(F) = Cn(RCl(F))$.
By Theorem 3.4.10(1), $C(F) = Cn(RCl(F)) = C_{\not\models}^R(F)$. But $C_{\not\models}^R$ is cumulative by Theorem 3.3.4(8) and so C is cumulative.

Case 2: For each set F of formulas, $C(F) = Cn(RCl(F)) - Taut$.
By Theorem 3.4.10(2), $C(F) = Cn(RCl(F)) - Taut = RFrom(F)$. But $RFrom$ is cumulative by Theorem 3.3.4(8) and so C is cumulative.

So in both cases C is cumulative. By Lemma 3.4.4(4), C is cautiously monotonic and satisfies cut.
EndProofLemma3.4.19

3.4.3.6 Left Equivalence, Right Absorption

C satisfies left equivalence iff for all $\{F,G\} \subseteq \mathscr{P}(Fml)$, if $F \equiv G$ then $C(F) = C(G)$.
C is right absorptive iff for all $F \subseteq Fml$, $C(Cn(F)) = C(F)$.

Recall that by Lemma 3.4.4(7), C satisfies left equivalence iff C is right absorptive. But, the following example indicates that left equivalence, and hence right absorption, are not desirable.

Example 3.4.20. Let $F = \{a, \neg a\}$ and $G = \{a, \neg a, b\}$. Then $F \equiv G$, $Cl(F) = F$, $Cl(G) = G$, $PCl(F) = \{\}$, and $PCl(G) = \{b\}$. Therefore $PCl(F) \not\equiv PCl(G)$. So by Theorem 3.3.4(6), $C_{\not\models}(F) = C_{\vdash}(PCl(F)) = Cn(PCl(F)) = Cn(\{\}) = Taut$, and $C_{\not\models}(G) = C_{\vdash}(PCl(G)) = Cn(PCl(G)) = Cn(\{b\})$. Thus $C_{\not\models}(F) \neq C_{\not\models}(G)$.

The equivalence relation $\equiv$ makes all unsatisfiable sets of formulas equivalent. But each rational clauses function extracts a satisfiable subset of the set of clauses of an unsatisfiable set of formulas. So one should expect that a rational clauses function would distinguish between unsatisfiable sets of formulas. Hence left equivalence is not desirable.

This motivates the following definition.

Definition 3.4.21.
C-Equivalence $\equiv_C$: Let C be a rational consequence function. Then there exists a rational clauses function, RCl, such that either for each set F of formulas, $C(F) = Cn(RCl(F))$; or for each set F of formulas, $C(F) = Cn(RCl(F)) - Taut$. For all $\{F,G\} \subseteq \mathscr{P}(Fml)$, $F \equiv_C G$ iff $RCl(F) \equiv RCl(G)$. We say F and G are C-equivalent, and F is C-equivalent to G, to mean $F \equiv_C G$.
Left C-Equivalence: C satisfies left C-equivalence iff for all $\{F,G\} \subseteq \mathscr{P}(Fml)$, if $F \equiv_C G$ then $C(F) = C(G)$.
Satisfiable Left Equivalence: C satisfies satisfiable left equivalence iff for all $\{F,G\} \subseteq \mathscr{P}(Fml)$, if $F \equiv G$ and F is satisfiable then $C(F) = C(G)$.

3.4 Consequence Functions

Satisfiable Right Absorption: C is *satisfiably right absorptive* iff for all $F \subseteq Fml$, if F is satisfiable then $C(Cn(F)) = C(F)$.

Lemma 3.4.22. Let C be a rational consequence function. Let C' be a function from $\mathscr{P}(Fml)$ to $\mathscr{P}(Fml)$.
1) C satisfies left C-equivalence.
2) If $\{F,G\} \subseteq \mathscr{P}(Fml)$ and $F \equiv G$ and F is satisfiable then $F \equiv_C G$.
3) C satisfies satisfiable left equivalence.
4) C' satisfies satisfiable left equivalence iff C' is satisfiably right absorptive.
5) C is satisfiably right absorptive.

Proof

Let C be a rational consequence function. Then there is a rational clauses function, RCl, such that either for each set F of formulas, $C(F) = Cn(RCl(F))$; or for each set F of formulas, $C(F) = Cn(RCl(F)) - Taut$.

Suppose $\{F,G\} \subseteq \mathscr{P}(Fml)$.

(1) Suppose $F \equiv_C G$. Then $RCl(F) \equiv RCl(G)$.
Case 1: For each set F of formulas, $C(F) = Cn(RCl(F))$.
Then $C(F) = Cn(RCl(F)) = Cn(RCl(G)) = C(G)$.
Case 2: For each set F of formulas, $C(F) = Cn(RCl(F)) - Taut$.
Then $C(F) = Cn(RCl(F)) - Taut = Cn(RCl(G)) - Taut = C(G)$.
Hence C satisfies left C-equivalence.

(2) Suppose $F \equiv G$ and F is satisfiable. Then G is satisfiable. So $RCl(F)$
$= Cl(F)$ by Definition 3.3.1(RC3),
$\equiv F$ by Lemma 2.8.5(4),
$\equiv G$
$\equiv Cl(G)$ by Lemma 2.8.5(4),
$= RCl(G)$ by Definition 3.3.1(RC3).
Thus $F \equiv_C G$.

(3) This follows from part (2) and part (1).

(4) Let C' be a function from $\mathscr{P}(Fml)$ to $\mathscr{P}(Fml)$.
Suppose C' satisfies satisfiable left equivalence. That is, for all $\{F,G\} \subseteq \mathscr{P}(Fml)$, if $F \equiv G$ and F is satisfiable then $C'(F) = C'(G)$. We must show that for all $F \subseteq Fml$, if F is satisfiable then $C'(Cn(F)) = C'(F)$. Take any satisfiable set F of formulas. By Lemma 2.2.21(3), $F \equiv Cn(F)$. Hence $C'(F) = C'(Cn(F))$. Therefore C' is satisfiably right absorptive.

Conversely, suppose C' is satisfiably right absorptive. That is, for all $F \subseteq Fml$, if F is satisfiable then $C'(Cn(F)) = C'(F)$. We must show that for all $\{F,G\} \subseteq \mathscr{P}(Fml)$, if $F \equiv G$ and F is satisfiable then $C'(F) = C'(G)$. Take any sets, F and G, of formulas. Suppose $F \equiv G$ and F is satisfiable. Then G is satisfiable. By Lemma 2.2.21(5), $Cn(F) = Cn(G)$. So $C'(F) = C'(Cn(F)) = C'(Cn(G)) = C'(G)$. Therefore C' satisfies satisfiable left equivalence.

Thus C' satisfies satisfiable left equivalence iff C' is satisfiably right absorptive.

(5) This follows from part (3) and part (4).

EndProofLemma3.4.22

3.4.3.7 Classical Closure, Left Absorption

C is classically closed iff for all $\{F,G\} \subseteq \mathscr{P}(Fml)$, if $F \subseteq C(G)$ then $Cn(F) \subseteq C(G)$.
C is left absorptive iff for all $F \subseteq Fml$, $Cn(C(F)) = C(F)$.

Recall that by Lemma 3.4.4(8), C is classically closed iff C is left absorptive.

Since $\{\} \subseteq C(G)$, by Classical Closure we have $Taut = Cn(\{\}) \subseteq C(G)$. So Classical Closure forces $C(G)$ to contain every tautology; which we do not want. Therefore the best we can hope for is that rational consequence functions satisfy Tautology-free Classical Closure, as defined in Definition 3.4.23. Definition 3.4.23 also defines some other related properties.

Definition 3.4.23.
Tautology-free Classical Closure: C is *tautology-free classically closed* iff for all $\{F,G\} \subseteq \mathscr{P}(Fml)$, if $F \subseteq C(G)$ then $Cn(F) - Taut \subseteq C(G) - Taut$.
Unit Classical Closure: C satisfies *unit classical closure* iff for any formula f and any set of formulas G, if $f \in C(G)$ then $Cn(f) \subseteq C(G)$.
Right Weakening: $\mathrel{|\!\!\sim}_{\overline{C}}$ satisfies *right weakening* iff for any formula f and any set of formulas G, if $G \mathrel{|\!\!\sim}_{\overline{C}} f$ then $G \mathrel{|\!\!\sim}_{\overline{C}} Cn(f)$.
Tautology-free Left Absorption: C is *tautology-free left absorptive* iff for all $F \subseteq Fml$, $Cn(C(F)) - Taut = C(F) - Taut$.

We note that what we have called Unit Classical Closure is a special case of Classical Closure, where the set F of formulas has been replaced by the set $\{f\}$ of just one formula. It is clear that Unit Classical Closure and Right Weakening are the same property expressed differently. Right Weakening is the usual way of expressing this special case of Classical Closure.

Lemma 3.4.24. Let C be a rational consequence function. Let C' be a function from $\mathscr{P}(Fml)$ to $\mathscr{P}(Fml)$.
1) C is tautology-free classically closed.
2) C' is tautology-free classically closed iff C' is tautology-free left absorptive.
3) C is tautology-free left absorptive.

Proof

Let C be a rational consequence function. Then there is a rational clauses function, RCl, such that either for each set F of formulas, $C(F) = Cn(RCl(F))$; or for each set F of formulas, $C(F) = Cn(RCl(F)) - Taut$.

(1) Suppose $\{F,G\} \subseteq \mathscr{P}(Fml)$ and $F \subseteq C(G)$.
Case 1: For each set F of formulas, $C(F) = Cn(RCl(F))$.
Then $F \subseteq C(G) = Cn(RCl(G))$. So
$Cn(F) \subseteq Cn(Cn(RCl(G)))$ by Lemma 2.2.21(6),
$= Cn(RCl(G))$ by Lemma 2.2.21(8),
$= C(G)$.
Hence $Cn(F) - Taut \subseteq C(G) - Taut$.

Case 2: For each set F of formulas, $C(F) = Cn(RCl(F)) - Taut$.
Then $F \subseteq C(G) = Cn(RCl(G)) - Taut \subseteq Cn(RCl(G))$. So by the reasoning in Case 1,

3.4 Consequence Functions

$Cn(F) \subseteq Cn(RCl(G))$. Hence $Cn(F) - Taut$
$\subseteq Cn(RCl(G)) - Taut$
$= C(G)$
$= C(G) - Taut$.

Thus C is tautology-free classically closed.

(2) Let C' be a function from $\mathscr{P}(Fml)$ to $\mathscr{P}(Fml)$.

Suppose C' is tautology-free classically closed. That is, for all $\{F,G\} \subseteq \mathscr{P}(Fml)$, if $F \subseteq C'(G)$ then $Cn(F) - Taut \subseteq C'(G) - Taut$. We must show that for all $G \subseteq Fml$, $Cn(C'(G)) - Taut = C'(G) - Taut$. Take any set G of formulas. By letting $F = C'(G)$ we have $Cn(C'(G)) - Taut \subseteq C'(G) - Taut$. By Lemma 2.2.21(2), $C'(G) \subseteq Cn(C'(G))$, and so $C'(G) - Taut \subseteq Cn(C'(G)) - Taut$. Hence $Cn(C'(G)) - Taut = C'(G) - Taut$. Therefore C' is tautology-free left absorptive.

Conversely, suppose C' is tautology-free left absorptive. That is, for all $F \subseteq Fml$, $Cn(C'(F)) - Taut = C'(F) - Taut$. We must show that for all $\{F,G\} \subseteq \mathscr{P}(Fml)$, if $F \subseteq C'(G)$ then $Cn(F) - Taut \subseteq C'(G) - Taut$. Take any sets, F and G, of formulas. Suppose $F \subseteq C'(G)$. By Lemma 2.2.21(6) we have, $Cn(F) \subseteq Cn(C'(G))$. Therefore $Cn(F) - Taut \subseteq Cn(C'(G)) - Taut$. So by tautology-free left absorption, $Cn(F) - Taut \subseteq C'(G) - Taut$. Therefore C' is tautology-free classically closed.

Thus C' is tautology-free classically closed iff C' is tautology-free left absorptive.

(3) This follows from part (1) and part (2).

EndProofLemma3.4.24

3.4.3.8 Divisibility

C is divisible iff whenever F is a set of formulas and g is a formula then we have $C(F \cup \{g\}) \cap C(F \cup \{\neg g\}) \subseteq C(F)$.

Lemma 3.4.25. Let C be a rational consequence function. Then C is divisible.

Proof

Let C be a rational consequence function. Then there is a rational clauses function, RCl, such that either for each set F of formulas, $C(F) = Cn(RCl(F))$; or for each set F of formulas, $C(F) = Cn(RCl(F)) - Taut$.

Case 1: For each set F of formulas, $C(F) = Cn(RCl(F))$.
By Theorem 3.4.10(1), $C(F) = Cn(RCl(F)) = C_{\models^R}(F)$. But $C_{\models^R}$ is divisible by Theorem 3.3.4(9) and so C is divisible.

Case 2: For each set F of formulas, $C(F) = Cn(RCl(F)) - Taut$.
By Theorem 3.4.10(2), $C(F) = Cn(RCl(F)) - Taut = RFrom(F)$. But $RFrom$ is divisible by Theorem 3.3.4(9) and so C is divisible.

So in both cases C is divisible.

EndProofLemma3.4.25

3.4.3.9 Subclassicality, Supraclassicality

Definition 3.4.26. Let C be a function from $\mathscr{P}(Fml)$ to $\mathscr{P}(Fml)$.
Subclassicality: C is subclassical iff for all $F \subseteq Fml$, $C(F) \subseteq Cn(F)$.
Supraclassicality: C is supraclassical iff for all $F \subseteq Fml$, $Cn(F) \subseteq C(F)$.

Supraclassicality forces $C(F)$ to contain every tautology, which we do not want. Moreover if F is unsatisfiable then Supraclassicality forces $C(F)$ to contain every formula, which is exactly what we are trying to avoid. The extent to which Supraclassicality holds is given by Essential Classicality (Definition 3.4.7). However, we shall have more to say about Supraclassicality in Section 4.6.

Contrastingly, subclassicality is satisfied by rational consequence functions, as we now show.

Lemma 3.4.27. Let C be a rational consequence function. Then C is subclassical.
Proof

Let C be a rational consequence function. Then there is a rational clauses function, RCl, such that either for each set F of formulas, $C(F) = Cn(RCl(F))$; or for each set F of formulas, $C(F) = Cn(RCl(F)) - Taut$.

Suppose F is any set of formulas. Then $C(F)$
$\subseteq Cn(RCl(F))$
$\subseteq Cn(Cl(F))$ by Definition 3.3.1(RC2) and Theorem 2.9.2(5),
$= Cn(F)$ by Lemma 2.8.5(4).

Hence C is subclassical.
EndProofLemma3.4.27

3.4.3.10 *C*-Partial Order

We conclude this list of properties by investigating whether the consequence relation $\models_{\overline{c}}$ is a partial order or not; but first some definitions.

Definition 3.4.28. Let C be a function from $\mathscr{P}(Fml)$ to $\mathscr{P}(Fml)$, and $\models_{\overline{c}}$ be a relation from $\mathscr{P}(Fml)$ to Fml. Recall from Definition 3.4.1(1 and 4), that $F \models_{\overline{c}} G$ iff for all g in G, $F \models_{\overline{c}} g$, and that $F \models_{\overline{c}} G$ iff $G \subseteq C(F)$.
C-Antisymmetry: $\models_{\overline{c}}$ is *C*-antisymmetric iff for all $\{F, G\} \subseteq \mathscr{P}(Fml)$, if $F \models_{\overline{c}} G$ and $G \models_{\overline{c}} F$ then $C(F) = C(G)$.
Transitivity: $\models_{\overline{c}}$ is transitive iff for all $\{F, G, H\} \subseteq \mathscr{P}(Fml)$, if $F \models_{\overline{c}} G$ and $G \models_{\overline{c}} H$ then $F \models_{\overline{c}} H$.
C-Partial Order: $\models_{\overline{c}}$ is a *C*-partial order iff $\models_{\overline{c}}$ is *C*-antisymmetric and transitive.

The following lemma shows that $\models_{\overline{c}}$ is transitive. Although it is not antisymmetric it is *C*-antisymmetric.

Lemma 3.4.29. Let C be a rational consequence function. Then $\models_{\overline{c}}$ is a *C*-partial order.
Proof

3.4 Consequence Functions

Let C be a rational consequence function. By Lemma 3.4.11(1), for all $F \subseteq Fml$, $C(F)$ is satisfiable. By Lemma 3.4.14, for all $F \subseteq Fml$, $C(C(F)) = C(F)$. By Lemma 3.4.16, for all $\{F, G\} \subseteq \mathscr{P}(Fml)$, if $F \subseteq G$ and G is satisfiable then $C(F) \subseteq C(G)$.

C-Antisymmetry.
Suppose $\{F, G\} \subseteq \mathscr{P}(Fml)$. Suppose $F \models_{\overline{C}} G$ and $G \models_{\overline{C}} F$. Then $G \subseteq C(F)$ and $F \subseteq C(G)$. $C(F)$ is satisfiable and $C(G)$ is satisfiable. So $C(G) \subseteq C(C(F))$ and $C(F) \subseteq C(C(G))$. But $C(C(F)) = C(F)$ and $C(C(G)) = C(G)$. Hence $C(G) \subseteq C(F)$ and $C(F) \subseteq C(G)$. Thus $C(F) = C(G)$, and so $\models_{\overline{C}}$ is *C*-antisymmetric.

Transitivity.
Suppose $\{F, G, H\} \subseteq \mathscr{P}(Fml)$. Suppose $F \models_{\overline{C}} G$ and $G \models_{\overline{C}} H$. Then $G \subseteq C(F)$ and $H \subseteq C(G)$. $C(F)$ is satisfiable. Hence $C(G) \subseteq C(C(F))$. But $C(C(F)) = C(F)$ and so $C(G) \subseteq C(F)$. Therefore $H \subseteq C(G) \subseteq C(F)$. Thus $F \models_{\overline{C}} H$, and so $\models_{\overline{C}}$ is transitive.

C-Partial Order.
$\models_{\overline{C}}$ is a *C*-partial order because $\models_{\overline{C}}$ is *C*-antisymmetric and transitive.
EndProofLemma3.4.29

Chapter 4
Principles of Plausible Reasoning

Abstract The sections of Chapter 4 Principles of Plausible Reasoning are: 4.1 Introduction, 4.2 Representation, 4.3 Evidence and Non-Monotonicity, 4.4 Conjunction, 4.5 Disjunction, 4.6 Supraclassicality, 4.7 Right Weakening, 4.8 Consistency, 4.9 Multiple Intuitions: Ambiguity, 4.10 Decisiveness, 4.11 Truth Values, 4.12 Correctness, and 4.13 Some Non-monotonic Logics.

Plausible reasoning concerns situations whose inherent lack of precision is not quantified; that is, there are no degrees or levels of precision, and hence no use of numbers like probabilities. A hopefully comprehensive set of principles that clarifies what it means for a formal logic to do plausible reasoning is presented in Sections 4.2 to 4.11. Some important examples are given that act as signposts to some of the principles. Some non-numeric non-monotonic logics are considered in the last section to see whether or not they satisfy all the principles of this chapter and can satisfactorily reason with the examples of this chapter.

4.1 Introduction

We are interested in reasoning about situations that
(a) have imprecisely defined parts, and
(b) this lack of precision is not quantified.
Point (b) means that there are no degrees or layers or levels of precision, and in particular there are no numbers like probabilities, that would quantify the lack of precision. These situations are often indicated by the ordinary, rather than technical, use of words such as 'mostly', 'usually', 'typically', 'normally', 'probably', 'likely', 'plausible', 'believable', and 'reasonable'. Although these words are not synonymous, they share the following common property: that the evidence for something outweighs the evidence against it. The evidence may be expressed by using frequency of occurrence, in which case the property becomes: that something is true more often than not. Two

examples of the kind of situations we have in mind are: 'Most mammals have fur' and 'After a roll of a fair standard die the uppermost side is usually not 1'.

We shall call these situations *plausible-reasoning situations* because we shall call the reasoning used in such situations *plausible reasoning*.

Plausible reasoning is inherently less accurate than reasoning with probabilities because the given information is less precise. Situations that are defined by using probabilities can also be described plausibly by words like those listed above. For example, 'After a roll of a fair standard die the probability that the uppermost side is not 1 is 5/6.' can be described plausibly by 'After a roll of a fair standard die the uppermost side is usually not 1.'. Such plausibly described situations can be reasoned with and the results should be consistent with the results from reasoning with the original situation using probabilities. So any logic that does plausible reasoning can be checked by comparing it with reasoning using probabilities.

This chapter introduces principles that give a much clearer understanding of what it means for a formal logic to do the kind of reasoning indicated above, that is, plausible reasoning. The hope is that this set of principles will characterise the formal logics that do plausible reasoning. Whether it does or not, this seems to be the first such set of principles even though plausible reasoning has been used for at least 2500 years, see [83]. However on page 114 of [83] there is a list of 11 characteristics of plausible reasoning, rather than characteristics of formal logics that do plausible reasoning.

We could limit our attention to propositional languages; but there is no need to be so restrictive. So we shall allow both propositional and first-order languages. (Rather than define here the first-order concepts that are needed, we shall refer the interested reader to, for example, [68].) Let us use the term *closed formula* to denote either a formula of a propositional language, or a formula of a first-order language that has no free variables, and hence, in any structure for the language, either it or its negation is true. If we are dealing with a first-order language then we shall use the term *Taut* to denote the set of all formulas of the language that are true in every structure for that language.

We shall only consider those plausible-reasoning situations that can be specified by a *plausible-structure* $S = (Fact(S), Plaus(S))$ where $Fact(S)$ is a satisfiable set of closed formulas representing the factual part of S, and $Plaus(S)$ is a set representing the plausible part of S. The elements of $Plaus(S)$ can have a variety of forms. For example: defaults are used in Reiter's Default Logic, see [77]; defeasible rules are used in ASPIC, see [27], and in ASPIC$^+$, see [67]; and defeasible and warning rules are used in Defeasible Logic, see [11]. The plausible-structure construct is very general, while being specific enough to permit the definition of concepts needed later.

Both classical propositional logic and classical first-order logic can be the logic underlying a degenerate plausible-structure $S = (Fact(S), Plaus(S))$ where $Fact(S)$ is a satisfiable set of closed formulas and $Plaus(S)$ is the empty set $\{\}$. Hence some forthcoming definitions involving plausible-structures can also apply to classical propositional logic and classical first-order logic.

Lists of postulates, properties, or principles that concern special types of reasoning are useful for at least the following reasons.

4.1 Introduction

1) They help *characterise* the intended special type of reasoning.
2) They provide a means of *evaluating* existing reasoning systems to see how well they perform the intended special type of reasoning.
3) They provide *guidelines* for creating new reasoning systems for the intended special type of reasoning.
4) They explicitly show a *difference* between the intended special type of reasoning and an existing form of reasoning.

Notable examples of such lists are: the AGM postulates for belief change, see [2] and [40]; the properties of non-monotonic consequence relations, see [63] and [58]; and the postulates that a rule-based argumentation system should satisfy, see [27].

We shall state the principles of this chapter by referring to the logic or proof algorithm directly; rather than the more common indirect approach of referring to consequence relations. A consequence relation, say $\models_c$, relates a set F of formulas to a formula f; where $F \models_c f$ means that f is a consequence of F. Consequence relations are appropriate if the reasoning situations under consideration can be characterised by a set of formulas. But the plausible-reasoning situations we consider are specified by a plausible-structure $S = (Fact(S), Plaus(S))$ where the elements of $Plaus(S)$ may be very different from the formulas in $Fact(S)$. For these situations consequence relations are not appropriate. For example consider two fundamental properties that consequences relations may have; namely cut and cautious monotonicity, which together are equivalent to cumulativity, also called lemma addition. If F and G are sets of formulas then let $F \models_c G$ mean for all g in G, $F \models_c g$. Then cumulativity is the following property. If $F \models_c G$ then for all formulas h, $F \models_c h$ iff $F \cup G \models_c h$. A straightforward translation of $F \models_c f$ into our situation is $S \models_c f$, where S is a plausible-structure. But then it is really hard to know what $F \cup G$ might mean. Essentially we are trying to add proved formulas to S. But the only set of formulas in S is $Fact(S)$. So we could try letting $F \cup G$ be $(Fact(S) \cup G, Plaus(S))$. But this is only sensible when the formulas in G have been proved using only $Fact(S)$. When the formulas in G have been proved using $Plaus(S)$ then it is no longer sensible to treat the formulas in G as facts; because they are not facts, they are only plausible conclusions.

Some of the following principles of plausible reasoning are regarded as necessary and so use the word 'must'; the other principles are regarded as desirable and so use the word 'should'.

As well as the principles of plausible reasoning, several plausible-reasoning examples will be presented. Some of these examples will guide the development of some of the principles, and so we shall call these examples signpost examples.

Several examples involve an *n-lottery* based on N; that is, randomly selecting a number from a set N of positive integers where $|N| = n$. In most cases $N = [1..n]$. We shall use s_i to denote that the number i was selected.

Several equivalents to an *n*-lottery occur in the literature; for example an *n*-sided die, or a (2-sided) coin, or an *n*-sided top, or an urn containing *n* balls appropriately numbered.

Our first signpost example is a 3-lottery example.

Example 4.1.1 (The 3-lottery example). Consider the 3-lottery based on $[1..3]$. Then we have the following.
1) Exactly one element of $\{s_1, s_2, s_3\}$ is true.
2) Each element of $\{\neg s_1, \neg s_2, \neg s_3\}$ is likely.
3) The disjunction of any pair of elements of $\{s_1, s_2, s_3\}$ is likely.

This example illustrates some important properties of plausible reasoning that will be considered in several of the following sections.

The following notation will be convenient. Let $Thm(\mathcal{L}, \alpha, \mathcal{S})$ denote the set of all formulas derivable from the plausible-structure $\mathcal{S}$ by using the proof algorithm α of the logic $\mathcal{L}$. If F is a set of closed formulas then $Thm(F)$ denotes all the formulas derivable from F by (the proof algorithm of) any classical propositional, or first-order, logic. This simpler notation is unambiguous because $Thm(F)$ is independent of the logic (for example Hilbert systems, natural deduction, or resolution systems) and its proof algorithm.

4.2 Representation

Plausible-reasoning situations may contain facts as well as plausible information. For instance in the context of Example 4.1.1, the 3-lottery example, the formula $\vee\{s_1, s_2, s_3\}$ is a fact and always true; whereas the formula $\neg s_1$ is not a fact but only a plausible statement. Semantically these are very different formulas and so they need to be differentiated syntactically. Hence the following necessary principle of plausible reasoning.

Principle 4.2.1 (The Representation Principle). A logic for plausible reasoning must be able to represent, and distinguish between, factual and plausible statements.

Although the inherent lack of precision of plausible-reasoning situations is not quantified, a logic could represent this lack of precision with undue accuracy, for instance by using probabilities. Forbidding this is too restrictive, because the logic may deduce a conclusion using the probabilities but then present that conclusion without using probabilities. All that is needed is that the conclusions are not unduly precise. In particular, if a formula is proved by using plausible information then it must not be presented as a fact. Hence following necessary principle of plausible reasoning.

Principle 4.2.2 (The Precision Principle). The formulas proved by a logic for plausible reasoning must not be more precise than the information used to derive them.

Recall from the third paragraph of the Introduction 4.1 that when a situation is precisely defined, perhaps using probabilities, a logic for plausible reasoning should be able to reason with the corresponding imprecisely described situation; Example 4.1.1 is such a situation.

We infer from Principle 4.2.1 and Principle 4.2.2 that we should be able to distinguish between conclusions that are factual and those that are merely plausible. One way of making this distinction is to have a *factual proof algorithm* that only uses facts and deduces only facts, and also a *plausible proof algorithm* that may use plausible statements and facts and deduces formulas that are only plausible. Of course, if a plausible proof algorithm deduces only facts when given just facts then it can be regarded as both a factual and a plausible proof algorithm. However, even if such a dual purpose proof algorithm exists, we shall see, in Section 4.9, that there is a need for at least two different plausible proof algorithms.

4.3 Evidence and Non-Monotonicity

Let us now see if we can establish some general guidelines concerning the provability of a given formula f. A plausible-reasoning situation will have evidence for and against f. So it seems reasonable to determine whether f is provable or not by just comparing these two sets of evidence, and declaring f provable iff the preponderance of evidence is for f. A consequence of this evidence criterion needs the following definition.

Definition 4.3.1. A proof algorithm α of a logic $\mathcal{L}$ is said to be *monotonic* iff for any two plausible-structures, $\mathcal{S}_1$ and $\mathcal{S}_2$, if $Fact(\mathcal{S}_1) \subseteq Fact(\mathcal{S}_2)$ and $Plaus(\mathcal{S}_1) \subseteq Plaus(\mathcal{S}_2)$ then $Thm(\mathcal{L}, \alpha, \mathcal{S}_1) \subseteq Thm(\mathcal{L}, \alpha, \mathcal{S}_2)$. A proof algorithm is *non-monotonic* iff it is not monotonic.

The proof algorithm of a classical propositional logic is monotonic. But a plausible proof algorithm is non-monotonic because the addition of evidence against a previously provable formula can cause it to be unprovable, as shown in our second signpost example.

Example 4.3.2 (The Non-Monotonicity example). Cephalopods are marine animals that have tentacles; octopuses, squids, cuttlefish, and nautiluses are all cephalopods. Consider the following four statements.
1) Nautiluses are cephalopods.
2) Cephalopods usually do not have external shells.
3) Nautiluses have external shells.
4) Nancy is a cephalopod.
From these four statements it is reasonable to conclude that Nancy probably does not have an external shell.

But suppose that later we discover the following statement.
5) Nancy is a nautilus.
Then it is not reasonable to conclude that Nancy probably does not have an external shell. Indeed from these five statements it is reasonable to conclude that Nancy has an external shell.

The discussion above justifies the following two necessary principles.

Principle 4.3.3 (The Evidence Principle).
A plausible proof algorithm proves a formula f iff all the evidence for f sufficiently outweighs all the evidence against f.

Principle 4.3.4 (The Non-Monotonicity Principle).
A plausible proof algorithm must be non-monotonic.

Exactly what constitutes evidence for or against f can only be determined when the particular logic for plausible reasoning is known. Also 'sufficiently outweighs' depends on the intuition that is being modelled, as well as the particular logic.

A proof algorithm that fails the Evidence Principle seems to be seriously flawed. So the Evidence Principle may be a principle that any sensible proof algorithm of any sensible logic should satisfy. The proof algorithm of a classical propositional logic fails the Evidence Principle because $\neg b$ can be proved from $\{a, \neg a, b\}$. Yet there is no evidence for $\neg b$ and there is evidence against $\neg b$, namely b.

A corollary of the Evidence Principle is what might be called the Relevant Information Principle: "Whether a plausible proof algorithm proves a formula f or not is determined only by all the information relevant to f".

4.4 Conjunction

Let us start by defining what we mean by a conjunctive proof algorithm.

Definition 4.4.1. A proof algorithm α of a logic $\mathcal{L}$ is said to be *conjunctive* iff for any plausible-structure, $\mathcal{S}$, and any two formulas f and g, if $\{f,g\} \subseteq Thm(\mathcal{L},\alpha,\mathcal{S})$ then $\wedge\{f,g\} \in Thm(\mathcal{L},\alpha,\mathcal{S})$. A proof algorithm is *non-conjunctive* iff it is not conjunctive.

The proof algorithm of any classical propositional logic is conjunctive.

Conjunctions of plausible formulas behave very differently from conjunctions of formulas that are certain. In Example 4.1.1, $\wedge\{\neg s_1, \neg s_2\}$ is equivalent to s_3. So although $\neg s_1$ is plausible and $\neg s_2$ is plausible, $\wedge\{\neg s_1, \neg s_2\}$ is not plausible. Therefore plausible proof algorithms are not conjunctive. Hence the following principle is necessary.

Principle 4.4.2 (The Non-Conjunctive Principle).
A plausible proof algorithm must not be conjunctive.

Although the conjunction of two plausible formulas is not necessarily plausible, the conjunction of two facts is a fact. So what about the conjunction of a fact and a plausible formula? Clearly it cannot be a fact, but is it always plausible? Intuitively, a fact f is always true, and a plausible formula g is true more often that not. So it seems necessary that their conjunction be true whenever g is true, and hence it is necessary that the conjunction is plausible.

This leads to the following definition.

4.5 Disjunction

Definition 4.4.3. A proof algorithm α of a logic $\mathcal{L}$ is said to be *plausibly conjunctive* iff for any plausible-structure, $\mathcal{S}$, and any formulas f and g, if $f \in Thm(Fact(\mathcal{S}))$ and $g \in Thm(\mathcal{L}, \alpha, \mathcal{S})$ then $\wedge \{f,g\} \in Thm(\mathcal{L}, \alpha, \mathcal{S})$.

The above discussion shows that our next principle is necessary.

Principle 4.4.4 (The Plausibly Conjunctive Principle).
A plausible proof algorithm must be plausibly conjunctive.

The Non-Conjunctive Principle is supported by the fact that the 'And' rule of [58], is not probabilistically sound, see [65](Section 2.1) where they call the 'And' rule the 'Right∧+' rule. Also Definition 2.4 of [53] defines a weak 'And' rule that is probabilistically sound and seems to have a similar intuition to our Plausibly Conjunctive Principle.

4.5 Disjunction

It could happen that whenever a disjunction is provable then at least one of its disjuncts is provable. When this happens we shall call the proof algorithm right disjunctive and the disjunction is often called point-wise disjunction. The formal definition follows.

Definition 4.5.1. A proof algorithm α of a logic $\mathcal{L}$ is said to be *right disjunctive* iff for any plausible-structure, $\mathcal{S}$, and any formulas f and g, if $\vee \{f,g\} \in Thm(\mathcal{L}, \alpha, \mathcal{S})$ then either $f \in Thm(\mathcal{L}, \alpha, \mathcal{S})$ or $g \in Thm(\mathcal{L}, \alpha, \mathcal{S})$.

The proof algorithm of any classical propositional logic is not right disjunctive.

The 3-lottery example (Example 4.1.1) shows that, although s_1 and s_2 are both unlikely their disjunction $\vee \{s_1, s_2\}$ is likely. Hence our next principle is necessary.

Principle 4.5.2 (The Not Right Disjunctive Principle).
A plausible proof algorithm must not be right disjunctive.

Recall that in Subsection 3.2.4 Disjunction, the left disjunction property was introduced. A first attempt at translating that property into our plausible reasoning setting might be the following. If $f \in Thm(\mathcal{L}, \alpha, \mathcal{S}+g)$ and $f \in Thm(\mathcal{L}, \alpha, \mathcal{S}+h)$ then $f \in Thm(\mathcal{L}, \alpha, \mathcal{S}+\vee\{g,h\})$. Of course the problem with this is, what is the meaning of $\mathcal{S}+g$ where g is a formula? Now $\mathcal{S} = (Fact(\mathcal{S}), Plaus(\mathcal{S}))$, where $Fact(\mathcal{S})$ is a satisfiable set of closed formulas representing the factual part of $\mathcal{S}$, and $Plaus(\mathcal{S})$ is a set representing the plausible part of $\mathcal{S}$. Since we do not want to further specify $Plaus(\mathcal{S})$, we shall let $\mathcal{S}+g$ mean $(Fact(\mathcal{S}) \cup \{g\}, Plaus(\mathcal{S}))$. Hence the following definition.

Definition 4.5.3. A proof algorithm α of a logic $\mathcal{L}$ is *left factually disjunctive* iff for any plausible-structure, $\mathcal{S} = (Fact(\mathcal{S}), Plaus(\mathcal{S}))$, and any formulas g and h, if $f \in Thm(\mathcal{L}, \alpha, (Fact(\mathcal{S}) \cup \{g\}, Plaus(\mathcal{S})))$ and $f \in Thm(\mathcal{L}, \alpha, (Fact(\mathcal{S}) \cup \{h\}, Plaus(\mathcal{S})))$ then $f \in Thm(\mathcal{L}, \alpha, (Fact(\mathcal{S}) \cup \{\vee\{g,h\}\}, Plaus(\mathcal{S})))$.

To help discover whether a plausible proof algorithm should be left factually disjunctive or not, consider our third signpost example, which is founded on a 7-lottery.

Example 4.5.4 (The Left Factual Disjunction example).
Let S be a plausible-structure that models the 7-lottery based on $[1..7]$.
Let g be $\wedge\{\neg s_1, \neg s_2\}$ and $S + g = (Fact(S) \cup \{g\}, Plaus(S))$.
Let h be $\wedge\{\neg s_3, \neg s_4\}$ and $S + h = (Fact(S) \cup \{g\}, Plaus(S))$.
Let $S + \vee\{g, h\} = (Fact(S) \cup \{\vee\{g, h\}\}, Plaus(S))$.
Let f be $\vee\{s_5, s_6, s_7\}$.
Then we have the following.
1) In S exactly one element of $\{s_1, s_2, s_3, s_4, s_5, s_6, s_7\}$ is true.
   Hence in S, f is unlikely.
2) In $S + g$ exactly one element of $\{s_3, s_4, s_5, s_6, s_7\}$ is true.
   Hence in $S + g$, f is likely.
3) In $S + h$ exactly one element of $\{s_1, s_2, s_5, s_6, s_7\}$ is true.
   Hence in $S + h$, f is likely.
But $Fact(S) \equiv Fact(S) \cup \{\vee\{g, h\}\}$. So in $S + \vee\{g, h\}$ exactly one element of $\{s_1, s_2, s_3, s_4, s_5, s_6, s_7\}$ is true. Hence in $S + \vee\{g, h\}$, f is unlikely.

Example 4.5.4 shows that a plausible proof algorithm must not be left factually disjunctive. Hence our next principle is necessary.

Principle 4.5.5 (The Not Left Factually Disjunctive Principle).
A plausible proof algorithm must not be left factually disjunctive.

4.6 Supraclassicality

Definition 4.6.1. A proof algorithm α of a logic $\mathcal{L}$ is said to be *supraclassical* iff for any plausible-structure, S, $Thm(Fact(S)) \subseteq Thm(\mathcal{L}, \alpha, S)$.

Supraclassicality could be phrased as 'what is true is usually true'. This is a reasonable requirement except that as we saw in Chapter 3, classical propositional logic proves all tautologies. We do not want to force logics for plausible reasoning to prove all tautologies. So we introduce the following notion of being plausibly supraclassical.

Definition 4.6.2. A proof algorithm α of a logic $\mathcal{L}$ has the *plausible supraclassicality property* and is said to be *plausibly supraclassical* iff for any plausible-structure, S, $Thm(Fact(S)) - Taut \subseteq Thm(\mathcal{L}, \alpha, S)$.

The corresponding principle is desirable.

Principle 4.6.3 (The Plausible Supraclassicality Principle).
Factual and plausible proof algorithms should be plausibly supraclassical.

Since $Thm(Fact(S)) - Taut \subseteq Thm(Fact(S))$, it is clear that if α is supraclassical then it is plausibly supraclassical.

4.7 Right Weakening

Definition 4.7.1. A proof algorithm α of a logic $\mathcal{L}$ has the *right weakening property* iff for any plausible-structure, $\mathcal{S}$, and any formula f, if $f \in Thm(\mathcal{L}, \alpha, \mathcal{S})$ and $f \models g$ then $g \in Thm(\mathcal{L}, \alpha, \mathcal{S})$.

Right weakening can be thought of as closure under classical inference. By replacing g with any tautology, we see that a consequence of the right weakening property is $Taut \subseteq Thm(\mathcal{L}, \alpha, \mathcal{S})$. But we do not want to force logics for plausible reasoning to prove all tautologies. Hence the next definition.

Definition 4.7.2. A proof algorithm α of a logic $\mathcal{L}$ has the *weak right weakening property* iff for any plausible-structure, $\mathcal{S}$, and any formula f, if $f \in Thm(\mathcal{L}, \alpha, \mathcal{S})$ then $Thm(\{f\}) - Taut \subseteq Thm(\mathcal{L}, \alpha, \mathcal{S})$.

However, suppose that whenever the facts of the plausible-structure $\mathcal{S}$ and a formula f are true then the formula g is also true; in symbols $Fact(\mathcal{S}) \cup \{f\} \models g$. Then in the situation defined by $\mathcal{S}$, g is true at least as often as f. So if f is usually true then g should also be usually true. Hence the next definition.

Definition 4.7.3. A proof algorithm α of a logic $\mathcal{L}$ has the *strong right weakening property* iff for any plausible-structure, $\mathcal{S}$, and any formula f, if $f \in Thm(\mathcal{L}, \alpha, \mathcal{S})$ and $Fact(\mathcal{S}) \cup \{f\} \models g$ then $g \in Thm(\mathcal{L}, \alpha, \mathcal{S})$.

Combining these ideas produces the following definition.

Definition 4.7.4. A proof algorithm α of a logic $\mathcal{L}$ has the *plausible right weakening property* iff for any plausible-structure, $\mathcal{S}$, and any formula f, if $f \in Thm(\mathcal{L}, \alpha, \mathcal{S})$ then $Thm(Fact(\mathcal{S}) \cup \{f\}) - Taut \subseteq Thm(\mathcal{L}, \alpha, \mathcal{S})$.

If α does not satisfy the plausible right weakening property then there are formulas that should be provable but are not. Such an algorithm is insufficient; but it can be easily augmented so that the plausible right weakening property does hold. Hence the corresponding principle is regarded as necessary.

Principle 4.7.5 (The Plausible Right Weakening Principle).
A plausible proof algorithm must have the plausible right weakening property.

It is straightforward to see that strong right weakening implies all the other right weakening properties, and also that weak right weakening is implied by all the other right weakening properties.

4.8 Consistency

There are 11 characteristics of plausible reasoning given on page 114 of [83]. Characteristic 8 is 'stability'; which seems to mean (bottom of page 97 of [83]) that plausible statements are consistent. However, as we shall show, where consistency is concerned the number of plausible statements is important.

Classical reasoning about facts always produces a satisfiable set of conclusions. So it is a fundamental shock that this is not true for plausible reasoning.

First we need the following two definitions that relate consistency and the number of statements.

Definition 4.8.1. A proof algorithm α of a logic $\mathcal{L}$ is *n-consistent* iff for any plausible-structure, $\mathcal{S}$, and any set of formulas, F, if $F \subseteq \mathit{Thm}(\mathcal{L}, \alpha, \mathcal{S})$, and $|F| \leq n$ then F is satisfiable.

Recall that if $\mathcal{S}$ be a plausible-structure then $\mathit{Fact}(\mathcal{S})$ is a satisfiable set of closed formulas.

Definition 4.8.2. A proof algorithm α of a logic $\mathcal{L}$ is *strongly n-consistent* iff for any plausible-structure, $\mathcal{S}$, and any set of formulas, F, if $F \subseteq \mathit{Thm}(\mathcal{L}, \alpha, \mathcal{S})$, and $|F| \leq n$ then $\mathit{Fact}(\mathcal{S}) \cup F$ is satisfiable.

So if a proof algorithm is strongly n-consistent then it is n-consistent. Also if a proof algorithm is strongly $(n+1)$-consistent then it is strongly n-consistent; and if a proof algorithm is $(n+1)$-consistent then it is n-consistent.

In classical propositional logic or classical first-order logic, if S is a satisfiable set of closed formulas then $S \cup \mathit{Thm}(S)$ is satisfiable. So classical propositional logic is strongly $\aleph_0$-consistent.

Contradictions are not plausible. Therefore plausible proof algorithms must be 1-consistent.

If $f \in \mathit{Thm}(\mathcal{L}, \alpha, \mathcal{S})$ then in the situation defined by $\mathcal{S}$, f is more likely to be true than not. Therefore $\mathit{Fact}(\mathcal{S}) \cup \{f\}$ is satisfiable. That is, strong 1-consistency holds.

Now consider strong 2-consistency. So suppose f and g are formulas such that $\{f,g\} \subseteq \mathit{Thm}(\mathcal{L}, \alpha, \mathcal{S})$. By strong 1-consistency, we have that both $\mathit{Fact}(\mathcal{S}) \cup \{f\}$ and $\mathit{Fact}(\mathcal{S}) \cup \{g\}$ are satisfiable. So f is not a contradiction and g is not a contradiction. If $\mathit{Fact}(\mathcal{S}) \cup \{f,g\}$ is unsatisfiable then f is not a tautology and g is not a tautology. Also $\mathit{Fact}(\mathcal{S}) \cup \{g\} \models \neg f$. By the plausible right weakening property, $\neg f \in \mathit{Thm}(\mathcal{L}, \alpha, \mathcal{S})$. Thus we have $\{f, \neg f\} \subseteq \mathit{Thm}(\mathcal{L}, \alpha, \mathcal{S})$. Although 'likely' is an imprecise concept, it has the property that for any formula f, at most one of f and $\neg f$ is likely. Therefore we do not have $\{f, \neg f\} \subseteq \mathit{Thm}(\mathcal{L}, \alpha, \mathcal{S})$. This contradiction shows that $\mathit{Fact}(\mathcal{S}) \cup \{f,g\}$ is satisfiable. So plausible proof algorithms are strongly 2-consistent. Hence our first consistency principle is necessary.

Principle 4.8.3 (The Strong 2-Consistency Principle).
A plausible proof algorithm must be strongly 2-consistent.

Consider the 3-lottery example (Example 4.1.1). Let $U = \{\neg s_1, \neg s_2, \vee\{s_1, s_2\}\}$. Then for each x in U, x is likely; and $\neg x$ is not likely. But U is unsatisfiable. Hence the necessity of our second consistency principle.

Principle 4.8.4 (The Non-3-Consistency Principle).
A plausible proof algorithm that can prove disjunctions must not be 3-consistent.

4.9 Multiple Intuitions: Ambiguity

With the possible exception of tautologies, classical propositional logic captures our intuition about what follows from a satisfiable set of facts. But there are different well-informed intuitions about what follows from a plausible-reasoning situation. For example, as early as 1987, see Section 4.1 of [82], it was recognised that a plausible-reasoning situation could elicit different sensible conclusions, depending on whether ambiguity was blocked or propagated. The essence of Figure 3 in [82] is our fourth signpost example.

Example 4.9.1 (The Ambiguity Puzzle).
1) a is a fact. b is a fact. c is a fact.
2) If a is true then e is likely.
3) If d is true then $\neg e$ is likely.
4) If b is true then d is likely.
5) If c is true then $\neg d$ is likely.

The Ambiguity Puzzle can be represented by the following diagram.

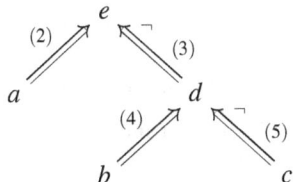

What can be concluded about e? The evidence for e is a and (2). The evidence against e comes from d and (3), and b and (4). If we knew that d was definitely true then the evidence for e and against e would be equal. Ignoring (5), d is only likely by (4), so the evidence against e is weaker than the evidence for e. But c and (5) means that d is even less likely, and so the evidence against e has been further weakened. Thus e is more likely than $\neg e$.

So reasoning that is based on the 'best bet' or the 'most likely' or the 'balance of probabilities' concludes e. However, d may be true, and so it is reasonable to have some doubts about the truth of e. So reasoning that is 'beyond reasonable doubt' does not conclude e.

A formula f is said to be *ambiguous* iff there is evidence for f and there is evidence against f and neither f nor $\neg f$ can be proved. Since b and (4) and c and (5) give equal evidence for and against d, d is ambiguous.

If the evidence against e has been weakened sufficiently to allow e to be concluded, then e is not ambiguous. So the ambiguity of d has been blocked from propagating to e. An algorithm that can prove e is said to be *ambiguity blocking*. This level of reasoning is appropriate if the benefit of being right outweighs the penalty for being wrong.

If the evidence against e has not been weakened sufficiently to allow e to be concluded, then e is ambiguous. So the ambiguity of d has been propagated to e. An algorithm that cannot prove e is said to be *ambiguity propagating*. This more cautious level of reasoning is appropriate if the penalty for being wrong outweighs the benefit of being right.

The Anglo-American legal system uses a hierarchy of proof levels. (Search for 'proof', 'burden of proof', or 'standard of proof' in any Anglo-American law dictionary, for example see page 584 of [66].) Two of these levels are the 'balance of probabilities' (used in civil cases) which, as noted above, is ambiguity blocking; and 'beyond reasonable doubt' (used in criminal cases) which, as noted above, is ambiguity propagating. So there is a need for a proof algorithm that blocks ambiguity and one that propagates ambiguity.

To avoid confusion, one should know which algorithm is used; unless it is irrelevant to the point being made. This, and our observation at the end of Section 4.2 Representation, that a logic for plausible reasoning should have a factual proof algorithm, leads to our next principle; which is clearly desirable.

Principle 4.9.2 (The Many Proof Algorithms Principle).
A logic for plausible reasoning should have at least
1) a factual proof algorithm,
2) an ambiguity blocking plausible proof algorithm, and
3) an ambiguity propagating plausible proof algorithm.
Also, the proof algorithm used to prove a formula should be explicit or irrelevant.

Clearly the algorithms in (2) and (3) must be different. But, as indicated at the end of Section 4.2, the factual proof algorithm could be the same as a plausible proof algorithm.

4.10 Decisiveness

For a formula, f, a proof algorithm, α, must satisfy exactly one of the following four conditions.
   i) α does not terminate.
  ii) α terminates in a state indicating that f is proved,
 iii) α terminates in a state indicating that f is not provable,
  iv) α terminates in some other state.

Definition 4.10.1. A proof algorithm α is *decisive* iff for every formula f, α terminates in either a state indicating that f is proved, or a state indicating that f is not provable.

Our next principle is clearly desirable.

Principle 4.10.2 (The Decisiveness Principle).
Factual and plausible proof algorithms should be decisive.

4.11 Truth Values

Let us change our focus from deduction to the more semantic notion of assigning truth values to statements. For classical propositional logic there are exactly two truth values: **T** for true and **F** for false. If v is a valuation and f and g are formulas then
1) Either $v(f) = \mathsf{T}$ or $v(\neg f) = \mathsf{T}$ but not both, (the Excluded Middle property) and
2) $v(\wedge\{f,g\}) = \mathsf{T}$ iff $v(f) = \mathsf{T} = v(g)$, and
3) $v(\vee\{f,g\}) = \mathsf{T}$ iff either $v(f) = \mathsf{T}$ or $v(g) = \mathsf{T}$ (or both).

The 3-lottery example (Example 4.1.1) shows that the closest plausible reasoning can get to (2) and (3) is (4) and (5) below.
4) If $v(\wedge\{f,g\}) = \mathsf{T}$ then $v(f) = \mathsf{T} = v(g)$.
5) If $v(f) = \mathsf{T}$ or $v(g) = \mathsf{T}$ then $v(\vee\{f,g\}) = \mathsf{T}$.

Moreover consider our fifth signpost example.

Example 4.11.1 (The 4-lottery example). Consider the 4-lottery based on $[1..4]$. Then we have the following four statements.
1) Exactly one element of $\{s_1, s_2, s_3, s_4\}$ is true.
2) Each element of $\{s_1, s_2, s_3, s_4\}$ is probably false.
3) The disjunction of any 2 different elements of $\{s_1, s_2, s_3, s_4\}$ is not probably true and not probably false.
4) The disjunction of any 3 elements of $\{s_1, s_2, s_3, s_4\}$ is probably true.

Intuitively some formulas concerning Example 4.11.1 have different truth values; for example $\vee\{s_1, s_2, s_3, s_4\}$ is definitely true, $\neg\vee\{s_1, s_2, s_3, s_4\}$ is definitely false, $\neg s_1$ is probably true, s_1 is probably false, and $\vee\{s_1, s_2\}$ is as likely to be true as false. So plausible reasoning distinguishes between at least 3 truth values:
one indicating that a formula is more likely to be true than false,
one indicating that a formula is as likely to be true as false, and
one indicating that a formula is more likely to be false than true.

Therefore a logic with only two truth values cannot fully express the different possibilities that exist in plausible reasoning. Hence a logic for plausible reasoning that cannot have a truth theory with at least three truth values is lacking an essential feature. Hence the following necessary principle of plausible reasoning.

Principle 4.11.2 (The Included Middle Principle).
A logic for plausible reasoning must have a truth theory with at least 3 truth values.

The Included Middle Principle (Principle 4.11.2) is our last formal principle of plausible reasoning. However, there is another informal, but necessary, principle of plausible reasoning that we shall consider in the next section.

4.12 Correctness

A logic that satisfies all the previous principles could nonetheless have a fatal flaw. It could give an unsatisfactory answer to a particular example. Some examples may well have no set of answers that are generally agreed upon. But some examples, in particular the signpost examples, do have a set of answers that are generally agreed upon. We might call these answers the correct answers. So it is tempting to state a principle of correctness similar to "When correct answers exist, a logic must give all the correct answers, and no incorrect answers.".

The problem with such a principle is that it is impossible to show that any logic satisfies it. The most that can be done is to produce a counter-example that shows a logic fails the principle, or demonstrate that for a chosen set of examples the logic gets the correct answers. But there might exist a counter-example that shows the logic fails the principle of correctness.

Thus we shall refrain from trying to formally state a Correctness Principle, even though such a principle is highly desirable. A verifiable correctness principle is likely to be very difficult to obtain, because it is likely to imply sufficient conditions for a logic to do plausible reasoning.

4.13 Some Non-monotonic Logics

We shall consider the relationship between some non-numeric non-monotonic logics and the principles and examples of this chapter.

There are three well-known non-monotonic logics, namely Default Logic, Circumscription, and Autoepistemic Logic; see [3] for an introduction. Answer Set Programming (ASP), see [7], is a well-known Knowledge Representation system.

Each of the proof algorithms of these four well-known systems is conjunctive and so fails the Non-Conjunctive Principle (Principle 4.4.2). Also for each of these four proof algorithms, the set of all provable formulas is either satisfiable or contains all formulas. So all four proof algorithms fail the Non-3-Consistency Principle (Principle 4.8.4). Hence none of these logics reasons correctly about the 3-lottery example (Example 4.1.1). Finally all four of these proof algorithms are ambiguity propagating but not ambiguity blocking. So each of these logics fails the Many Proof Algorithms Principle (Principle 4.9.2). Hence when ambiguity blocking is required — for instance in civil cases — these logics do not get the right answers.

Logics that deal with only literals are incapable of the reasoning required by the 3-lottery example (Example 4.1.1). Logics in this category include inheritance networks, see [54]; the DeLP system of [39]; the ASPIC system mentioned in [27]; the logic in [76]; Ordered logic, see [42]; and most Defeasible Logics, see [11].

Propositional Plausible Logic (PPL), which is defined in the next chapter, is a member of the family of Defeasible Logics. The only Defeasible Logics that deal with conjunction and disjunction, besides PPL, are the logic in [20], let us call it DL1, and

4.13 Some Non-monotonic Logics

the logic in [11], let us call it DL8. But the plausible proof algorithms of both DL1 and DL8 are conjunctive and so do not satisfy the Non-Conjunctive Principle (Principle 4.4.2). Also the Decisiveness Principle (Principle 4.10.2) fails for the plausible proof algorithms that define the Defeasible Logics in: [10], [20], [62], [11], and [12]. Since all Defeasible Logics apart from PPL are closely related to a Defeasible Logic in these five citations, all Defeasible Logics apart from PPL fail the Decisiveness Principle. As we shall show, PPL is the only Defeasible Logic that satisfies all the principles in this chapter.

Argumentation systems, see [35], are well-known non-monotonic reasoning systems that can use rules, for example ASPIC, see [27], and ASPIC$^+$, see [67]. Let $E \in \{$admissible, complete, preferred, grounded, ideal, semi-stable, stable$\}$. Then the semantics of ASPIC$^+$ defined by intersecting all E-extensions is ambiguity propagating and so fails the Many Proof Algorithms Principle (Principle 4.9.2(2)). An early argumentation system is given in [79] and it also is ambiguity propagating and so fails the Many Proof Algorithms Principle (Principle 4.9.2(2)). It also has other problems mentioned in [41].

Three postulates that a rule-based argumentation system should satisfy are given in [27]. Postulate 1 is closure under strict rules; that is Modus Ponens for strict rules (see Theorem 6.1.2(3)). It is a kind of right weakening property (Section 4.7). Postulate 2 requires the set of all proved literals to be consistent. If only literals can be proved, as in [27], then this is implied by the Strong 2-Consistency Principle (Principle 4.8.3). Postulates 1 and 2 jointly imply Postulate 3.

It is not surprising that Conditional Logics, see [72] and [6], have been used to analyse non-monotonic reasoning. Let $\dashrightarrow$ denote a weak conditional; so that for formulas f and g, $f \dashrightarrow g$ means 'if f then ... g' where '...' could be 'normally', 'typically', 'probably', or any other similar word or phrase. A set of such weak conditionals is called a 'conditional knowledge base'. The following two rules are particularly important for differentiating our plausible reasoning from other kinds of reasoning.

And-rule: If $f \dashrightarrow g$ and $f \dashrightarrow h$ then $f \dashrightarrow \wedge\{g,h\}$.
Or-rule: If $g \dashrightarrow f$ and $h \dashrightarrow f$ then $\vee\{g,h\} \dashrightarrow f$.

The And-rule is also called the CC-rule, and the Or-rule is also called the CA-rule.

Let $Ax3 = \wedge\{\vee\{s_1,s_2,s_3\}, \neg\wedge\{s_1,s_2\}, \neg\wedge\{s_1,s_3\}, \neg\wedge\{s_2,s_3\}\}$ be the formula that characterises the 3-lottery example, Example 4.1.1(1). As noted in Section 4.4, we have $Ax3 \dashrightarrow \neg s_1$ and $Ax3 \dashrightarrow \neg s_2$ but not $Ax3 \dashrightarrow \wedge\{\neg s_1, \neg s_2\}$. So reasoning systems that satisfy the And-rule do not do plausible reasoning.

We shall adapt the Left Factual Disjunction example, Example 4.5.4, as follows. The formula that characterises the 7-lottery based on [1..7] is the conjunction of the following 22 formulas: $\vee\{s_1,s_2,s_3,s_4,s_5,s_6,s_7\}$, and $\neg\wedge\{s_i,s_j\}$, where $1 \leq i < j \leq 7$; which will be denoted by $Ax7$. Let g be $\wedge\{Ax7, \neg s_1, \neg s_2\}$, let h be $\wedge\{Ax7, \neg s_3, \neg s_4\}$, and let f be $\vee\{s_5,s_6,s_7\}$. Then g is equivalent to exactly one of s_3 or s_4 or s_5 or s_6 or s_7, and h is equivalent to exactly one of s_1 or s_2 or s_5 or s_6 or s_7. So $g \dashrightarrow f$ and $h \dashrightarrow f$. But $\vee\{g,h\}$ does not restrict the selected number at all, and f is not a usual result of a 7-lottery. So we do not have $\vee\{g,h\} \dashrightarrow f$. Therefore reasoning systems that satisfy the Or-rule do not do plausible reasoning.

In [32] it is observed that the following reasoning systems satisfy both the And-rule and the Or-rule and hence do not do our plausible reasoning: systems based on intuitions from probability theory such as [1] and [73]; systems based on intuitions from qualitative possibilistic logic such as [33]; systems based on C4, see [60], and on CT4, see [23]; and system S of [26].

A logic called 'conditional entailment' is defined in [43]. The second paragraph on page 235 of [43] contains the following sentence. "In the propositional case, the only difference between conditional entailment and prioritized circumscription is the source of the priorities: while prioritized circumscription relies on the user, conditional entailment extracts the priorities from the knowledge base itself." As noted near the beginning of this section, circumscription fails the Non-Conjunctive Principle (Principle 4.4.2), the Non-3-Consistency Principle (Principle 4.8.4), and the Many Proof Algorithms Principle (Principle 4.9.2). Hence conditional entailment also fails these principles.

The consequence function of [63](pages 14 and 15) and the cumulative conditional knowledge bases of [58](page 13) satisfy both the And-rule and the Or-rule. The rational conditional knowledge bases of [61] are preferential conditional knowledge bases, see [58]; and preferential conditional knowledge bases are cumulative. Therefore both the preferential and rational closures of a conditional knowledge base satisfy both the And-rule and the Or-rule. Thus both the preferential and rational closures of a conditional knowledge base do not do the plausible reasoning we are trying to characterise.

As noted in [32] the following systems are 'essentially the same as' rational closure and hence do not do our plausible reasoning: System Z, see [74]; systems based on conditional logic, see [28]; on modal logic, see [24]; on possibilistic logic, see [9]; and on conditional objects, see [34].

The Propositional Typicality Logic (PTL) of [22] and [21] has several semantics. Each semantics is at least preferential and so satisfies both the And-rule and the Or-rule. Hence PTL does not do the plausible reasoning we are trying to characterise.

The conditional logic C of [32] does not satisfy the Plausible Right Weakening Principle (Principle 4.7.5). Also C and the extensions of C considered in [32] have only one proof algorithm and so fail the Many Proof Algorithms Principle (Principle 4.9.2).

Apart from the problems mentioned in Section 5 of [44], System Z, see [74], and System Z^+, see [44], are ambiguity propagating but not ambiguity blocking. Hence they fail the Many Proof Algorithms Principle (Principle 4.9.2). Moreover, although they can represent the 3-lottery example (Example 4.1.1), they cannot prove anything about the example because the set of rules is not 'consistent' as defined in [74] and [44].

The logic implemented by THEORIST, see [75], and the Preferred Subtheories logic in [25], both generate consistent extensions and so fail the Non-3-Consistency Principle (Principle 4.8.4).

Although many principles of plausible reasoning have been stated in this chapter, there are four principles that distinguish plausible reasoning from the reasoning done

4.13 Some Non-monotonic Logics

by the logics reviewed above. More precisely, every logic reviewed above fails at least one of the following four principles:

the Non-Conjunctive Principle (Principle 4.4.2),

the Non-3-Consistency Principle (Principle 4.8.4),

the Many Proof Algorithms Principle (Principle 4.9.2), and

the correctness principle as instanced by the 3-lottery example (Example 4.1.1).

As far as we know, Propositional Plausible Logic (PPL) is the only non-numeric non-monotonic propositionally adequate logic that satisfies all the principles in this chapter. Moreover PPL can represent and reason with all the signpost examples in this chapter. The next three chapters are devoted to the definition of PPL and the verification of the claims made about PPL.

Chapter 5
Propositional Plausible Logic (PPL)

Abstract The sections of Chapter 5 Propositional Plausible Logic (PPL) are: 5.1 Plausible Descriptions, 5.2 The Proof Relation and the Proof Algorithms, 5.3 The Proof Theory, and 5.4 A Truth Theory.

A new propositional logic called Propositional Plausible Logic (PPL) is defined. As far as we know this is the first non-numeric non-monotonic propositionally adequate logic that satisfies all the principles in Chapter 4.

PPL has four different proof algorithms that are defined by defining a proof relation. Only one of these proof algorithms mimics classical propositional logic. A proof in PPL is a tree. A proof function is also defined because it is easier to use than either the proof relation or a proof tree. The chapter concludes by defining four truth values.

5.1 Plausible Descriptions

The purpose of this chapter is to define a propositional logic, called Propositional Plausible Logic (PPL), that satisfies all the principles in Chapter 4. PPL reasons about plausible-reasoning situations that may contain facts, like definitions and membership of categories. For example 'Whales are mammals'. These facts are represented by formulas that are converted into clauses called axioms and these axioms are then converted into strict rules. The plausible information is represented by defeasible rules, warning rules, and a priority relation, $>$, on rules.

Intuitively the various kinds of rules have the following meanings. The strict rule $A \rightarrow c$ means if every formula in A is accepted then c is acceptable. So strict rules are like material implication except that A is a finite set of formulas rather than a single formula. (We have already seen that A and $\wedge A$ behave differently.) For example, 'nautiluses are cephalopods' could be written as $\{n\} \rightarrow c$, and 'cephalopods are molluscs' could be written as $\{c\} \rightarrow m$.

The defeasible rule $A \Rightarrow c$ means if every formula in A is accepted and there is no evidence against c then c is likely. For example, 'molluscs usually have shells' could be written as $\{m\} \Rightarrow s$, and 'cephalopods usually have no shells' could be written as $\{c\} \Rightarrow \neg s$.

The warning rule $A \rightsquigarrow c$ roughly means if every formula in A is accepted and there is no evidence against c then c might be acceptable. So $A \rightsquigarrow \neg c$ warns against concluding usually c, but does not support usually $\neg c$. Warning rules can be used to prevent conclusions that are too risky. For example, 'objects that look red in red light might not be red' could be written as $\{\text{looks-red-in-red-light}\} \rightsquigarrow \neg r$. The idea is that looking red in a red light is not evidence against the object being red; rather it is not sufficient evidence to conclude that the object is probably red. Warning rules can be used to prevent unwanted chaining. For example, suppose we have 'if a then usually b' ($\{a\} \Rightarrow b$) and 'if b then usually c' ($\{b\} \Rightarrow c$). Then it may be too risky to conclude 'usually c' from a. Without introducing evidence for $\neg c$, the conclusion of 'usually c' from a can be prevented by the warning rule $\{a\} \rightsquigarrow \neg c$. An instance of this example can be created by letting a be $x \in \{1,2,3,4\}$, b be $x \in \{2,3,4\}$, and c be $x \in \{3,4,5\}$. Warning rules have also been called 'defeaters' and 'interfering rules'.

Let us start the formalities by taking the alphabet of PRL and adding three more connectives, namely the three kinds of arrows mentioned above.

Definition 5.1.1. An *alphabet* for PPL consists of the following three parts.
1) A countable set, *Atm*, of (propositional) *atoms*.
2) The set $\{\neg, \wedge, \vee, \rightarrow, \Rightarrow, \rightsquigarrow\}$ of *connectives* for PPL denoting negation, conjunction, disjunction, strict implication, defeasible implication, and warning implication respectively.
3) The set of punctuation symbols consisting of the comma and both braces (curly brackets).

We also need to make sure that *Atm* does not contain any misleading, or potentially ambiguous, symbols. In particular, *Atm* must not contain any sequence of connectives or punctuation symbols.

The formulas and quasi-formulas of PPL are the same as those of PRL. The formal definition of a rule and its associated terms follow.

Definition 5.1.2. A *generalised-rule*, r, is any triple $(A(r), arrow(r), c(r))$ such that $A(r)$, called the *set of antecedents* of r, is a finite (possibly empty) set of formulas; $arrow(r) \in \{\rightarrow, \Rightarrow, \rightsquigarrow\}$; and $c(r)$, called the *consequent* of r, is a quasi-formula.

A generalised-rule r is a *strict rule*, written $A(r) \rightarrow c(r)$, iff $arrow(r)$ is the strict arrow, $\rightarrow$, and $c(r)$ is a formula or a conjunction of a countable set of formulas.

A generalised-rule r is a *defeasible rule*, written $A(r) \Rightarrow c(r)$, iff $arrow(r)$ is the defeasible arrow, $\Rightarrow$, and $c(r)$ is a formula.

A generalised-rule r is a *warning rule*, written $A(r) \rightsquigarrow c(r)$, iff $arrow(r)$ is the warning arrow, $\rightsquigarrow$, and $c(r)$ is a formula.

A generalised-rule is a *rule* iff it is a strict rule or a defeasible rule or a warning rule.

5.1 Plausible Descriptions

A priority relation, $>$, on rules is used to indicate the more relevant of two rules. For instance, the specific rule 'cephalopods usually have no shells', $\{c\} \Rightarrow \neg s$, is more relevant than the general rule 'molluscs usually have shells', $\{m\} \Rightarrow s$, when reasoning about the external appearance of cephalopods. Hence $\{c\} \Rightarrow \neg s\ >\ \{m\} \Rightarrow s$. More generally, some common policies for defining $>$ are the following. Prefer specific rules over general rules; prefer authoritative rules, (for instance national laws override state laws); prefer recent rules (because they are more up-to-date); and prefer more reliable rules. If r and s are rules and $r > s$ then we often say r is *superior* to s and s is *inferior* to r.

Although the priority relation does not have to be a partial order or even transitive, it does have to be acyclic.

Let us now consider the conversion of the facts of a plausible-reasoning situation represented by a set F of formulas into strict rules. First we transform F into a suitable set of clauses. A detailed investigation of how to do this is given in Subsection 3.3.1. Definition 3.3.1 defines a family of functions called rational clauses functions. Any rational clauses function, RCl, transforms F into a suitable set of clauses $RCl(F)$. If F is satisfiable then $RCl(F) = Cl(F)$. But if F is unsatisfiable then there are at least four different rational clauses functions to choose from, depending on one's intuition and tolerance of risk. So it is best, but not necessary, to make sure that the facts are consistent.

Next we generate the set of axioms, Ax by letting $Ax = CorRes(RCl(F))$. Finally we convert a contingent clause with n literals into $2^n - 1$ strict rules. The conversion is done by the function $Rul(.)$ in the usual way as shown by the following example.
$Rul(\vee\{a,b,c\}) = \{\ \{\} \to \vee\{a,b,c\},$
$\{\wedge\{\neg b, \neg c\}\} \to a,\quad \{\wedge\{\neg a, \neg c\}\} \to b,\quad \{\wedge\{\neg a, \neg b\}\} \to c,$
$\{\neg a\} \to \vee\{b,c\},\quad \{\neg b\} \to \vee\{a,c\},\quad \{\neg c\} \to \vee\{a,b\}\ \}.$

The full definition of $Rul(.)$ and $Rul(.,.)$, as well as some useful notation, is given in the next definition.

Definition 5.1.3. Let R be a set of rules, F be a finite set of formulas, and C be a set of contingent clauses.
1) R_s is the set of strict rules in R. That is, $R_s = \{r \in R : r \text{ is a strict rule}\}$.
2) R_d is the set of defeasible rules in R. That is, $R_d = \{r \in R : r \text{ is a defeasible rule}\}$.
3) R_w is the set of warning rules in R. That is, $R_w = \{r \in R : r \text{ is a warning rule}\}$.
4) $c(R)$ is the set of consequents of the rules in R. That is, $c(R) = \{c(r) : r \in R\}$.
5) If $c \in C$ then
$Rul(c) = \{\ \{\} \to ns(c)\} \cup \{\ \{ns(\wedge \sim (L-K))\} \to ns(\vee K) : c = \vee L \text{ and } \{\} \subset K \subset L\}$.
6) $Rul(C) = \bigcup \{Rul(c) : c \in C\}$.
7) $Rul(C,F)$ is the set of rules in $Rul(C)$ whose set of antecedents is F. That is, $Rul(C,F) = \{r \in Rul(C) : A(r) = F\}$.

We note that the set of antecedents of any strict rule formed by $Rul(.)$ has at most one element.

Although $Rul(Ax)$ gives us the strict rules that characterise the set F of facts we started with, we can reduce the number of these strict rules by 'anding' all those that have the same antecedent. For example, the 'anding' of $\{a\} \to c_1$, $\{a\} \to c_2$, and $\{a\} \to c_3$ is $\{a\} \to \wedge\{c_1,c_2,c_3\}$. We now have the set of strict rules that we want. This set is formally defined by PD2 below. The formal structure used for describing plausible-reasoning situations is called a plausible description and is defined below.

Definition 5.1.4. If F is a set of formulas, RCl is a rational clauses function, R is a set of rules, and $\mathcal{P} = (R,>)$ then $\mathcal{P}$ is a *plausible description* iff PD1, PD2, PD3, and PD4 all hold.

PD1) $Ax(\mathcal{P}) = Ax(R) = CorRes(RCl(F))$.

The set of *axioms of* $\mathcal{P}$, $Ax(\mathcal{P})$, and the set of *axioms of* R, $Ax(R)$, are usually denoted by Ax provided there is no ambiguity.

PD2) $R_s = \{A \to ns(\wedge c(Rul(Ax,A))) : A \in \{A(r) : r \in Rul(Ax)\}\}$.

PD3) If $Ax \neq \{\}$ then r_{ax} denotes the strict rule $\{\} \to \wedge Ax$.

PD4) $>$ is a *priority relation* on R; that is, $> \,\subseteq R \times (R - \{r_{ax}\})$ and $>$ is not cyclic.

Suppose $(R,>)$ is a plausible description. Then Ax is empty iff R_s is empty. If $Ax \neq \{\}$ then $r_{ax} \in R_s$. If R_s is not empty we can extract Ax from the consequent of r_{ax}. This shows that $Ax(R)$ is indeed dependent on R. Different strict rules in R have different sets of antecedents, and no rule is superior ($>$) to r_{ax}.

For PPL the plausible-structure is a plausible description $(R,>)$ and the factual part is $Ax(R)$, which by PD2 is equivalent to the strict rules in R, R_s. The plausible part consists of the non-strict rules in R and the priority relation $>$.

It is occasionally useful to write $s < t$ instead of $t > s$; hence the next definition.

Definition 5.1.5. Let $(R,>)$ be a plausible description. Define the relation $<$ on R by $s < t$ iff $t > s$. We read $s < t$ as s is *inferior* to t, or t is *superior* to s.

Some basic results about the set of axioms is the content of the following lemma.

Lemma 5.1.6. Suppose $(R,>)$ is a plausible description and $Ax = Ax(R)$.
1) Ax is satisfiable.
2) Each axiom in Ax is contingent.
3) Each axiom in Ax is either a literal or $\vee L$ where L is a finite set of literals such that $|L| \geq 2$.

Proof

Suppose $(R,>)$ is a plausible description and $Ax = Ax(R)$.

(1) By Definition 5.1.4(PD1), $Ax = CorRes(C)$, where C is a set of clauses. By Definition 3.3.1(RC2), C is satisfiable. By Lemma 2.6.8(1), $CorRes(C)$ is satisfiable and so Ax is satisfiable.

(2) By Definition 5.1.4(PD1), $Ax = CorRes(C)$, where C is a set of clauses. So $Ax = CorRes(C) = nsMinCtge(Res(C))$; so each axiom in Ax is either contingent or empty. But Ax is satisfiable, so each axiom in Ax is contingent.

(3) This follows from part (2) and the fact that $Ax = ns(C)$ where C is a set of clauses.

EndProofLemma5.1.6

5.1 Plausible Descriptions

By Lemma 5.1.6(1), Ax is satisfiable. So in PPL we extend the meaning of 'fact' from just being an element of Ax to a formula that is implied by Ax. Explicitly, a formula f is said to be a *fact* iff $Ax \models f$.

We conclude this section with a lemma about strict rules.

Lemma 5.1.7. Let $(R,>)$ be a plausible description, Ax be its set of axioms, and f be a formula.
1) $Rul(Ax) = \{\{\} \to c : c \in Ax\} \cup$
   $\{\{ns(\wedge\sim(L-K))\} \to ns(\vee K) : \vee L \in Ax$ and $\{\} \subset K \subset L\}$.
2) If $Ax \neq \{\}$ then $r_{ax} \in R_s$.
3) If $r \in R_s$ then either $A(r) = \{\}$;
   or $A(r) = \{l\}$, where l is a literal;
   or $A(r) = \{\wedge L\}$, where $|L| \geq 2$ and L is contingent.
4) If $r \in Rul(Ax)$ then $\bigcup Lit(A(r) \cup \{c(r)\})$ is contingent,
   $Lit(c(r)) \cap \bigcup Lit(\sim A(r)) = \{\}$, and
   $Lit(c(r)) \cap \bigcup Lit(A(r)) = \{\}$.
5) If $r \in R_s$ then $Ax \cup A(r) \models c(r)$.
6) If $r \in R_s$ then $c(r)$ and every formula in $A(r)$ is contingent.

Proof

Let $(R,>)$ be a plausible description and Ax be its set of axioms.

(1) By Definitions 5.1.4 and 5.1.3(6) we have, $Rul(Ax) = \{\{\} \to ns(c) : c \in Ax\} \cup \{\{ns(\wedge\sim(L-K))\} \to ns(\vee K) : \vee L \in Ax$ and $\{\} \subset K \subset L\}$. By Lemma 5.1.6(3), $R_s = \{\{\} \to c : c \in Ax\} \cup \{\{ns(\wedge\sim(L-K))\} \to ns(\vee K) : \vee L \in Ax$ and $\{\} \subset K \subset L\}$.

(2) This follows from part 1.

(3) By Lemma 5.1.6(2), if $\vee L \in Ax$ then $\vee L$ is contingent and so L is contingent. By Lemma 2.2.16(4), if $\{\} \subset K \subset L$ then $L-K$ is contingent. Hence $\sim(L-K)$ is contingent. Therefore the result holds for all r in $Rul(Ax)$. So by Definition 5.1.4(PD2), the result holds for all r in R_s.

(4) Suppose $r \in Rul(Ax)$.

If $A(r) = \{\}$ then by part 1, $c(r) \in Ax$ and hence by Lemma 5.1.6(2), $c(r)$ is contingent. So $Lit(c(r))$ is contingent. But $\bigcup Lit(A(r) \cup \{c(r)\}) = \bigcup Lit(\{c(r)\}) = \bigcup \{Lit(c(r))\} = Lit(c(r))$. So $\bigcup Lit(A(r) \cup \{c(r)\})$ is contingent. Since $A(r) = \{\}$, $\sim A(r) = \{\}$, and so $\bigcup Lit(A(r)) = \{\}$ and $\bigcup Lit(\sim A(r)) = \{\}$. So part (4) holds in this case.

So suppose $A(r)$ is not empty. Then r is $\{ns(\wedge\sim(L-K))\} \to ns(\vee K)$ where $\vee L \in Ax$ and $\{\} \subset K \subset L$. By Lemma 5.1.6(2), $\vee L$ is contingent and hence L is contingent. Therefore $\bigcup Lit(A(r) \cup \{c(r)\})$
$= \bigcup Lit(\{ns(\wedge\sim(L-K)), ns(\vee K)\})$
$= \bigcup \{Lit(ns(\wedge\sim(L-K))), Lit(ns(\vee K))\}$
$= \bigcup \{\sim(L-K), K\}$
$= K \cup \sim(L-K)$.
Since L is contingent, $K \cup (L-K)$ is contingent, and so $K \cup \sim(L-K)$ is contingent. Hence $\bigcup Lit(A(r) \cup \{c(r)\})$ is contingent.

By Lemma 2.5.7(5), $Lit(c(r)) = Lit(ns(\vee K)) = Lit(\vee K) = K$.

By Lemma 2.5.7(5), $\bigcup Lit(\sim A(r))$
$= \bigcup Lit(\sim\{ns(\wedge\sim(L-K))\})$
$= \bigcup Lit(\{\sim ns(\wedge\sim(L-K))\})$
$= \bigcup Lit(\{ns(\sim\wedge\sim(L-K))\})$
$= \bigcup Lit(\{ns(\vee\sim\sim(L-K))\})$
$= \bigcup Lit(\{ns(\vee(L-K))\})$
$= \bigcup \{Lit(ns(\vee(L-K)))\}$
$= \bigcup \{Lit(\vee(L-K))\}$
$= \bigcup \{L-K\}$
$= L-K$.

Hence $Lit(c(r)) \cap \bigcup Lit(\sim A(r)) = K \cap (L-K) = \{\}$.

By Lemma 2.5.7(5), $\bigcup Lit(A(r))$
$= \bigcup Lit(\{ns(\wedge\sim(L-K))\})$
$= \bigcup \{Lit(ns(\wedge\sim(L-K)))\}$
$= \bigcup \{Lit(\wedge\sim(L-K))\}$
$= \bigcup \{\sim(L-K)\}$
$= \sim(L-K)$.

Hence $Lit(c(r)) \cap \bigcup Lit(A(r)) = K \cap \sim(L-K)$ which is empty since L is contingent.

(5) Take any r in $Rul(Ax)$, and suppose $v \models Ax \cup A(r)$. Then either r is $\{\} \to c$ where $c \in Ax$, or r is $\{ns(\wedge\sim(L-K))\} \to ns(\vee K)$, where $\{\} \subset K \subset L$ and $\vee L \in Ax$. In the first case $v \models c$, so $v \models c(r)$. In the second case $v \models \vee L$, and $v \models \wedge\sim(L-K)$. Hence for all $l \in L-K$, $v \models \sim l$ and so $v \not\models l$. But $L = K \cup (L-K)$. So $v \models \vee(K \cup (L-K))$ and hence $v \models \vee K$. Thus $v \models c(r)$ in the second case too. So the lemma holds for all r in $Rul(Ax)$.

Take any r_0 in $R_s - Rul(Ax)$, and suppose $v \models Ax \cup A(r_0)$. Then r_0 is $A(r_0) \to \wedge c(Rul(Ax,A(r_0)))$. For each r in $Rul(Ax,A(r_0))$, r is $A(r_0) \to c(r)$. Therefore, by the previous paragraph, $v \models c(r)$. But this is true for every r in $Rul(Ax,A(r_0))$, and so $v \models \wedge c(Rul(Ax,A(r_0)))$.

(6) Suppose $r \in R_s$ and $r = A \to c(r)$. Then $c(r) = ns(\wedge c(Rul(Ax,A)))$. By part (3) of this lemma, either $A(r) = \{\}$; or $A(r) = \{l\}$, where l is a literal; or $A(r) = \{\wedge L\}$, where $|L| \geq 2$ and L is contingent. If $r_i \in Rul(Ax,A)$ then $c(r_i)$ is a non-empty subclause of a clause in Ax. By Lemma 5.1.6(2), each clause in Ax is contingent. So each non-empty subclause of a clause in Ax is contingent. Hence each $c(r_i)$ is contingent, and so $c(r)$ is not a tautology. We now show that $c(r)$ is not a contradiction.

Case 1: $A = \{\}$.

Then for each r_i in $Rul(Ax,A)$, $c(r_i) \in Ax$. By Lemma 5.1.6(1), Ax is satisfiable and so $c(r)$ is satisfiable. Hence $c(r)$ is not a contradiction, and so is contingent.

Case 2: $A \neq \{\}$.

Then $A = \{ns(\wedge\sim M)\}$ where M is contingent; and for each r_i in $Rul(Ax,A)$, $c(r_i) = ns(\vee K_i)$ where K_i is not empty and $\vee(K_i \cup M) \in Ax$.

Assume $c(r)$ is a contradiction. If $c(r) = ns(\vee K_0)$ then K_0 must be empty; which contradicts each K_i being non-empty. So suppose $c(r) = \wedge\{c(r_i) : r_i \in Rul(Ax,A)\}$ $= \wedge\{ns(\vee K_i) : r_i \in Rul(Ax,A)\}$ where $|Rul(Ax,A)| \geq 2$. By using Lemma 2.4.12(5) we get, $\vee\{\} \in Res\{\vee K_i : r_i \in Rul(Ax,A)\}$. By using Lemma 2.4.11(11) we get, $\vee M \in Res\{\vee(K_i \cup M) : r_i \in Rul(Ax,A)\} \subseteq Res(Ax)$. Since M is contingent, $\vee M \in CtgeRes(Ax)$.

So there is a non-empty subset M' of M such that $\vee M' \in MinCtgeRes(Ax)$. Hence $ns(\vee M') \in CorRes(Ax)$. Let $Ax = CorRes(C)$. By Lemma 2.6.8(5), $CorRes(Ax) = CorRes(CorRes(C)) = CorRes(C) = Ax$. So $ns(\vee M') \in Ax$. But for each i, $M' \subset K_i \cup M$. So there are two elements of Ax, one of which is a strict subclause of the other. This contradicts the *Min* part of the definition of *Cor*, Definition 2.6.1. Therefore $c(r)$ is not a contradiction, and so is contingent.
EndProofLemma5.1.7

5.2 The Proof Relation and the Proof Algorithms

In this section we define the proof relation and the proof algorithms of PPL given a plausible description. This complex task will be accomplished by stating the top level plan, and then progressively refining this plan until all the terms used have been defined.

Any method of demonstrating that $Ax \models f$ will do as an algorithm for proving facts; so there is no need to specify a particular one. Hence the top level plan for proving a formula is the following.

Plan Distinguish between proving facts and proving formulas that are not facts.

Lower case Greek letters will be used to denote the proof algorithms that will be defined eventually. A general proof algorithm will be denoted by α (a for alpha and algorithm). We shall use φ (f for phi and fact) to denote our factual proof algorithm. Until a further refinement is needed we shall use the notation $\alpha \vdash f$ to denote that a formula f is proved by the proof algorithm α.

Since facts are always true they are (at least) probably true. So we shall decree that facts are provable by all proof algorithms. That is, if $Ax \models f$ then $\alpha \vdash f$. The factual algorithm proves a formula iff it is a fact. In symbols, $\varphi \vdash f$ iff $Ax \models f$. (Recall that by Theorem 2.9.2(1) and Definition 2.9.1 we have, $Ax \models f$ iff $Ax \vdash f$ iff $\vee \{\} \in Res(Cl(Ax \cup \{\sim f\}))$.)

Now consider formulas f that are not facts, that is, $Ax \not\models f$. To (plausibly) prove f we need to do two things. First, establish some evidence for f. Second, defeat all the evidence against f. This will satisfy the requirements of the Evidence Principle, Principle 4.3.3. So our first refinement of the plan is the following.

Refinement 5.2.1. Suppose $(R, >)$ is a plausible description, $Ax = Ax(R)$, and f is a formula.
1) If $Ax \models f$ then $\alpha \vdash f$. Also $\varphi \vdash f$ iff $Ax \models f$.
2) If $Ax \not\models f$ and $\alpha \neq \varphi$ then $\alpha \vdash f$ iff (2.1) and (2.2) hold.
   2.1) Some evidence for f is established.
   2.2) All the evidence against f is defeated.

In Refinement 5.2.1(2.1) the evidence for f consists of strict or defeasible rules that have a consequent that implies f. However since the axioms are always true, we

can weaken this to requiring that $Ax \cup \{c(r)\}$ implies f, provided that $Ax \cup \{c(r)\}$ is satisfiable. So if $R' \subseteq R$ it will be convenient to let $R'[f] = \{r \in R' : Ax \cup \{c(r)\}$ is satisfiable and $Ax \cup \{c(r)\} \models f\}$. (We can check whether r is in $R'[f]$ or not as follows. $Ax \cup \{c(r)\}$ is satisfiable
iff $Cl(Ax \cup \{c(r)\})$ is satisfiable by Lemma 2.8.5(4),
iff $\vee\{\} \notin Res(Cl(Ax \cup \{c(r)\}))$ by Lemma 2.4.12(5); and
$Ax \cup \{c(r)\} \models f$ iff $\vee\{\} \in Res(Cl(Ax \cup \{c(r), \sim f\}))$ by Theorem 2.9.2(1) and Definition 2.9.1.)

In the case of Refinement 5.2.1(2.1) we have $Ax \not\models f$ and so r_{ax} cannot support f. Hence the following notation is convenient: $R_d^s = (R_s \cup R_d) - \{r_{ax}\}$. So the evidence for f is all the rules in R that support f, that is $R_d^s[f]$.

To establish some evidence for f we must prove the set of antecedents of a rule supporting f. So we need to find a rule r in $R_d^s[f]$ and prove $A(r)$.

But $A(r)$ is a finite set of formulas, not a formula. By proving a finite set F of formulas we shall mean proving every formula in F. In symbols, $\alpha \vdash F$ iff $\forall f \in F$, $\alpha \vdash f$. So if F is empty we have $\alpha \vdash \{\}$.

Collecting these ideas together gives the following formal definition and our second refinement.

Definition 5.2.2. Suppose $(R, >)$ is a plausible description, $Ax = Ax(R)$, $R' \subseteq R$, and f is a formula.
1) $R_d^s = (R_s \cup R_d) - \{r_{ax}\}$.
2) $R'[f] = \{r \in R' : Ax \cup \{c(r)\}$ is satisfiable and $Ax \cup \{c(r)\} \models f\}$.

Refinement 5.2.3. Suppose $(R, >)$ is a plausible description, $Ax = Ax(R)$, and f is a formula.
1) If F is a finite set of formulas then $\alpha \vdash F$ iff $\forall f \in F$, $\alpha \vdash f$.
2) If $Ax \models f$ then $\alpha \vdash f$. Also $\varphi \vdash f$ iff $Ax \models f$.
3) If $Ax \not\models f$ and $\alpha \neq \varphi$ then $\alpha \vdash f$ iff $\exists r \in R_d^s[f]$ such that (3.1) and (3.2) hold.
   3.1) $\alpha \vdash A(r)$.
   3.2) All the evidence against f is defeated.

Each rule whose consequent implies $\neg f$ is evidence against f. The set of such rules is $R[\neg f]$. In Refinement 5.2.3(3) we have a rule r in $R_d^s[f]$. So any rule in $R[\neg f]$ that is inferior to r has already been defeated by r and hence need not be explicitly considered. This reduces the set of evidence against f that must be considered to the set of rules in $R[\neg f]$ that are not inferior to r; in symbols, $\{s \in R[\neg f] : s \not< r\}$.

A rule s in $\{s \in R[\neg f] : s \not< r\}$ is defeated either by team defeat or by disabling s. The team of rules for f is $R_d^s[f]$. The rule s is defeated by *team defeat* iff there is a rule t in the team of rules for f, $R_d^s[f]$, such that t is superior to s, $t > s$, and the set of antecedents of t, $A(t)$, is proved $\alpha \vdash A(t)$. So if $R' \subseteq R$ it will be convenient to let $R'[f; s]$ denote the set of all rules in $R'[f]$ that are superior to s. In symbols, $R'[f; s] = \{t \in R'[f] : t > s\}$.

Alternatively s is *disabled* iff the set of antecedents of s, $A(s)$, cannot be proved; that is, $\alpha \not\vdash A(s)$.

5.2 The Proof Relation and the Proof Algorithms

Collecting these ideas together gives the following formal definition and our third refinement.

Definition 5.2.4. Suppose $(R, >)$ is a plausible description, $R' \subseteq R$, f is a formula, and $s \in R$. Then define $R'[f; s] = \{t \in R'[f] : t > s\}$.

Refinement 5.2.5. Suppose $(R, >)$ is a plausible description, $Ax = Ax(R)$, and f is a formula.
1) If F is a finite set of formulas then $\alpha \vdash F$ iff $\forall f \in F$, $\alpha \vdash f$.
2) If $Ax \models f$ then $\alpha \vdash f$. Also $\varphi \vdash f$ iff $Ax \models f$.
3) If $Ax \not\models f$ and $\alpha \neq \varphi$ then $\alpha \vdash f$ iff $\exists r \in R_d^s[f]$ such that (3.1) and (3.2) hold.
    3.1) $\alpha \vdash A(r)$.
    3.2) $\forall s \in \{s \in R[\neg f] : s \not< r\}$ either
        3.2.1) $\exists t \in R_d^s[f; s]$ such that $\alpha \vdash A(t)$; or
        3.2.2) $\alpha \not\vdash A(s)$.

We now have a refinement in which there are no undefined terms. Unfortunately Refinement 5.2.5 has two failings. Apart from the factual proof algorithm φ, there is only one other proof algorithm (denoted by α). Hence the Many Proof Algorithms Principle, Principle 4.9.2, fails. Also, a proof may get into a loop, and hence the Decisiveness Principle, Principle 4.10.2, fails.

Before we consider looping, let us invent the other proof algorithms. The α in Refinement 5.2.5(3.2.2) evaluates evidence against f; and this need not be the same α as in (3.1) and (3.2.1) which evaluates evidence for f. To avoid confusion let us call the α in (3.2.2), α'. Replacing α by α' in (3.2.2) creates the need to decide what $(\alpha')'$ is. Let us simplify $(\alpha')'$ to α''. Some obvious choices are: $\alpha'' = \alpha$, or $\alpha'' = \alpha'$, or α'' is some other proof algorithm. The third choice postpones and complicates the choice that must eventually be made. Experimentation shows that the second choice has some properties that we would rather avoid. So we let $\alpha'' = \alpha$.

Another change that can be made is to the set $\{s \in R[\neg f] : s \not< r\}$ of rules that a proof algorithm regards as evidence against f. Let $Foe(\alpha, f, r)$ denote the set of rules that α regards as the evidence against f that is not inferior to r.

These ideas gives us our fourth refinement.

Refinement 5.2.6. Suppose $(R, >)$ is a plausible description, $Ax = Ax(R)$, and f is a formula.
1) If F is a finite set of formulas then $\alpha \vdash F$ iff $\forall f \in F$, $\alpha \vdash f$.
2) If $Ax \models f$ then $\alpha \vdash f$. Also $\varphi \vdash f$ iff $Ax \models f$.
3) If $Ax \not\models f$ and $\alpha \neq \varphi$ then $\alpha \vdash f$ iff $\exists r \in R_d^s[f]$ such that (3.1) and (3.2) hold.
    3.1) $\alpha \vdash A(r)$.
    3.2) $\forall s \in Foe(\alpha, f, r)$ either
        3.2.1) $\exists t \in R_d^s[f; s]$ such that $\alpha \vdash A(t)$; or
        3.2.2) $\alpha' \not\vdash A(s)$.

Let us create our first non-factual proof algorithm, β, by changing Refinement 5.2.6 as little as possible. So let $Foe(\beta, f, r) = \{s \in R[\neg f] : s \not< r\}$. Let β be defined by replacing each α in Refinement 5.2.6 with β. Of course now β' must be defined. First let $Foe(\beta', f, r) = \{s \in R[\neg f] : s \not< r\}$. Then let β' be defined by replacing each α in Refinement 5.2.6 with β'. (Recall that $\beta'' = \beta$.) Later we shall show that β is ambiguity blocking (b for beta and blocking). We are not really concerned with any primed algorithm as they only assist with the definition of their non-primed co-algorithm. But later we shall show that β and β' prove exactly the same formulas. So why is β' needed? Without β' it is exceedingly difficult to prove the relationship between β and the other algorithms we are about to define.

Our next algorithm, π, will be shown to be ambiguity propagating (p for pi and propagating). We want to make π as strong as possible; that is, π proves f if there is no evidence against f. This can be done by making its co-algorithm π' as weak as possible; that is, π' ignores all evidence against f; hence $Foe(\pi', f, r) = \{\}$. This is the only change we make to Refinement 5.2.6. Explicitly, let $Foe(\pi, f, r) = \{s \in R[\neg f] : s \not< r\}$. Let π be defined by replacing each α in Refinement 5.2.6 with π. Let π' be defined by replacing each α in Refinement 5.2.6 with π'. (Recall that $\pi'' = \pi$.)

Our last algorithm, ψ, will also be shown to be ambiguity propagating (p for psi and propagating). We want to make ψ weaker than π. This can be done by making its co-algorithm ψ' regard those rules that imply $\neg f$ and are superior to r as evidence against f; hence $Foe(\psi', f, r) = \{s \in R[\neg f] : s > r\}$. This is the only change we make to Refinement 5.2.6. Explicitly, let $Foe(\psi, f, r) = \{s \in R[\neg f] : s \not< r\}$. Let ψ be defined by replacing each α in Refinement 5.2.6 with ψ. Let ψ' be defined by replacing each α in Refinement 5.2.6 with ψ'. (Recall that $\psi'' = \psi$.)

To emphasise that we are only interested in the non-primed proof algorithms we note that there are examples in which both π' and ψ' can prove both f and $\neg f$. This is fine as both π' and ψ' only assess the evidence against f, rather than try to defeasibly justify accepting f, as the non-primed algorithms do.

The following two formal definitions collect together for easy reference the above notations concerning algorithms and $Foe(.,.,.)$.

Definition 5.2.7. Define the set, Alg, of names of the proof algorithms by $Alg = \{\varphi, \pi, \psi, \beta, \beta', \psi', \pi'\}$. Define $\varphi' = \varphi$. If $\alpha \in \{\pi, \psi, \beta\}$ then define $(\alpha')' = \alpha'' = \alpha$. If $\alpha \in Alg$ then the co-algorithm of α is α'.

Definition 5.2.8. Suppose $(R, >)$ is a plausible description, f is a formula, and $r \in R$.
1) If $\alpha \in \{\pi, \psi, \beta, \beta'\}$ and $r \neq r_{ax}$ then $Foe(\alpha, f, r) = \{s \in R[\neg f] : s \not< r\}$.
2) If $\alpha \in \{\varphi, \pi'\}$ or $r = r_{ax}$ then $Foe(\alpha, f, r) = \{\}$.
3) $Foe(\psi', f, r) = \{s \in R[\neg f] : s > r\} = R[\neg f; r]$.

Finally, let us consider looping. To prove f we use α and a rule r. While proving f we may have to prove other formulas. During a proof of one of these other formulas, if we choose to use α and r again then we will be in a loop and so this choice should fail. To prevent such a looping choice we need to record that α and r have been used previously. We shall call such a record of used algorithms and rules a history. Its formal definition follows.

5.2 The Proof Relation and the Proof Algorithms

Definition 5.2.9. Suppose $(R,>)$ is a plausible description and $\alpha \in Alg$. Define $\alpha R = \{\alpha r : r \in R\}$. Then H is an α-*history* iff H is a finite sequence of elements of $\alpha R \cup \alpha' R$ that has no repeated elements. H is a *history* iff there is a proof algorithm α such that H is an α-history.

Unfortunately using a history complicates Refinement 5.2.6 because we now no longer have just an algorithm proving a formula, but an algorithm and a history proving a formula. Therefore in (1), (2), and (3), $\alpha \vdash x$ becomes $(\alpha, H) \vdash x$. In (3.1) α and r have now been used so H must be updated to $H + \alpha r$ and therefore $\alpha \vdash A(r)$ becomes $(\alpha, H + \alpha r) \vdash A(r)$. Also in (3.2.1) α and t have been used so H must be updated to $H + \alpha t$ and hence $\alpha \vdash A(t)$ becomes $(\alpha, H + \alpha t) \vdash A(t)$. Similarly in (3.2.2) α' and s have been used and so H must be updated to $H + \alpha' s$ and hence $\alpha' \nvdash A(s)$ becomes $(\alpha', H + \alpha' s) \nvdash A(s)$. Finally to prevent looping we must be sure that $\alpha r \notin H$ in (3.1), $\alpha t \notin H$ in (3.2.1), and $\alpha' s \notin H$ in (3.2.2).

Incorporating these changes into Refinement 5.2.6 gives our formal definition of the proof algorithms and proof relation $\vdash$. The letter I is attached to these final inference conditions.

Definition 5.2.10. Suppose $\mathcal{P} = (R,>)$ is a plausible description, $Ax = Ax(R)$, f is a formula, $\alpha \in Alg$, and H is an α-history. The *proof relation* for $\mathcal{P}$, $\vdash$, and the *proof algorithms* are defined by I1 to I3.
I1) If F is a finite set of formulas then $(\alpha, H) \vdash F$ iff $\forall f \in F$, $(\alpha, H) \vdash f$.
I2) If $Ax \models f$ then $(\alpha, H) \vdash f$. Also $(\varphi, H) \vdash f$ iff $Ax \models f$.
I3) If $Ax \nvDash f$ and $\alpha \neq \varphi$ then $(\alpha, H) \vdash f$ iff $\exists r \in R_d^s[f]$ such that I3.1 and I3.2 hold.
   I3.1) $\alpha r \notin H$ and $(\alpha, H + \alpha r) \vdash A(r)$.
   I3.2) $\forall s \in Foe(\alpha, f, r)$ either
       I3.2.1) $\exists t \in R_d^s[f;s]$ such that $\alpha t \notin H$ and $(\alpha, H + \alpha t) \vdash A(t)$; or
       I3.2.2) $\alpha' s \notin H$ and $(\alpha', H + \alpha' s) \nvdash A(s)$.

As Definition 5.2.10 shows, the proof relation, $\vdash$, is a subset of $\{(\mathcal{P}, \alpha, H, x) : \mathcal{P}$ is a plausible description, $\alpha \in Alg$, H is an α-history, and x is either a formula or a finite set of formulas$\}$. But, rather than write $(\mathcal{P}, \alpha, H, x) \in \vdash$ we shall, as usual, write $(\mathcal{P}, \alpha, H) \vdash x$; and of course, $(\mathcal{P}, \alpha, H) \nvdash x$ means $\text{not}[(\mathcal{P}, \alpha, H) \vdash x]$. Providing there is no ambiguity we often omit the reference to $\mathcal{P}$, so that $(\mathcal{P}, \alpha, H) \vdash x$ becomes $(\alpha, H) \vdash x$.

The following notation is useful.

Definition 5.2.11. Suppose $\mathcal{P}$ is a plausible description, $\alpha \in Alg$, and f is a formula.
1) We say f is α-*provable*, and write $\alpha \vdash f$, iff $(\alpha, ()) \vdash f$.
2) The α-*consequences* of $\mathcal{P}$, $\mathcal{P}(\alpha)$, is the set of all α-provable formulas. That is, $\mathcal{P}(\alpha) = \{f \in Fml : \alpha \vdash f\}$.

A semantic remark

Section 5.2, and its culmination in Definition 5.2.10, can be given a semantic interpretation. By Theorem 6.1.2(3), the meaning of the strict rule $A \rightarrow f$ is that for any proof algorithm α, if $\alpha \vdash A$ then $\alpha \vdash f$. The meaning of the defeasible rule $A \Rightarrow f$ is that for any non-factual proof algorithm α, if $\alpha \vdash A$ and the evidence against f is defeated then $\alpha \vdash f$. Exactly what the evidence against f is and how it is defeated is given by I3.2. By I3.2 the warning rule $s = A \rightsquigarrow \neg f$ can only be used as evidence against f; and exactly how s can be defeated is also given by I3.2. Thus Definition 5.2.10 can be seen as giving a meaning to each of the three kinds of rules.

Similarly, Definition 5.2.10, and the explanations preceding it, can be seen as giving a meaning to each of the proof algorithms. By I2, we see that $\varphi \vdash f$ means $Ax \models f$. Each of the non-factual proof algorithms, α, regards $R_d^s[f]$ as the set of potential evidence for f; and how α establishes that there is actual evidence for f is given by I3.1. Given some actual evidence r for f, the set that α regards as evidence against f is $Foe(\alpha, f, r)$. Exactly how this evidence can be defeated is given by I3.2.

End remark

We finish this section by proving some results that are used later and also give a better understanding of some of the previous definitions. The first result is a simple observation about the factual proof algorithm φ.

Lemma 5.2.12. Let $(R, >)$ be a plausible description, $Ax = Ax(R)$, H be a φ-history, f be a formula, and F be a finite set of formulas.
1) $(\varphi, H) \vdash f$ iff $Ax \models f$.
2) $(\varphi, H) \vdash F$ iff $Ax \models F$.
Proof

Let $(R, >)$ be a plausible description, $Ax = Ax(R)$, H be a φ-history, f be a formula, and F be a finite set of formulas.

(1) This is part of Definition 5.2.10(I2).

(2) Suppose $(\varphi, H) \vdash F$. By Definition 5.2.10(I1), for all f in F, $(\varphi, H) \vdash f$. By part (1), for all f in F, $Ax \models f$. Therefore $Ax \models F$.

Conversely, suppose $Ax \models F$. Then for each f in F, $Ax \models f$, and so by part (1), $(\varphi, H) \vdash f$. Therefore $(\varphi, H) \vdash F$.
EndProofLemma5.2.12

The next result also makes some simple observations.

Lemma 5.2.13. Suppose $(R, >)$ is a plausible description, $Ax = Ax(R)$, $\alpha \in Alg$, H is an α-history, and f is a formula.
1) If $R[f] \neq \{\}$ then $Ax \cup \{f\}$ is satisfiable and $Ax \not\models \neg f$.
2) If $(\alpha, H) \vdash f$ then $Ax \cup \{f\}$ is satisfiable and $Ax \not\models \neg f$.
Proof

Suppose $(R, >)$ is a plausible description, $Ax = Ax(R)$, $\alpha \in Alg$, H is an α-history, and f is a formula.

(1) Suppose $R[f] \neq \{\}$. By Definition 5.2.2(2) we see that, there exists r in R such that $Ax \cup \{c(r)\}$ is satisfiable and $Ax \cup \{c(r)\} \models f$. Hence $Ax \cup \{f\}$ is satisfiable, and so $Ax \not\models \neg f$.

5.2 The Proof Relation and the Proof Algorithms

(2) Suppose $(\alpha, H) \vdash f$. By Definition 5.2.10(I2,I3), either $Ax \models f$ or $R_d^s[f] \neq \{\}$. If $Ax \models f$ then by part 1, $Ax \cup \{f\}$ is satisfiable. If $R_d^s[f] \neq \{\}$ then by part 1, $Ax \cup \{f\}$ is satisfiable. But $Ax \cup \{f\}$ is satisfiable implies $Ax \not\models \neg f$.

EndProofLemma5.2.13

The last result in this section is not as obvious as the previous two results, but is more useful.

Lemma 5.2.14. Suppose $(R_0, >)$ is a plausible description, Ax is its set of axioms, $R \subseteq R_0$, f and g are formulas, $\alpha \in Alg$, and $\{r, s\} \subseteq R_0$.
1) $R[f] \subseteq R$.
2) If $R' \subseteq R$ then $R'[f] \subseteq R[f]$.
3) If $Ax \cup \{f\} \equiv Ax \cup \{g\}$ then $R[f] = R[g]$.
   If $f \equiv g$ then $R[f] = R[g]$.
4) If $Ax \models f$ then (a) $r_{ax} \in R[f]$ and so $r_{ax} \in R_s[f]$,
       (b) $R[\wedge\{f, g\}] = R[g]$, and
       (c) $Foe(\alpha, \wedge\{f, g\}, r) \subseteq Foe(\alpha, g, r)$.
5) If $Ax \cup \{f\} \models g$ then (a) $Ax \cup \{\neg g\} \models \neg f$,
       (b) $R[f] \subseteq R[g]$,
       (c) $R[f; s] \subseteq R[g; s]$,
       (d) $R[\neg g] \subseteq R[\neg f]$, and
       (e) $Foe(\alpha, g, r) \subseteq Foe(\alpha, f, r)$.
6) If $Ax \cup \{f, g\}$ is unsatisfiable then $R[f] \subseteq R[\neg g]$ and $R[g] \subseteq R[\neg f]$.
7) If $r \in (R_0)_s[f]$ and $a \in A(r)$ and $r_1 \in (R_0)_s[a]$ then $r_1 \in (R_0)_s[f]$.

Proof

Suppose $(R_0, >)$ is a plausible description, Ax is its set of axioms, $R \subseteq R_0$, f and g are formulas, $\alpha \in Alg$, and $\{r, s\} \subseteq R_0$.

(1) By Definition 5.2.2(2), $R[f] \subseteq R$.

(2) Suppose $R' \subseteq R$. Then $R'[f]$
$= \{r \in R' : Ax \cup \{c(r)\}$ is satisfiable and $Ax \cup \{c(r)\} \models f\}$
$\subseteq \{r \in R : Ax \cup \{c(r)\}$ is satisfiable and $Ax \cup \{c(r)\} \models f\}$
$= R[f]$.

(3) Suppose $Ax \cup \{f\} \equiv Ax \cup \{g\}$. By Definition 5.2.2(2), $R[f]$
$= \{r \in R : Ax \cup \{c(r)\}$ is satisfiable and $Ax \cup \{c(r)\} \models f\}$
$= \{r \in R : Ax \cup \{c(r)\}$ is satisfiable and $Ax \cup \{c(r)\} \models g\}$
$= R[g]$.

If $f \equiv g$ then $Ax \cup \{f\} \equiv Ax \cup \{g\}$ and so $R[f] = R[g]$ by the previous paragraph.

(4) Suppose $Ax \models f$.

(a) Then $Ax \cup \{c(r_{ax})\} \models f$. Also $Ax \models c(r_{ax})$. By Lemma 5.1.6(1), Ax is satisfiable and so $Ax \cup \{c(r_{ax})\}$ is satisfiable. Hence by Definition 5.2.2(2), $r_{ax} \in R[f]$, and so $r_{ax} \in R_s[f]$.

(b) By Definition 5.2.2(2), $R[g]$
$= \{r \in R : Ax \cup \{c(r)\}$ is satisfiable and $Ax \cup \{c(r)\} \models g\}$
$= \{r \in R : Ax \cup \{c(r)\}$ is satisfiable and $Ax \cup \{c(r)\} \models f$ and $Ax \cup \{c(r)\} \models g\}$

$= \{r \in R : Ax \cup \{c(r)\}$ is satisfiable and $Ax \cup \{c(r)\} \models \wedge\{f,g\}\}$
$= R[\wedge\{f,g\}]$.

(c) Let Claim 1 be: $Foe(\alpha, \wedge\{f,g\}, r) \subseteq Foe(\alpha, g, r)$.

If $\alpha \in \{\varphi, \pi'\}$ or $r = r_{ax}$ then $Foe(\alpha, \wedge\{f,g\}, r) = \{\}$. Hence Claim 1 holds.

Suppose $\alpha = \psi'$. Then we have $Foe(\psi', \wedge\{f,g\}, r) = \{s \in R[\neg \wedge\{f,g\}] : s > r\}$ and $Foe(\psi', g, r) = \{s \in R[\neg g] : s > r\}$. Take any s in $Foe(\psi', \wedge\{f,g\}, r)$. Then $s \in R$, $Ax \cup \{c(s)\}$ is satisfiable, $Ax \cup \{c(s)\} \models \neg \wedge\{f,g\}$, and $s > r$. But $Ax \models f$, therefore $Ax \cup \{c(s)\} \models \neg g$ and so $s \in Foe(\psi', g, r)$. Hence Claim 1 holds.

So suppose $\alpha \in \{\pi, \psi, \beta, \beta'\}$ and $r \neq r_{ax}$. Then we have $Foe(\alpha, \wedge\{f,g\}, r) = \{s \in R[\neg \wedge\{f,g\}] : s \not< r\}$ and $Foe(\alpha, g, r) = \{s \in R[\neg g] : s \not< r\}$. Take any s in $Foe(\alpha, \wedge\{f,g\}, r)$. Then $s \in R$, $Ax \cup \{c(s)\}$ is satisfiable, $Ax \cup \{c(s)\} \models \neg \wedge\{f,g\}$, and $s \not< r$. But $Ax \models f$, so $Ax \cup \{c(s)\} \models \neg g$ and therefore $s \in Foe(\alpha, g, r)$. Hence Claim 1 holds.

Thus Claim 1 is proved.

(5) Suppose $Ax \cup \{f\} \models g$. From Definitions 5.2.2(2) and 5.2.4 we have, $R[f] = \{r \in R : Ax \cup \{c(r)\}$ is satisfiable and $Ax \cup \{c(r)\} \models f\}$ and $R[f; s] = \{t \in R[f] : t > s\}$.

(a) Every valuation satisfies exactly one of f or $\neg f$. Hence $Ax \cup \{\neg g\} \models \neg f$.

(b) Take any r in $R[f]$. Then $Ax \cup \{c(r)\}$ is satisfiable and $Ax \cup \{c(r)\} \models f$. Hence $Ax \cup \{c(r)\} \models g$ and so $r \in R[g]$. Thus $R[f] \subseteq R[g]$.

(c) Take any t in $R[f; s]$. Then $t \in R[f]$ and $t > s$. By part (b), $t \in R[g]$ and so $t \in R[g; s]$. Thus $R[f; s] \subseteq R[g; s]$.

(d) This follows from parts (a) and (b).

(e) This follows from parts (a) and (c).

(6) Suppose $Ax \cup \{f, g\}$ is unsatisfiable. Take any r in $R[f]$. Then $Ax \cup \{c(r)\}$ is satisfiable and $Ax \cup \{c(r)\} \models f$. Hence $Ax \cup \{c(r)\} \models \neg g$. Therefore $r \in R[\neg g]$ and so $R[f] \subseteq R[\neg g]$. By swapping f and g we get $R[g] \subseteq R[\neg f]$.

(7) Let $R = R_0$ and suppose $r \in R_s[f]$, $a \in A(r)$, and $r_1 \in R_s[a]$. By Lemma 5.1.7(3), $A(r) = \{a\}$. Since $r_1 \in R_s[a]$, $Ax \cup \{c(r_1)\}$ is satisfiable and $Ax \cup \{c(r_1)\} \models a$. Also $Ax \cup \{c(r)\} \models f$. By Lemma 5.1.7(5), $Ax \cup A(r) \models c(r)$, that is, $Ax \cup \{a\} \models c(r)$. So $Ax \cup \{c(r_1)\} \models f$ and hence $r_1 \in R_s[f]$.

EndProofLemma5.2.14

5.3 The Proof Theory

The previous section defined the proof algorithms and a proof relation. But so far there is no concept of a proof or a derivation. Such a concept is useful because it enables results to be proved by induction on a derivation or proof. In this section we shall define the concept of a proof. In classical propositional logic a derivation is a sequence of formulas. But in Propositional Plausible Logic, although it is possible to define a derivation to be a sequence, it is more intuitive to define it to be a tree. The nodes of such a tree will have special tags which we now define.

5.3 The Proof Theory

Definition 5.3.1. Suppose $(R,>)$ is a plausible description, $\alpha \in Alg$, F is a finite set of formulas, H is an α-history, f is a formula, $r \in R_d^s[f]$, $s \in R[\neg f]$, and p is a node of a tree.
1) The *tag*, $t(p)$, of p is a triple $t(p) = (Subj(p), op(p), pv(p))$.
2) The *subject* of p, $Subj(p)$, has one of the following forms: (α, H, F), $-(\alpha, H, F)$, (α, H, f), (α, H, f, r), or (α, H, f, r, s).
3) The *operation* of p, $op(p)$, is either min (for minimum), max (for maximum), or $-$. If $op(p)$ is min [resp. max, $-$] then p is referred to as a min [resp. max, minus] node.
4) The *proof value* of p, $pv(p)$, is either $+1$ or -1.

Roughly a proof value of $+1$ means that the formula or set of formulas associated with the node has been proved. Whereas a proof value of -1 means that the formula or set of formulas associated with the node has been disproved; that is, it has been proved that there is no proof.

The arithmetic properties of the proof values $+1$ and -1 are given in the next definition.

Definition 5.3.2. Suppose $S \subseteq \{+1, -1\}$.
(1) $\min S = -1$ iff $-1 \in S$.   (2) $\min S = +1$ iff $-1 \notin S$.
(3) $\max S = +1$ iff $+1 \in S$.   (4) $\max S = -1$ iff $+1 \notin S$.
(5) $--1 = +1$.   (6) $-+1 = -1$.

A proof tree will be a subtree of an evaluation tree, which is defined below. An evaluation tree records all the possibilities that may arise from using the proof relation $\vdash$ of Definition 5.2.10.

Definition 5.3.3. Let $\mathcal{P} = (R, >)$ be a plausible description. Then T is an *evaluation tree* of $\mathcal{P}$ iff T is a rooted tree constructed as follows. Each node, p, of T has exactly one tag, $t(p)$. For each node p of T there is exactly one number, $\#_p$, in $[1..6]$ such that p satisfies T$\#_p$ and T7.

T1) $Subj(p) = (\alpha, H, F)$, $\alpha \in Alg$, H is an α-history, and F is a finite set of formulas. Define $S(p) = \{(\alpha, H, f) : f \in F\}$. Then $op(p) = \min$, p has $|S(p)|$ children, and each element of $S(p)$ is the subject of exactly one child of p. If $S(p) = \{\}$ then $pv(p) = +1$.

T2) $Subj(p) = (\alpha, H, f)$, $\alpha \in Alg$, H is an α-history, f is a formula, and $Ax \models f$. Then p has no children and $t(p) = ((\alpha, H, f), \min, +1)$.

T3) $Subj(p) = (\alpha, H, f)$, $\alpha \in Alg - \{\varphi\}$, H is any α-history, f is any formula, and $Ax \not\models f$. Define $S(p) = \{(\alpha, H, f, r) : \alpha r \notin H$ and $r \in R_d^s[f]\}$. Then $op(p) = \max$, p has $|S(p)|$ children, and each element of $S(p)$ is the subject of exactly one child of p. If $S(p) = \{\}$ then $pv(p) = -1$.

T4) $Subj(p) = (\alpha, H, f, r)$, $\alpha \in Alg - \{\varphi\}$, H is any α-history, f is any formula, $Ax \not\models f$, $\alpha r \notin H$, and $r \in R_d^s[f]$. Define $S(p) = \{(\alpha, H+\alpha r, A(r))\} \cup \{(\alpha, H, f, r, s) : s \in Foe(\alpha, f, r)\}$. Then $op(p) = \min$, p has $|S(p)|$ children, and each element of $S(p)$ is the subject of exactly one child of p.

T5) $Subj(p) = (\alpha, H, f, r, s)$, $\alpha \in Alg - \{\varphi, \pi'\}$, H is any α-history, f is any formula, $Ax \not\models f$, $\alpha r \notin H$, $r \in R_d^s[f]$, and $s \in Foe(\alpha, f, r)$. Define $S(p) = \{(\alpha, H + \alpha t, A(t)) : \alpha t \notin H$ and $t \in R_d^s[f;s]\} \cup \{-(\alpha', H + \alpha' s, A(s)) : \alpha' s \notin H\}$. Then $op(p) = $ max, p has $|S(p)|$ children, and each element of $S(p)$ is the subject of exactly one child of p. If $S(p) = \{\}$ then $pv(p) = -1$.

T6) $Subj(p) = -(\alpha', H, F)$, $\alpha \in \{\pi, \psi, \beta, \beta'\} \cup \{\psi' : >$ is not empty$\}$, H is any α-history, and F is any finite set of formulas. Then $op(p) = -$; p has exactly one child, say p_1; and $Subj(p_1) = (\alpha', H, F)$.

T7) If $op(p) = $ min then $pv(p) = \min\{pv(c) : c$ is a child of $p\}$.
If $op(p) = $ max then $pv(p) = \max\{pv(c) : c$ is a child of $p\}$.
If $op(p) = -$ and c is the child of p then $pv(p) = -pv(c)$.

As usual we would like our proofs to be finite. But it is possible for an evaluation tree T of $(R, >)$ to be infinite. So we make the following distinction.

Definition 5.3.4. A plausible description $\mathcal{P}$ is a *plausible theory* iff every evaluation tree of $\mathcal{P}$ is finite. A *propositional plausible logic* (PPL) consists of a plausible theory and its proof relation.

We could force an evaluation tree of $(R, >)$ to be finite by allowing only a finite number of rules; that is, forcing R to be finite. But this is not necessary. So we shall now search for necessary and sufficient conditions that will make every evaluation tree of $(R, >)$ finite. We start with three definitions and a lemma that establishes some useful relationships.

Definition 5.3.5. Suppose $\mathcal{P} = (R, >)$ is a plausible description, $\alpha \in Alg$, H is an α-history, F is a finite set of formulas, f is a formula, r and s are any rules, T is an evaluation tree of $\mathcal{P}$, and p is any node of T. Suppose $Subj(p) \in \{(\alpha, H, F), (\alpha, H, f), (\alpha, H, f, r), (\alpha, H, f, r, s), -(\alpha, H, F)\}$.
1) The *history of p*, $Hist(p)$, is defined by $Hist(p) = H$.
2) The *algorithm of p*, $alg(p)$, is defined by $alg(p) = \alpha$.
3) If $H = (\alpha_1 r_1, \alpha_2 r_2, ..., \alpha_n r_n)$ then define $Rul(H) = (r_1, r_2, ..., r_n)$.

Definition 5.3.6. Suppose $(R, >)$ is a plausible description, $R' \subseteq R$, and F is a set of formulas. Then define $R'[F] = \bigcup \{R'[f] : f \in F\}$.

Definition 5.3.7. Let $(R, >)$ be a plausible description.
1) A *train of rules* of R is a 2to1 sequence $\mathbf{r} = (r_1, r_2, ...)$ of rules of R such that if $r_i \in \mathbf{r}$ and $r_{i+1} \in \mathbf{r}$ then $r_{i+1} \in R_d^s[A(r_i)] \cup R[\neg A(r_i)]$.
2) A *positive train of rules* of R is a 1to1 sequence $\mathbf{r} = (r_1, r_2, ...)$ of rules R such that if $r_i \in \mathbf{r}$ and $r_{i+1} \in \mathbf{r}$ then $r_{i+1} \in R_d^s[A(r_i)]$.

Lemma 5.3.8. Suppose $\mathcal{P} = (R, >)$ is a plausible description, T is an evaluation tree of $\mathcal{P}$, p_0 is the root of T, and p and q are any nodes of T.
1) If q is any child of p then $Hist(p)$ is a prefix of $Hist(q)$ and $|Hist(p)| \leq |Hist(q)| \leq |Hist(p)| + 1$.

5.3 The Proof Theory

2) If p is an ancestor of q then $Hist(p)$ is a prefix of $Hist(q)$.
3) If p_1, p_2, p_3, p_4, p_5, and p_6 are any 6 nodes of T such that for each i in $[1..5]$, p_{i+1} is a child of p_i, then $Hist(p_1)$ is a proper prefix of $Hist(p_6)$.
4) The level of p is at most $5|Hist(p)-Hist(p_0)|+4$.
5) If R is finite and H is a history then $|H| \leq 2|R|$.
6) If R is finite then the height of T is at most $10|R|+4$.
7) $Rul(Hist(p)-Hist(p_0))$ is a train of rules of R.

Proof

Suppose $\mathcal{P} = (R, >)$ is a plausible description, T is an evaluation tree of $\mathcal{P}$, p_0 is the root of T, and p and q are any nodes of T. It will be convenient to denote the relation 'is a proper prefix of' by '<'.

(1) Let q be any child of p. If p satisfies T1 or T3 or T6 then $Hist(p) = Hist(q)$. Since p has a child, p does not satisfy T2. If p satisfies T4 then either $Hist(p) < Hist(q)$ and $|Hist(q)| = |Hist(p)|+1$ or $Hist(p) = Hist(q)$. If p satisfies T5 then $Hist(p) < Hist(q)$ and $|Hist(q)| = |Hist(p)|+1$.

(2) Part (2) follows from part (1).

(3) In this part whenever we write p_{j+1} we shall mean that the node p_{j+1} is a child of the node p_j. If p_i satisfies T6 then $Hist(p_i) = Hist(p_{i+1})$ and p_{i+1} satisfies T1. So either p_{i+1} is a leaf, or $Hist(p_i) = Hist(p_{i+1}) = Hist(p_{i+2})$. If p_{i+2} satisfies T2 then p_{i+2} is a leaf. If p_{i+2} satisfies T3 then $Hist(p_i) = Hist(p_{i+2}) = Hist(p_{i+3})$ and p_{i+3} satisfies T4. So either $Hist(p_i) = Hist(p_{i+3}) < Hist(p_{i+4})$, or $Hist(p_i) = Hist(p_{i+3}) = Hist(p_{i+4})$ and p_{i+4} satisfies T5. So either p_{i+4} is a leaf, or $Hist(p_i) = Hist(p_{i+4}) < Hist(p_{i+5})$. Thus part (3) is proved.

(4) By part (3), the level of p is at most $(6-1)|Hist(p)-Hist(p_0)|+(6-2)$; that is, $5|Hist(p)-Hist(p_0)|+4$.

(5) This follows directly from Definition 5.2.9.

(6) This follows from parts 4 and 5.

(7) If $Rul(Hist(p)-Hist(p_0))$ has less than 2 rules then it is a train of rules of R. So suppose $Rul(Hist(p)-Hist(p_0))$ has at least 2 rules. Let $Rul(Hist(p)-Hist(p_0)) = \mathbf{r} = (r_1, r_2, ...)$. Let q_1 be the first node on the path from p_0 to p such that $Hist(q_1) = Hist(p_0)+\alpha r_1$. Then the parent of q_1 satisfies either T4 or T5. In both cases the third component of $Subj(q_1)$ is $A(r_1)$. Applying T6, if necessary, and then T1, T3, T4, and if necessary T5, we see that for some $f \in A(r_1)$, $r_2 \in R_d^s[f] \cup R[\neg f]$. Hence $r_2 \in R_d^s[A(r_1)] \cup R[\neg A(r_1)]$.

Continuing this line of reasoning we see that $\mathbf{r}$ is a train of rules of R.

EndProofLemma5.3.8

As well as some useful relationships, we now show that if R is finite then $(R, >)$ is a plausible theory. Recall that $|p|$ is the number of children that a node p has.

Lemma 5.3.9. Let $\mathcal{P} = (R, >)$ be a plausible description, T be an evaluation tree of $\mathcal{P}$, p be any node of T, and $\{r, s\} \subseteq R$. Suppose either T or R is finite.
1) If p satisfies T1 and $Subj(p) = (\alpha, H, A(r))$ then $|p| = |A(r)|$.
2) If p satisfies T2 then $|p| = 0$.
3) If p satisfies T3 and $Subj(p) = (\alpha, H, f)$ then $|p| \leq |R_d^s[f]|$.

4) If p satisfies T4 and $Subj(p) = (\alpha, H, f, r)$ then $|p| = 1 + |Foe(\alpha, f, r)|$.
5) If p satisfies T5 and $Subj(p) = (\alpha, H, f, r, s)$ then $|p| \leq |R_d^s[f;s]| + 1$.
6) If p satisfies T6 and $Subj(p) = -(\alpha', H, A(s))$ then $|p| = 1$.
7) If R is finite then T is finite, and so $\mathcal{P}$ is a plausible theory.
Proof

Parts (1) to (6) follow directly from Definition 5.3.3.

(7) Suppose R is finite. By parts (1) to (6) every node of T has finitely many children. In fact if $c = \max[\{|A(r)| : r \in R\} \cup \{|R|+1\}]$ then $|p| \leq c$. By Lemma 5.3.8(6), if h is the height of T then $h \leq 10|R| + 4$. If R is empty then $c = 1$ and so by Lemma 1.2.5, $|T| = h + 1 \leq 5$. If R is not empty then $c \geq 2$ and so by Lemma 1.2.5, $|T| \leq (c^{h+1} - 1)/(c - 1)$.
EndProofLemma5.3.9

The next two lemmas give some sufficient and necessary conditions for a plausible description to be a plausible theory.

Lemma 5.3.10. Suppose $(R, >)$ is a plausible description and Ax is its set of axioms. If (1), (2), and (3) all hold then $(R, >)$ is a plausible theory.
1) If f is satisfiable and $Ax \not\models f$ then $R_d^s[f]$ is finite.
2) If $\alpha \in Alg$ and f is satisfiable and $Ax \not\models f$ and $r \in R_d^s[f]$ then $Foe(\alpha, f, r)$ is finite.
3) Each train of rules of R is finite.
Proof

Suppose $(R, >)$ is a plausible description and Ax is its set of axioms. Also suppose that (1), (2), and (3) all hold. We must show that every evaluation tree of $(R, >)$ is finite. Let T be any evaluation tree of $(R, >)$, and suppose p is any node of T.

If p satisfies T1 and $Subj(p) = (\alpha, H, F)$ then F is finite and $|p| = |F|$. So $|p|$ is finite. If p satisfies T2 then $|p| = 0$. So $|p|$ is finite. If p satisfies T3 and $Subj(p) = (\alpha, H, f)$ then f is satisfiable and $Ax \not\models f$ and $|p| \leq |R_d^s[f]|$. So by (1), $|p|$ is finite. If p satisfies T4 and $Subj(p) = (\alpha, H, f, r)$ then $\alpha \in Alg$, f is satisfiable, $Ax \not\models f$, $r \in R_d^s[f]$, and $|p| = 1 + |Foe(\alpha, f, r)|$. So by (2), $|p|$ is finite. If p satisfies T5 and $Subj(p) = (\alpha, H, f, r, s)$ then f is satisfiable and $Ax \not\models f$ and $|p| \leq |R_d^s[f;s]| + 1$. So by (1), $|p|$ is finite. If p satisfies T6 then $|p| = 1$. So $|p|$ is finite. Thus in all cases p has only finitely many children.

Consider a path $P = (p_0, p_1, ...)$ in T, where p_0 is the root of T. By Lemma 5.3.8(3), as i increase so does $Hist(p_i)$, and hence so does $Rul(Hist(p_i) - Hist(p_0))$ which, by Lemma 5.3.8(7), is a train of rules of R. Thus if P has no finite length then there is a train of rules of R that has no finite length. But this contradicts (3), so P must be finite. Thus every path in T is finite.

So by Lemma 1.2.6, T is finite and hence $(R, >)$ is a plausible theory.
EndProofLemma5.3.10

Lemma 5.3.11. Suppose $(R, >)$ is a plausible theory, $Ax = Ax(R)$, and f is a formula.
1) If f is satisfiable and $Ax \not\models f$ then $R_d^s[f]$ is finite.
2) If $\alpha \in Alg$ and f is satisfiable and $Ax \not\models f$ and $r \in R_d^s[f]$ then $Foe(\alpha, f, r)$ is finite.
3) Each positive train of rules of R is finite.

5.3 The Proof Theory 163

Proof

Suppose $(R,>)$ is a plausible theory, $Ax = Ax(R)$, and f is a formula.

(1) Suppose f is satisfiable and $Ax \not\models f$. Take any α in $Alg - \{\varphi\}$. By Definition 5.3.4, $T[\alpha,(),f]$ is finite, and so its root has only finitely many children. The root of $T[\alpha,(),f]$ satisfies T3 of Definition 5.3.3. Therefore $R_d^s[f]$ is finite.

(2) Suppose $\alpha \in Alg$ and f is satisfiable and $Ax \not\models f$ and $r \in R_d^s[f]$. By Definition 5.3.4, $T[\alpha,(),f]$ is finite, and so its root, say p_0, has only finitely many children. If $\alpha \in \{\varphi, \pi'\}$ then $Foe(\alpha,f,r) = \{\}$ and so is finite. So suppose $\alpha \in Alg - \{\varphi, \pi'\}$. Then the root of $T[\alpha,(),f]$ satisfies T3 of Definition 5.3.3. By T3, $(\alpha,(),f,r)$ is the subject of a child, say p_1, of p_0. By T4, for each s in $Foe(\alpha,f,r)$, $(\alpha,(),f,r,s)$ is the subject of a child of p_1. But p_1 has only finitely many children, so $Foe(\alpha,f,r)$ is finite.

(3) Let $\mathbf{r} = (r_1, r_2, ...)$ be any positive train of rules of R. Take any α in $Alg - \{\varphi\}$. Then $r_2 \in R_d^s[A(r_1)]$. So there exists a_1 in $A(r_1)$ such that $r_2 \in R_d^s[a_1]$. We shall construct a path $P = (p_0, p_1, ...)$ in $T[\alpha,(),a_1]$ such that if r_i is in $\mathbf{r}$ then p_{3i-4} is in P.

Let p_0 be the root of $T[\alpha,(),a_1]$. Then p_0 satisfies T3 and $Subj(p_0) = (\alpha,(),a_1)$. By T3, p_0 has a child p_1 such that $Subj(p_1) = (\alpha,(),a_1,r_2)$ and p_1 satisfies T4. By T4, p_1 has a child p_2 such that $Subj(p_2) = (\alpha,(\alpha r_2), A(r_2))$ and p_2 satisfies T1.

If $r_3 \in \mathbf{r}$ then $r_3 \in R_d^s[A(r_2)]$ and so there exists a_2 in $A(r_2)$ such that $r_3 \in R_d^s[a_2]$. By T1, p_2 has a child p_3 such that $Subj(p_3) = (\alpha,(\alpha r_2), a_2)$ and p_3 satisfies T3. By T3, p_3 has a child p_4 such that $Subj(p_4) = (\alpha,(\alpha r_2), a_2, r_3)$ and p_4 satisfies T4. By T4, p_4 has a child p_5 such that $Subj(p_5) = (\alpha,(\alpha r_2, \alpha r_3), A(r_3))$ and p_5 satisfies T1.

If $i \geq 3$ then the reasoning in the above paragraph can be repeated for each r_i in $\mathbf{r}$ as follows.

If $r_i \in \mathbf{r}$ then $r_i \in R_d^s[A(r_{i-1})]$. So there exists a_{i-1} in $A(r_{i-1})$ such that $r_i \in R_d^s[a_{i-1}]$. By T1, $p_{3(i-1)-4}$ has a child p_{3i-6} such that $Subj(p_{3i-6}) = (\alpha,(\alpha r_2, ..., \alpha r_{i-1}), a_{i-1})$ and p_{3i-6} satisfies T3. By T3, p_{3i-6} has a child p_{3i-5} such that $Subj(p_{3i-5}) = (\alpha,(\alpha r_2, ..., \alpha r_{i-1}), a_{i-1}, r_i)$ and p_{3i-5} satisfies T4. By T4, p_{3i-5} has a child p_{3i-4} such that $Subj(p_{3i-4}) = (\alpha,(\alpha r_2, ..., \alpha r_i), A(r_i))$ and p_{3i-4} satisfies T1.

But $(R,>)$ is a plausible theory so P must be finite. Hence $\mathbf{r}$ must be finite.

EndProofLemma5.3.11

We still have not defined what a proof tree in Propositional Plausible Logic should be. Roughly a proof tree will be an evaluation tree with some unnecessary parts removed. The formalities are contained in the next three definitions.

Definition 5.3.12. Suppose $\mathcal{P} = (R,>)$ is a plausible theory, $\alpha \in Alg$, H is any α-history, F is a finite set of formulas, f is a formula, $r \in R_d^s[f]$, and $s \in R[\neg f]$. Let (X) be an element of $\{(\alpha, H, F), (\alpha, H, f), (\alpha, H, f, r), (\alpha, H, f, r, s)\}$.
1) $T[X]$ is the evaluation tree of $\mathcal{P}$ whose root has the subject (X).
2) $T(X)$ is the proof value of the root of $T[X]$.
3) $T[-(\alpha, H, F)]$ is the evaluation tree of $\mathcal{P}$ whose root has the subject $-(\alpha, H, F)$.
4) $T(-(\alpha, H, F))$ is the proof value of the root of $T[-(\alpha, H, F)]$.

Definition 5.3.13. Let $\mathcal{P}$ be a plausible theory and T be an evaluation tree of $\mathcal{P}$. Then T_r is a *reduct* [*lean reduct*] of T iff whenever p is a node of T_r the following eight conditions all hold.
red1) T_r is a subtree of T.
red2) The root of T_r is the root of T.
red3) The tag of p in T_r is the same as the tag of p in T.
red4) If $op(p) = \max$ and $pv(p) = +1$ then, in T_r, p has a [exactly one] child c, and $pv(c) = +1$.
red5) If $op(p) = \max$ and $pv(p) = -1$ and c is a child of p in T then c is a child of p in T_r.
red6) If $op(p) = \min$ and $pv(p) = -1$ then, in T_r, p has a [exactly one] child c, and $pv(c) = -1$.
red7) If $op(p) = \min$ and $pv(p) = +1$ and c is a child of p in T then c is a child of p in T_r.
red8) If $op(p) = -$ and c is the child of p in T then c is the child of p in T_r.

All the nodes of T_r satisfy T7 of Definition 5.3.3.

Definition 5.3.14. Suppose $\mathcal{P}$ is a plausible theory, $\alpha \in Alg$, H is an α-history, and x is either a formula or a finite set of formulas.
1) If $T(\alpha,(),x) = +1$ then an α-*proof* of x is any reduct of $T[\alpha,(),x]$.
2) If $T(\alpha,(),x) = -1$ then an α-*disproof* of x is any reduct of $T[\alpha,(),x]$.

Recall from Definition 4.10.1 that a proof algorithm α is *decisive* iff for every formula f, α terminates in either a state indicating that f is proved, or a state indicating that f is not provable. We now show that Propositional Plausible Logic is decisive, and so decidable.

Theorem 5.3.15 (Decisiveness). Suppose $\mathcal{P}$ is a plausible theory, $\alpha \in Alg$, H is any α-history, and x is either a formula or a finite set of formulas.
1) $T[\alpha,H,x]$ has finitely many nodes.
2) Either $T(\alpha,H,x) = +1$ or $T(\alpha,H,x) = -1$ but not both.
3) Either x is α-proved or x is α-disproved.
Proof
   (1) follows from Definition 5.3.4.
   (2) follows from part (1) of this lemma and Definition 5.3.3.
   (3) follows from part (2) of this lemma and Definition 5.3.14.
EndProofTheorem5.3.15

The proof relation, $\vdash$, and evaluation trees are both cumbersome to use for manual derivations. This is the motivation for the following definition of a proof function, P, that is easier to use. It is a straightforward translation of the proof relation $\vdash$ of Definition 5.2.10 into the function P with the following property:
$P(\alpha,H,x) = +1$ iff $(\alpha,H) \vdash x$, and
$P(\alpha,H,x) = -1$ iff $(\alpha,H) \not\vdash x$.
The definition of P is facilitated by the following two auxiliary functions:

5.3 The Proof Theory

For (evidence for), and
Dftd (defeated).

Definition 5.3.16. Suppose $\mathcal{P} = (R, >)$ is any plausible theory, $\alpha \in Alg$, H is any α-history, and f is any formula. The *proof function for* $\mathcal{P}$, P, and its auxiliary functions *For* and *Dftd* are defined by P1 to P5.
P1) If F is a finite set of formulas, then $P(\alpha, H, F) = \min\{P(\alpha, H, f) : f \in F\}$.
P2) If $Ax \models f$ then $P(\alpha, H, f) = +1$. Also $P(\varphi, H, f) = +1$ iff $Ax \models f$.
P3) If $Ax \not\models f$ and $\alpha \neq \varphi$ then
$$P(\alpha, H, f) = \max\{For(\alpha, H, f, r) : \alpha r \notin H \text{ and } r \in R_d^s[f]\}.$$
P4) If $Ax \not\models f$, $\alpha \neq \varphi$, $\alpha r \notin H$, and $r \in R_d^s[f]$ then
$$For(\alpha, H, f, r) = \min[\{P(\alpha, H+\alpha r, A(r))\} \cup \{Dftd(\alpha, H, f, r, s) : s \in Foe(\alpha, f, r)\}].$$
P5) If $Ax \not\models f$, $\alpha \in Alg - \{\varphi, \pi'\}$, $\alpha r \notin H$, $r \in R_d^s[f]$, and $s \in Foe(\alpha, f, r)$ then
$$Dftd(\alpha, H, f, r, s) = \max[\{P(\alpha, H+\alpha t, A(t)) : \alpha t \notin H \text{ and } t \in R_d^s[f;s]\} \cup \{-P(\alpha', H+\alpha' s, A(s)) : \alpha' s \notin H\}].$$

Now we have three different representations of the proof status of a formula, or finite set of formulas. The next result shows that these representations do not contradict each other.

Theorem 5.3.17. Suppose $(R, >)$ is any plausible theory, $\alpha \in Alg$, H is any α-history, F is any finite set of formulas, and f is any formula.
1) $P(\alpha, H, F) = +1$ iff $(\alpha, H) \vdash F$ iff $T(\alpha, H, F) = +1$.
2) $P(\alpha, H, f) = +1$ iff $(\alpha, H) \vdash f$ iff $T(\alpha, H, f) = +1$.
3) If $Ax \not\models f$, $\alpha \neq \varphi$, $\alpha r \notin H$, $r \in R_d^s[f]$, and f is satisfiable then $For(\alpha, H, f, r) = +1$
   iff $T(\alpha, H, f, r) = +1$
   iff $(\alpha, H+\alpha r) \vdash A(r)$ and $\forall s \in Foe(\alpha, f, r)$, $Dftd(\alpha, H, f, r, s) = +1$.
4) If $Ax \not\models f$, $\alpha \in Alg - \{\varphi, \pi'\}$, $\alpha r \notin H$, $r \in R_d^s[f]$, f is satisfiable, and $s \in Foe(\alpha, f, r)$
   then $Dftd(\alpha, H, f, r, s) = +1$
   iff $T(\alpha, H, f, r, s) = +1$
   iff either $\exists t \in R_d^s[f; s]$ such that $\alpha t \notin H$ and $(\alpha, H+\alpha t) \vdash A(t)$;
   or $\alpha' s \notin H$ and $(\alpha', H+\alpha' s) \not\vdash A(s)$.

Proof

Let $\mathcal{P} = (R, >)$ be a plausible theory. The proof is by induction on the number of nodes in an evaluation tree of $\mathcal{P}$. Let p be the only node of an evaluation tree of $\mathcal{P}$.

If p satisfies T1 then the subject of p is $(\alpha, H, \{\})$ and the proof value of p is $+1$. So $T(\alpha, H, \{\}) = +1$. By P1, $P(\alpha, H, \{\}) = \min\{\} = +1$. By I1, $(\alpha, H) \vdash \{\}$.

If p satisfies T2 then the subject of p is (α, H, f) and the proof value of p is $+1$. So $T(\alpha, H, f) = +1$. By P2, $P(\alpha, H, f) = +1$. By I2, $(\alpha, H) \vdash f$.

If p satisfies T3 then the subject of p is (α, H, f) and the proof value of p is -1. So $T(\alpha, H, f) = -1$. Let $S(p) = \{(\alpha, H, f, r) : \alpha r \notin H \text{ and } r \in R_d^s[f]\}$. By T3, $S(p)$ is empty. By P3, $P(\alpha, H, f) = \max\{For(\alpha, H, f, r) : (\alpha, H, f, r) \in S(p)\} = \max\{\} = -1$. Since $S(p)$ is empty, for all r in $R_d^s[f]$, $\alpha r \in H$. Hence I3.1 fails and so $(\alpha, H) \not\vdash f$.

Since p has no children, p does not satisfy T4 or T6.

If p satisfies T5 then the subject of p is (α,H,f,r,s) and the proof value of p is -1. So $T(\alpha,H,f,r,s) = -1$. Let
$S(p) = \{(\alpha,H+\alpha t,A(t)) : \alpha t \notin H$ and $t \in R_d^s[f;s]\} \cup \{-(\alpha',H+\alpha's,A(s)) : \alpha's \notin H\}$.
Also let
$S' = \{P(\alpha,H+\alpha t,A(t)) : \alpha t \notin H$ and $t \in R_d^s[f;s]\} \cup \{-P(\alpha',H+\alpha's,A(s)) : \alpha's \notin H\}$.
By T5, $S(p)$ is empty, and so S' is also empty. By P5, $Dftd(\alpha,H,f,r,s) = \max S' = \max\{\} = -1$. Since $S(p) = \{\}$, we have for each t in $R_d^s[f;s]$, $\alpha t \in H$; and $\alpha's \in H$. Hence the last characterisation of $Dftd(\alpha,H,f,r,s) = +1$ in part (4) is false.

Thus the result holds for all evaluation trees of $\mathcal{P}$ that have only the root node.

Take any positive integer n. We shall denote the following inductive hypothesis by IndHyp. Suppose the result holds for all evaluation trees of $\mathcal{P}$ that have less than $n+1$ nodes. Let T be an evaluation tree of $\mathcal{P}$ that has $n+1$ nodes and let p be the root of T. Then p has at least one child.

If p satisfies T1 then the subject of p is (α,H,F). By T1, IndHyp, P1, and I1,
$T(\alpha,H,F) = +1$ iff for all f in F, $T(\alpha,H,f) = +1$
   iff for all f in F, $P(\alpha,H,f) = +1$ iff $P(\alpha,H,F) = +1$
   iff for all f in F, $(\alpha,H) \vdash f$ iff $(\alpha,H) \vdash F$.
Since p has a child, p does not satisfy T2.

If p satisfies T3 then the subject of p is (α,H,f).
Let $S(p) = \{(\alpha,H,f,r) : \alpha r \notin H$ and $r \in R_d^s[f]\}$. By T3, IndHyp, P3, and I3,
$T(\alpha,H,f) = +1$
iff there exists (α,H,f,r) in $S(p)$ such that $T(\alpha,H,f,r) = +1$
iff there exists (α,H,f,r) in $S(p)$ such that $For(\alpha,H,f,r) = +1$
iff $P(\alpha,H,f) = +1$.
iff there exists (α,H,f,r) in $S(p)$ such that $For(\alpha,H,f,r) = +1$
iff there exists (α,H,f,r) in $S(p)$ such that $(\alpha,H+\alpha r) \vdash A(r)$ and
   $\forall s \in Foe(\alpha,f,r)$, $Dftd(\alpha,H,f,r,s) = +1$
iff there exists (α,H,f,r) in $S(p)$ such that $(\alpha,H+\alpha r) \vdash A(r)$ and
   $\forall s \in Foe(\alpha,f,r)$, either $\exists t \in R_d^s[f;s]$ such that $\alpha t \notin H$ and $(\alpha,H+\alpha t) \vdash A(t)$;
      or $\alpha's \notin H$ and $(\alpha',H+\alpha's) \not\vdash A(s)$
iff $\exists r \in R_d^s[f]$ such that I3.1 and I3.2
iff $(\alpha,H) \vdash f$.

If p satisfies T4 then the subject of p is (α,H,f,r). By T4, IndHyp, and P4,
$T(\alpha,H,f,r) = +1$
iff $T(\alpha,H+\alpha r,A(r)) = +1$ and $\forall s \in Foe(\alpha,f,r)$, $T(\alpha,H,f,r,s) = +1$
iff $(\alpha,H+\alpha r) \vdash A(r)$ and $\forall s \in Foe(\alpha,f,r)$, $Dftd(\alpha,H,f,r,s) = +1$
iff $P(\alpha,H+\alpha r,A(r)) = +1$ and $\forall s \in Foe(\alpha,f,r)$, $Dftd(\alpha,H,f,r,s) = +1$
iff $For(\alpha,H,f,r) = +1$.

If p satisfies T5 then the subject of p is (α,H,f,r,s). By T5, IndHyp, T6, T7, and P5, $T(\alpha,H,f,r,s) = +1$
iff either $\exists t \in R_d^s[f;s]$ such that $\alpha t \notin H$ and $T(\alpha,H+\alpha t,A(t)) = +1$;
   or $\alpha's \notin H$ and $T(-(\alpha',H+\alpha's,A(s))) = +1$
iff either $\exists t \in R_d^s[f;s]$ such that $\alpha t \notin H$ and $P(\alpha,H+\alpha t,A(t)) = +1$;
   or $\alpha's \notin H$ and $T(\alpha',H+\alpha's,A(s)) = -1$

5.3 The Proof Theory

iff either $\exists t \in R_d^s[f;s]$ such that $\alpha t \notin H$ and $P(\alpha, H+\alpha t, A(t)) = +1$;
or $\alpha's \notin H$ and $P(\alpha', H+\alpha's, A(s)) = -1$
iff $Dftd(\alpha, H, f, r, s) = +1$.

If p satisfies T6 then the subject of p is $-(\alpha', H, F)$. By T6, T7, and IndHyp,
$T(-(\alpha', H, F)) = +1$
iff $T(\alpha', H, F) = -1$
iff $P(\alpha', H, F) = -1$
iff $(\alpha', H) \not\vdash F$.

Thus the theorem is proved by induction.
EndProofTheorem5.3.17

We close this section by noting the conditions under which changing the history has no effect on what is provable.

Definition 5.3.18. Suppose $(R, >)$ is a plausible theory, $\alpha \in Alg$, H is an α-history, and F is a finite set of formulas.
1) If $T_l[\alpha, H, ...]$ is a lean reduct of $T[\alpha, H, ...]$ then $T_l(\alpha, H, ...)$ denotes the proof value of the root of $T_l[\alpha, H, ...]$.
2) If $T_l[-(\alpha, H, F)]$ is a lean reduct of $T[-(\alpha, H, F)]$ then $T_l(-(\alpha, H, F))$ denotes the proof value of the root of $T_l[-(\alpha, H, F)]$.

Definition 5.3.19. Suppose H is any sequence and J is any set. Then define $H-J$ to be the subsequence of H formed by only deleting from H all the occurrences of each element in J.

Suppose $\mathcal{P}$ is any plausible theory, T is any evaluation tree of $\mathcal{P}$, and J is any set. Then define $T-J$ to be the tree formed from T by, for each node p of T, only replacing $Hist(p)$ by $Hist(p)-J$.

It is clear that $|T-J| = |T|$ and if H is an α-history then $H-J$ is also an α-history.

Lemma 5.3.20. Suppose $(R, >)$ is a plausible theory, $\alpha \in \{\varphi, \pi'\}$, H is an α-history, $J \subseteq Set(H)$, and x is either a formula or a finite set of formulas.
1) If $T_l[\alpha, H, ...]$ is a lean reduct of $T[\alpha, H, ...]$ and $T(\alpha, H, ...) = +1$
   then $T_l[\alpha, H, ...] - J$ is a lean reduct of $T[\alpha, H-J, ...]$.
2) If $(\alpha, H) \vdash x$ then $(\alpha, H-J) \vdash x$.
3) Suppose H is a π'-history and $I \subseteq \pi R$. If $(\pi', H-I) \vdash x$ then $(\pi', H) \vdash x$.

Proof

Suppose $(R, >)$ is a plausible theory. It will sometimes be convenient to denote $T[\alpha, H, f, r]$ by $T[\alpha, H, (f, r)]$; $T(\alpha, H, f, r)$ by $T(\alpha, H, (f, r))$; $T[\alpha, H, f, r, s]$ by $T[\alpha, H, (f, r, s)]$; and $T(\alpha, H, f, r, s)$ by $T(\alpha, H, (f, r, s))$.

(1) Let $Y(n)$ denote the following conditional statement. "If $\alpha \in \{\varphi, \pi'\}$, H is an α-history, $J \subseteq Set(H)$, F is a finite set of formulas, f is a formula, $r \in R_d^s[f]$, $s \in R[\neg f]$, $x \in \{F, f, (f, r), (f, r, s)\}$, $T(\alpha, H, x) = +1$, $T_l[\alpha, H, x]$ is a lean reduct of $T[\alpha, H, x]$, and $|T_l[\alpha, H, x]| \leq n$ then $T_l[\alpha, H, x] - J$ is a lean reduct of $T[\alpha, H-J, x]$."

By Theorem 5.3.17(1,2), and Theorem 5.3.15(1), it suffices to prove $Y(n)$ by induction on n.

Suppose $n=1$. Let the antecedent of $Y(1)$ hold. Let the only node of $T_l[\alpha,H,x]$ be denoted by p_0.

If p_0 satisfies T1 then $x=\{\}$ and so $T[\alpha,H-J,\{\}]$ has only one node. Hence $T_l[\alpha,H,\{\}]-J$ is a lean reduct of $T[\alpha,H-J,\{\}]$.

Since $T(\alpha,H,x) = +1$ and p_0 has no children, p_0 does not satisfy T3 or T4 or T5 or T6.

If p_0 satisfies T2 then x is a satisfiable formula, say f, and $Ax \models f$. So $T[\alpha,H-J,f]$ has only one node. Hence $T_l[\alpha,H,f]-J$ is a lean reduct of $T[\alpha,H-J,f]$.

Thus $Y(1)$ is proved.

Take any positive integer n. Suppose $Y(n)$ is true. We shall prove $Y(n+1)$.

Suppose the antecedent of $Y(n+1)$ holds and that $|T_l[\alpha,H,x]| = n+1$. Then $T_l(\alpha,H,x) = T(\alpha,H,x) = +1$. Let the root of $T[\alpha,H,x]$, and hence of $T_l[\alpha,H,x]$, be p_0. Then, in $T_l[\alpha,H,x]$, p_0 has a child.

If p_0 satisfies T1 then let $x=F$, hence $|T_l[\alpha,H,F]| = n+1$. Since p_0 is a min node and $T(\alpha,H,F) = +1$, by Definition 5.3.13(red7), every child of p_0 in $T[\alpha,H,F]$ is a child of p_0 in $T_l[\alpha,H,F]$. For each f in F, let p_f be the root of $T[\alpha,H,f]$. Then the subtree of $T_l[\alpha,H,F]$ generated by p_f is a lean reduct of $T[\alpha,H,f]$ which we shall denoted by $T_l[\alpha,H,f]$. So $T(\alpha,H,f) = +1$ and $|T_l[\alpha,H,f]| \leq n$. By $Y(n)$, $T_l[\alpha,H,f]-J$ is a lean reduct of $T[\alpha,H-J,f]$. Since $T(\alpha,H,f) = +1$, the proof value of the root of $T_l[\alpha,H,f]-J$ is $+1$ and so $T(\alpha,H-J,f) = +1$. But this is true for all f in F, so $T(\alpha,H-J,F) = +1$ and $T_l[\alpha,H,F]-J$ is a lean reduct of $T[\alpha,H-J,F]$.

Since p_0 has a child, p_0 does not satisfy T2.

If p_0 satisfies T3, let $x=f$. Then $\alpha = \pi'$. Hence $|T_l[\pi',H,f]| = n+1$. Since p_0 is a max node and $T(\pi',H,f) = +1$, by Definition 5.3.13(red4), in $T_l[\pi',H,f]$, there is exactly one child, p_1, of p_0 such that $pv(p_1) = +1$. Let the subject of p_1 be (π',H,f,r). Then $T(\pi',H,f,r) = +1$, and $|T_l[\pi',H,f,r]| \leq n$. By $Y(n)$, $T_l[\pi',H,f,r]-J$ is a lean reduct of $T[\pi',H-J,f,r]$. Since $T(\pi',H,f,r) = +1$, the proof value of the root of $T_l[\pi',H,f,r]-J$ is $+1$; hence $T(\pi',H-J,f,r) = +1$. Therefore $T_l[\pi',H,f]-J$ is a lean reduct of $T[\pi',H-J,f]$.

If p_0 satisfies T4, let $x = (f,r)$. Then $\alpha = \pi'$. Hence $|T_l[\pi',H,f,r]| = n+1$. Since p_0 is a min node and $T(\pi',H,f,r) = +1$, by Definition 5.3.13(red7), every child of p_0 in $T[\pi',H,f,r]$ is a child of p_0 in $T_l[\pi',H,f,r]$. Each child, $p_1(y)$, of p_0 has a subject of the form $(\pi',H+\pi'r,A(r))$, and also $T(\pi',H+\pi'r,A(r)) = +1$. By T4, $\pi'r \notin H$ and so $\pi'r \notin J$. For each child $p_1(y)$ of p_0, the subtree of $T_l[\pi',H,f,r]$ generated by $p_1(y)$ is a lean reduct of $T[\pi',H+\pi'r,A(r)]$ which we shall denote by $T_l[\pi',H+\pi'r,A(r)]$. So $|T_l[\pi',H+\pi'r,A(r)]| \leq n$. By $Y(n)$, $T_l[\pi',H+\pi'r,A(r)]-J$ is a lean reduct of $T[\pi',H+\pi'r-J,A(r)]$. Since $T(\pi',H+\pi'r,A(r)) = +1$, the proof value of the root of $T_l[\pi',H+\pi'r,A(r)]-J$ is $+1$ and so $T(\pi',H+\pi'r-J,A(r)) = +1$. But this is true for all the children of $T[\pi',H-J,f,r]$, so $T(\pi',H-J,f,r) = +1$. Hence $T_l[\pi',H,f,r]-J$ is a lean reduct of $T[\pi',H-J,f,r]$.

Since $\alpha \in \{\varphi,\pi'\}$, p_0 does not satisfies T5.

Since the subject of p_0 is (α,H,x), p_0 does not satisfy T6.

Thus $Y(n)$ is proved by induction.

5.3 The Proof Theory

(2) Suppose $\alpha \in \{\varphi, \pi'\}$, H is an α-history, $J \subseteq Set(H)$, x is either a formula or a finite set of formulas, and $(\alpha, H) \vdash x$. By Theorem 5.3.17(1,2), $T(\alpha, H, x) = +1$. Let $T_l[\alpha, H, x]$ be a lean reduct of $T[\alpha, H, x]$. By part (1), $T_l[\alpha, H, x] - J$ is a lean reduct of $T[\alpha, H-J, x]$. So $T(\alpha, H-J, x)$
= the proof value of the root of $T[\alpha, H-J, x]$
= the proof value of the root of $T_l[\alpha, H, x] - J$
= the proof value of the root of $T_l[\alpha, H, x]$
= $T(\alpha, H, x)$
= $+1$.
Hence by Theorem 5.3.17(1,2), $(\alpha, H-J) \vdash x$.

(3) Let Ax be the set of axioms of the plausible theory $(R, >)$. Let $Y(n)$ denote the following conditional statement. "If H is a π'-history, $I \subseteq \pi R$, x is either a formula or a finite set of formulas, $T(\pi', H-I, x) = +1$, and $|T[\pi', H-I, x]| \leq n$ then $T(\pi', H, x) = +1$." By Theorem 5.3.17(1,2), and Theorem 5.3.15(1), it suffices to prove $Y(n)$ by induction on n.

Suppose $n = 1$. Also suppose the antecedent of $Y(1)$ holds. Let p_0 be the root of $T[\pi', H-I, x]$ and q_0 be the root of $T[\pi', H, x]$. Then p_0 has no children. If p_0 satisfies T1 then $x = \{\}$ and so q_0 satisfies T1. So by T1, $T[\pi', H, \{\}]$ has only one node and $T(\pi', H, \{\}) = +1$. If p_0 satisfies T2 then $x = f$ and $Ax \models f$. So q_0 satisfies T2 and hence $T(\pi', H, f) = +1$. Since p_0 has no children and the proof value of p_0 is $+1$, p_0 does not satisfy T3. Since the subject of p_0 is $(\pi', H-I, x)$, p_0 does not satisfy T4, or T5, or T6. Thus the base case holds.

Take any positive integer n. Suppose that $Y(n)$ is true. We shall prove $Y(n+1)$.

Suppose the antecedent of $Y(n+1)$ holds and that $|T[\pi', H-I, x]| = n+1$. Let p_0 be the root of $T[\pi', H-I, x]$ and q_0 be the root of $T[\pi', H, x]$. Then p_0 has a child.

If p_0 satisfies T1 then let x be F. We see that q_0 also satisfies T1. So $t(p_0) = ((\pi', H-I, F), \min, +1)$ and $t(q_0) = ((\pi', H, F), \min, w_0)$, where $w_0 = T(\pi', H, F) \in \{+1, -1\}$. Let $\{p_f : f \in F\}$ be the set of children of p_0 and $\{q_f : f \in F\}$ be the set of children of q_0. Let f be any formula in F. Then the subject of p_f is $(\pi', H-I, f)$, and the subject of q_f is (π', H, f). Also $|T[\pi', H-I, f]| \leq n$ and the proof value of p_f is $+1$ because p_0 is a min node with proof value $+1$. So by $Y(n)$ the proof value of q_f is $+1$. But this is true for each f, so $w_0 = +1$, as required.

Since p_0 has a child, p_0 does not satisfy T2.

If p_0 satisfies T3 then let x be f. We see that q_0 also satisfies T3. So $t(p_0) = ((\pi', H-I, f), \max, +1)$ and $t(q_0) = ((\pi', H, f), \max, w_0)$, where $w_0 = T(\pi', H, f) \in \{+1, -1\}$.

It is convenient to adopt the following naming conventions. Each non-root node of $T[\pi', H-I, f]$ is denoted by $p_l(\#, y)$ where l is the level of the node, # is the number in [1..6] such that the node satisfies T#, and y is a rule, or a formula, or a set, which distinguishes siblings. The proof value of $p_l(\#, y)$ will be denoted by $v_l(\#, y)$. For non-root nodes in $T[\pi', H, f]$ we shall use $q_l(\#, y)$, and its proof value will be denoted by $w_l(\#, y)$.

Let the set of children of p_0 be $\{p_1(4,r) : \pi'r \notin H-I$ and $r \in R_d^s[f]\}$, where the tags of these children are: $t(p_1(4,r)) = ((\pi', H-I, f, r), \min, v_1(4,r))$. So $+1 = \max\{v_1(4,r) : \pi'r \notin H-I$ and $r \in R_d^s[f]\}$.

Let the set of children of q_0 be $\{q_1(4,r) : \pi'r \notin H$ and $r \in R_d^s[f]\}$, where the tags of these children are: $t(q_1(4,r)) = ((\pi', H, f, r), \min, w_1(4,r))$. So $w_0 = \max\{w_1(4,r) : \pi'r \notin H$ and $r \in R_d^s[f]\}$.

From above there exists r_0 in $R_d^s[f]$ such that $\pi'r_0 \notin H-I$ and $v_1(4,r_0) = +1$. So $t(p_1(4,r_0)) = ((\pi', H-I, f, r_0), \min, +1)$.

Let the only child of $p_1(4,r_0)$ be $p_2(1,r_0)$, where $t(p_2(1,r_0)) = ((\pi', (H-I) + \pi'r_0, A(r_0)), \min, v_2(1,r_0))$. So $+1 = v_1(4,r_0) = v_2(1,r_0)$.

Let the only child of $q_1(4,r_0)$ be $q_2(1,r_0)$, where $t(q_2(1,r_0)) = ((\pi', H+\pi'r_0, A(r_0)), \min, w_2(1,r_0))$. So $w_1(4,r_0) = w_2(1,r_0)$.

Since $(H-I) + \pi'r_0 = (H+\pi'r_0) - I$, and $|T[\pi', (H-I) + \pi'r_0, A(r_0)]| \le n$, and $T(\pi', (H-I) + \pi'r_0, A(r_0)) = v_2(1,r_0) = +1$, by $Y(n)$, $+1 = T(\pi', H+\pi'r_0, A(r_0)) = w_2(1,r_0) = w_1(4,r_0)$. Hence $w_0 = +1$, as required.

Since the subject of p_0 is $(\pi', H-I, x)$, p_0 does not satisfy T4, or T5, or T6.

Thus $Y(n+1)$ holds, and so the lemma is proved by induction.
EndProofLemma5.3.20

5.4 A Truth Theory

Logics often have a function from the set of all formulas to a set of truth values with the following two properties.
a) The truth value of a formula is related to its proof value.
b) The truth value of a formula is related to the truth values of its parts.
Section 4.11 deals with (b), while this section is concerned with (a).

Consider the possibilities that could occur when the proof algorithm α evaluates the evidence for and against the formula f.

If there is sufficient evidence for both f and $\neg f$ then, as far as α is concerned, f is ambiguous and $\neg f$ is ambiguous. Therefore both f and $\neg f$ should be assigned the **ambiguous** truth value **a**.

If there is insufficient evidence for both f and $\neg f$ then α does not know enough about f or about $\neg f$. Therefore both f and $\neg f$ should be assigned the **undetermined** truth value **u**.

If there is sufficient evidence for f but insufficient evidence for $\neg f$ then, as far as α is concerned, f cannot be ambiguous and $\neg f$ cannot be undetermined. Therefore f should be assigned the **usually true** truth value **t**, and $\neg f$ should be assigned the **usually false** truth value **f**.

Since the truth value of a formula, f, depends on the proof algorithm, α, evaluating its evidence, we need a truth function V (for veracity) such that $V(\alpha, f)$ is in the set of plausible truth values $\{\mathbf{a}, \mathbf{t}, \mathbf{f}, \mathbf{u}\}$. This is accomplished by the next definition.

5.4 A Truth Theory

Definition 5.4.1. Suppose $\mathcal{P} = (R, >)$ is a plausible theory, $\alpha \in Alg$, and f is any formula. The *truth function for* $\mathcal{P}$, V, from $Alg \times Fml$ to the *set of plausible truth values* $\{\mathbf{a}, \mathbf{t}, \mathbf{f}, \mathbf{u}\}$ is defined by V1 to V4.
V1) $V(\alpha, f) = \mathbf{a}$ iff $\alpha \vdash f$ and $\alpha \vdash \neg f$.
V2) $V(\alpha, f) = \mathbf{t}$ iff $\alpha \vdash f$ and $\alpha \not\vdash \neg f$.
V3) $V(\alpha, f) = \mathbf{f}$ iff $\alpha \not\vdash f$ and $\alpha \vdash \neg f$.
V4) $V(\alpha, f) = \mathbf{u}$ iff $\alpha \not\vdash f$ and $\alpha \not\vdash \neg f$.

Results about these truth values are in Section 6.3 Truth Values.

Now that Propositional Plausible Logic is defined, we should check that it satisfies the principles given in Chapter 4 Principles of Plausible Reasoning. This will be done by proving results in Chapter 6 Properties of Propositional Plausible Logic, and investigating examples in Chapter 7 Examples.

Chapter 6
Properties of Propositional Plausible Logic

Abstract The sections of Chapter 6 Properties of Propositional Plausible Logic are: 6.1 Conjunction and Right Weakening, 6.2 Consistency, 6.3 Truth Values, and 6.4 Proof Algorithm Hierarchy.

We prove that Propositional Plausible Logic satisfies many important properties concerning: Conjunction; Right Weakening; Modus Ponens for strict rules; Consistency; Truth Values, Soundness and Completeness, Conjunction and Disjunction; and the relationship between the proof algorithms.

6.1 Conjunction and Right Weakening

The conjunction of two proved facts should be provable. But as we saw in Section 4.4 this does not hold for plausible formulas. However, Principle 4.4.4 (The Plausibly Conjunctive Principle) says that "a plausible proof algorithm must be plausibly conjunctive". For Propositional Plausible Logic this means that whenever $\alpha \in \{\pi, \psi, \beta\}$, and $Ax \models f$, and $\alpha \vdash g$ then $\alpha \vdash \wedge\{f,g\}$. This is a special case of the following theorem.

Theorem 6.1.1 (Plausible Conjunction). Suppose $(R,>)$ is a plausible description, $Ax = Ax(R)$, $\alpha \in Alg$, H is an α-history, and f and g are both formulas. If $Ax \models f$ and $(\alpha, H) \vdash g$ then $(\alpha, H) \vdash \wedge\{f,g\}$.
Proof
  Suppose $(R,>)$ is a plausible description, $Ax = Ax(R)$, $\alpha \in Alg$, H is an α-history, and f and g are both formulas. Further suppose that $Ax \models f$ and $(\alpha, H) \vdash g$. We shall use Definition 5.2.10.
  Since $(\alpha, H) \vdash g$, by Definition 5.2.10(I3), there exists r in $R^s_d[g]$ such that I3.1 and I3.2 both hold. By Lemma 5.2.14(4), there exists r in $R^s_d[\wedge\{f,g\}]$ such that I3.1 and I3.2 both hold. Thus $(\alpha, H) \vdash \wedge\{f,g\}$.
EndProofTheorem6.1.1

Principle 4.7.5 (The Plausible Right Weakening Principle) says that "a plausible proof algorithm must have the plausible right weakening property". For Propositional Plausible Logic this means that if $\alpha \in \{\pi, \psi, \beta\}$, and $\alpha \vdash f$, and $Ax \cup \{f\} \models g$, and g is not a tautology then $\alpha \vdash g$. This is a special case of the following theorem.

Theorem 6.1.2 (Right Weakening). Suppose that $(R, >)$ is a plausible description, $Ax = Ax(R)$, $\alpha \in Alg$, H is an α-history, and f and g are both formulas.
1) If $(\alpha, H) \vdash f$ and $Ax \cup \{f\} \models g$ then $(\alpha, H) \vdash g$. [Strong Right Weakening]
2) If $(\alpha, H) \vdash f$ and $f \models g$ then $(\alpha, H) \vdash g$. [Right Weakening]
3) If $A \to g \in R_s$ and $(\alpha, H) \vdash A$ then $(\alpha, H) \vdash g$. [Modus Ponens for strict rules]
Proof

Suppose $(R, >)$ is a plausible description, $Ax = Ax(R)$, $\alpha \in Alg$, H is an α-history, and f and g are both formulas. Further suppose that $(\alpha, H) \vdash f$. We shall use Definition 5.2.10.

(1) Suppose $Ax \cup \{f\} \models g$.
If $Ax \models f$ then $Ax \models g$ and so by I2, $(\alpha, H) \vdash g$. So suppose $Ax \not\models f$. Then $\alpha \neq \varphi$.

Since $(\alpha, H) \vdash f$, by I3.1 for f, $\exists r_0 \in R_d^s[f]$ such that $\alpha r_0 \notin H$ and $(\alpha, H + \alpha r_0) \vdash A(r_0)$. By Lemma 5.2.14(5)(b), $R_d^s[f] \subseteq R_d^s[g]$. Hence $r_0 \in R_d^s[g]$. So I3.1 holds for g.

By Lemma 5.2.14(5)(d,e), $R[\neg g] \subseteq R[\neg f]$ and $Foe(\alpha, g, r_0) \subseteq Foe(\alpha, f, r_0)$. By Lemma 5.2.14(5)(b,c), $R_d^s[f] \subseteq R_d^s[g]$ and $R_d^s[f;s] \subseteq R_d^s[g;s]$.

Now take any s_0 in $Foe(\alpha, g, r_0)$. Then $s_0 \in Foe(\alpha, f, r_0)$. If I3.2.1 holds for f then $\exists t_0 \in R_d^s[f;s_0]$ such that $\alpha t_0 \notin H$ and $(\alpha, H + \alpha t_0) \vdash A(t_0)$. Hence $t_0 \in R_d^s[g;s_0]$ and so I3.2.1 holds for g. If I3.2.2 holds for f then $\alpha' s_0 \notin H$ and $(\alpha', H + \alpha' s_0) \not\vdash A(s_0)$. Hence I3.2.2 holds for g.

Thus I3.2 holds for g and so $(\alpha, H) \vdash g$.

(2) Suppose $f \models g$.
Since $f \models g$, we have $Ax \cup \{f\} \models g$. So by part (1), $(\alpha, H) \vdash g$.

(3) Suppose $A \to g \in R_s$ and $(\alpha, H) \vdash A$. By Lemma 5.1.7(3), either $A = \{\}$, or $A = \{a\}$ where a is formula.

Case 1: $A = \{\}$.
By Definition 5.1.4, $g = \wedge Ax$ and so $Ax \models g$. By Definition 5.2.10(I2), $(\alpha, H) \vdash g$.

Case 2: $A = \{a\}$ where a is formula.
By Definition 5.2.10(I1), $(\alpha, H) \vdash a$. By Lemma 5.1.7(5), $Ax \cup \{a\} \models g$. So by part (1), $(\alpha, H) \vdash g$.
EndProofTheorem6.1.2

6.2 Consistency

When reasoning with facts, which classical propositional logic does, it is highly desirable that the set of all formulas derivable from a consistent set of facts is consistent. However, as we saw in Section 4.8 Consistency, when reasoning plausibly, consistency is much less desirable. The Strong 2-Consistency Principle (Principle 4.8.3) says that "a plausible proof algorithm must be strongly 2-consistent".

6.2 Consistency

To prove that the plausible proof algorithms of Propositional Plausible Logic, namely π, ψ, and β, are strongly 2-consistent we shall need to compare their "sizes". Roughly an algorithm λ is bigger than, or encompasses, a smaller algorithm α if and only if every formula proved by the smaller algorithm α can be proved by the bigger algorithm λ. The notation and terms used are defined below.

Definition 6.2.1. Suppose $\alpha \in Alg$ and $\lambda \in Alg$. α is *encompassed* by λ, symbolised by $\alpha \lesssim \lambda$, and λ *encompasses* α, symbolised by $\lambda \gtrsim \alpha$, iff for every plausible theory, $\mathcal{P}$, $\mathcal{P}(\alpha) \subseteq \mathcal{P}(\lambda)$.

We start by showing that φ is the smallest algorithm.

Lemma 6.2.2. Suppose $(R, >)$ is a plausible theory, $\alpha \in Alg$, I is a φ-history, H is an α-history, and x is either a formula or a finite set of formulas. If $(\varphi, I) \vdash x$ then $(\alpha, H) \vdash x$. Hence $\varphi \lesssim \alpha$.
Proof

Suppose $(R, >)$ is a plausible theory, Ax is its set of axioms, $\alpha \in Alg$, H is an α-history, and x is either a formula or a finite set of formulas. Let $(\varphi, I) \vdash x$. We shall use Definition 5.2.10.

Case 1: x is a formula.
Let $x = f$. Then $(\varphi, I) \vdash f$. By Definition 5.2.10(I2), $Ax \models f$ and $(\alpha, H) \vdash f$.

Case 2: x is a finite set of formulas.
Let $x = F$. Then $(\varphi, I) \vdash F$. By I1, for all f in F, $(\varphi, I) \vdash f$. By Case 1, $(\alpha, H) \vdash f$. So by I1, $(\alpha, H) \vdash F$.
EndProofLemma6.2.2

In general, when comparing algorithms, we shall need to change the histories that are associated with them.

Definition 6.2.3. Suppose H is an α-history.
1) If $\alpha = \varphi$ then define $H(\varphi := \pi')$ to be the sequence formed from H by only replacing each φ by π'.
2) If $\alpha \in \{\pi, \pi', \psi, \psi', \beta, \beta'\}$ then define $H(\alpha := \pi')$ to be the sequence formed from H by only replacing each α by π', and each α' by π.

It is clear that $H(\alpha := \pi')$ is a π'-history.
We can now show that π' is the biggest algorithm.

Lemma 6.2.4. Suppose $(R, >)$ is a plausible theory, $\alpha \in Alg$, H is an α-history, and x is either a formula or a finite set of formulas. If $(\alpha, H) \vdash x$ then $(\pi', H(\alpha := \pi')) \vdash x$. Hence $\alpha \lesssim \pi'$.
Proof

Suppose $(R, >)$ is a plausible theory. Let $Y(n)$ denote the following conditional statement. "If $\alpha \in Alg$, H is an α-history, f is a formula, F is a finite set of formulas, $x \in \{F, f\}$, $T(\alpha, H, x) = +1$, and $|T[\alpha, H, x]| \leq n$ then $T(\pi', H(\alpha := \pi'), x) = +1$."

By Theorem 5.3.17(1,2), and Theorem 5.3.15(1), it suffices to prove $Y(n)$ by induction on n.

Suppose $n = 1$. Let the antecedent of $Y(1)$ hold. Let the only node of $T[\alpha, H, x]$ be p_0. Let the root of $T[\pi', H(\alpha := \pi'), x]$ be q_0.

If p_0 satisfies T1 then $x = \{\}$ and so q_0 satisfies T1. So by T1, $T[\pi', H(\alpha := \pi'), \{\}]$ has only one node and $T(\pi', H(\alpha := \pi'), \{\}) = +1$. If p_0 satisfies T2 then $x = f$ and $Ax \models f$. So q_0 satisfies T2 and hence $T(\pi', H(\alpha := \pi'), f) = +1$. Since p_0 has no children and the proof value of p_0 is $+1$, p_0 does not satisfy T3. Since the subject of p_0 is (α, H, x), p_0 does not satisfy T4, or T5, or T6. Thus the base case holds.

Take any positive integer n. Suppose $Y(n)$ is true. We shall prove $Y(n+1)$.

Suppose the antecedent of $Y(n+1)$ holds and that $|T[\alpha, H, x]| = n+1$. Then $T(\alpha, H, x) = +1$. Let p_0 denote the root of $T[\alpha, H, x]$ and let q_0 denote the root of $T[\pi', H(\alpha := \pi'), x]$.

If p_0 satisfies T1 then $x = F$. We see that q_0 also satisfies T1. Therefore we have $t(p_0) = ((\alpha, H, F), \min, +1)$ and $t(q_0) = ((\pi', H(\alpha := \pi'), F), \min, w_0)$, where $w_0 = T(\pi', H(\alpha := \pi'), F) \in \{+1, -1\}$. Let $\{p_f : f \in F\}$ be the set of children of p_0 and $\{q_f : f \in F\}$ be the set of children of q_0. Let f be any formula in F. Then the subject of p_f is (α, H, f), and the subject of q_f is $(\pi', H(\alpha := \pi'), f)$. Also $|T[\alpha, H, f]| \leq n$ and the proof value of p_f is $+1$ because p_0 is a min node with proof value $+1$. So by $Y(n)$ the proof value of q_f is $+1$. But this is true for each f, so $w_0 = +1$, as required.

Since p_0 has a child, p_0 does not satisfy T2.

If p_0 satisfies T3 then $x = f$. We see that q_0 also satisfies T3. Therefore we have $t(p_0) = ((\alpha, H, f), \max, +1)$ and $t(q_0) = ((\pi', H(\alpha := \pi'), f), \max, w_0)$, where $w_0 = T(\pi', H(\alpha := \pi'), f) \in \{+1, -1\}$.

We shall adopt the following naming conventions. Each non-root node of $T[\alpha, H, f]$ is denoted by $p_l(\#, y)$ where l is the level of the node, $\#$ is the number in $[1..6]$ such that the node satisfies T#, and y is a rule, or a formula, or a set, which distinguishes siblings. The proof value of $p_l(\#, y)$ will be denoted by $v_l(\#, y)$. For non-root nodes in $T[\pi', H(\alpha := \pi'), f]$ we shall use $q_l(\#, y)$, and its proof value will be denoted by $w_l(\#, y)$.

Let the set of children of p_0 be $\{p_1(4, r) : \alpha r \notin H$ and $r \in R_d^s[f]\}$, where the tags of these children are: $t(p_1(4, r)) = ((\alpha, H, f, r), \min, v_1(4, r))$. So $+1 = \max\{v_1(4, r) : \alpha r \notin H$ and $r \in R_d^s[f]\}$.

Let the set of children of q_0 be $\{q_1(4, r) : \pi' r \notin H(\alpha := \pi')$ and $r \in R_d^s[f]\}$, where the tags of these children are: $t(q_1(4, r)) = ((\pi', H(\alpha := \pi'), f, r), \min, w_1(4, r))$. So $w_0 = \max\{w_1(4, r) : \pi' r \notin H(\alpha := \pi')$ and $r \in R_d^s[f]\}$.

From above there exists r_0 in $R_d^s[f]$ such that $\alpha r_0 \notin H$ and $v_1(4, r_0) = +1$. By using Definition 6.2.3, if $\pi' r_0 \in H(\alpha := \pi')$ then $\alpha r_0 \in H$. So if $\alpha r_0 \notin H$ then $\pi' r_0 \notin H(\alpha := \pi')$. Hence $q_1(4, r_0)$ exists. We shall show that $w_1(4, r_0) = +1$ and hence that $w_0 = +1$, as required.

Because $p_1(4, r_0)$ is a min node whose proof value is $+1$, the proof value of every child of $p_1(4, r_0)$ must be $+1$. $p_2(1, r_0)$ is a child of $p_1(4, r_0)$ such that $t(p_2(1, r_0)) = ((\alpha, H + \alpha r_0, A(r_0)), \min, +1)$.

The only child of $q_1(4, r_0)$ is $q_2(1, r_0)$ where $t(q_2(1, r_0)) = ((\pi', H(\alpha := \pi') + \pi' r_0, A(r_0)), \min, w_2(1, r_0))$. Therefore $w_1(4, r_0) = w_2(1, r_0)$.

6.2 Consistency

Since $|T[\alpha, H+\alpha r_0, A(r_0)]| \leq n$ and $T(\alpha, H+\alpha r_0, A(r_0)) = +1$, by $Y(n)$ we have $T(\pi', H(\alpha := \pi') + \pi' r_0, A(r_0)) = w_2(1, r_0) = +1$. Hence $w_1(4, r_0) = +1$ and so $w_0 = +1$, as required.

Since the subject of p_0 is (α, H, x), p_0 does not satisfy T4, or T5, or T6.

Thus $Y(n)$, and hence the lemma, is proved by induction.
EndProofLemma6.2.4

If α is an algorithm, we shall sometimes need to interchange α and α'; as defined below.

Definition 6.2.5. Suppose $\{\alpha, \lambda\} \subseteq Alg$, H is a λ-history, and T is an evaluation tree of some plausible theory.
1) If $\lambda \notin \{\alpha, \alpha'\}$ then define $\lambda(\alpha : \alpha') = \lambda$; otherwise define $\alpha(\alpha : \alpha') = \alpha'$ and $\alpha'(\alpha : \alpha') = \alpha$.
2) If $H = (\lambda_1 r_1, ..., \lambda_n r_n)$ then define $H(\alpha : \alpha') = (\lambda_1(\alpha : \alpha') r_1, ..., \lambda_n(\alpha : \alpha') r_n)$.
3) Define $T(\alpha : \alpha')$ to be the tree formed from T by only changing the subject of each node as follows. For each node p of T replace $alg(p)$ by $alg(p)(\alpha : \alpha')$, and replace $Hist(p)$ by $Hist(p)(\alpha : \alpha')$.

We shall generalise Definition 6.2.5 in Definition 6.4.5.

After Refinement 5.2.6 we indicated that β and β' prove exactly the same formulas. We shall call such algorithms equivalent. However, β and β' are even more closely related. If β and β' are interchanged in an evaluation tree then the result is an evaluation tree. We shall call such algorithms isomorphic. These terms and their notation are defined below.

Definition 6.2.6. Suppose $\{\alpha, \lambda\} \subseteq Alg$.
1) α is *equivalent* to λ, symbolised by $\alpha \approx \lambda$, iff $\alpha \lesssim \lambda$ and $\lambda \lesssim \alpha$.
2) α is *isomorphic* to α', symbolised by $\alpha \simeq \alpha'$, iff for each plausible theory $\mathcal{P}$, if T is an evaluation tree of $\mathcal{P}$ then $T(\alpha : \alpha')$ is an evaluation tree of $\mathcal{P}$.

It should be clear that if $\alpha \simeq \alpha'$ then $\alpha \approx \alpha'$.
We can now prove the above claims for β and β'.

Lemma 6.2.7 (Isomorphic proof algorithms). β and β' are isomorphic, $\beta \simeq \beta'$, and so β and β' are equivalent, $\beta \approx \beta'$.
Proof

Suppose $\alpha \in \{\beta, \beta'\}$ and $\mathcal{P} = (R, >)$ is a plausible theory. Let $Y(n)$ denote the following conditional statement. "If T is an evaluation tree of $\mathcal{P}$ and $|T| \leq n$ then $T(\alpha : \alpha')$ is an evaluation tree of $\mathcal{P}$." By Definition 5.3.4, it suffices to prove $Y(n)$ by induction on n. By Definitions 5.3.3 and 5.3.12, if T is an evaluation tree of $\mathcal{P}$ then either $T = T[\alpha, H, ...]$ or $T = T[-(\alpha', H, F)]$. So if $T(\alpha : \alpha')$ is an evaluation tree of $\mathcal{P}$ then either $T(\alpha : \alpha') = T[\alpha', H(\alpha : \alpha'), ...]$ or $T(\alpha : \alpha') = T[-(\alpha, H(\alpha : \alpha'), F)]$.

Suppose $n = 1$ and the antecedent of $Y(1)$ holds. Let the root of T be p_0. Since $\alpha \in \{\beta, \beta'\}$, without loss of generality we can suppose $alg(p_0) = \alpha$. Let the root of $T(\alpha : \alpha')$ be q_0. Since p_0 has no children, q_0 has no children.

If p_0 satisfies T1 then let $Subj(p_0) = (\alpha, H, \{\})$. So $Subj(q_0) = (\alpha', H(\alpha:\alpha'), \{\})$ and hence q_0 satisfies T1. Thus $T(\alpha:\alpha')$ is an evaluation tree of $\mathcal{P}$.

If p_0 satisfies T2 then let $Subj(p_0) = (\alpha, H, f)$. Hence $Subj(q_0) = (\alpha', H(\alpha:\alpha'), f)$ and so q_0 satisfies T2. Thus $T(\alpha:\alpha')$ is an evaluation tree of $\mathcal{P}$.

If p_0 satisfies T3 then let $Subj(p_0) = (\alpha, H, f)$. So $Subj(q_0) = (\alpha', H(\alpha:\alpha'), f)$. Since p_0 has no children, $S(p_0) = \{\}$. So if $r \in R_d^s[f]$ then $\alpha r \in H$. Now $\alpha r \in H$ iff $\alpha' r \in H(\alpha:\alpha')$. Hence $S(q_0) = \{\}$ and so q_0 satisfies T3. Thus $T(\alpha:\alpha')$ is an evaluation tree of $\mathcal{P}$.

Since p_0 and q_0 have no children, p_0 and q_0 satisfy neither T4 nor T6.

If p_0 satisfies T5 then let $Subj(p_0) = (\alpha, H, f, r, s)$. Therefore we have $Subj(q_0) = (\alpha', H(\alpha:\alpha'), f, r, s)$. Since p_0 has no children, $S(p_0) = \{\}$ and so $S(p_0, \alpha) = \{\}$. Hence if $t \in R_d^s[f;s]$ then $\alpha t \in H$. Now $\alpha t \in H$ iff $\alpha' t \in H(\alpha:\alpha')$. Also $\alpha' s \in H$. Hence $\alpha s \in H(\alpha:\alpha')$. Therefore $S(q_0, \alpha') = \{\}$ and so $S(q_0) = \{\}$. Thus q_0 satisfies T5 and so $T(\alpha:\alpha')$ is an evaluation tree of $\mathcal{P}$.

All cases have been considered and so $Y(1)$ holds.

If T is any tree and p is any node of T for which $Subj(p)$ is defined, then define the set $S(p,T)$ of subjects of the children of p in T by $S(p,T) = \{Subj(c) : c$ is a child of p in $T\}$.

Take any integer n such that $n \geq 1$. Suppose that $Y(n)$ is true. We shall prove $Y(n+1)$. Suppose the antecedent of $Y(n+1)$ holds and that $|T| = n+1$. Let p_0 be the root of T. If $alg(p_0) \notin \{\alpha, \alpha'\}$ then $T(\alpha:\alpha')$ is T and so $T(\alpha:\alpha')$ is an evaluation tree of $\mathcal{P}$. So suppose $alg(p_0) \in \{\alpha, \alpha'\}$. Let q_0 be the root of $T(\alpha:\alpha')$.

If p_0 satisfies T1 then let $Subj(p_0) = (\alpha, H, F)$. So $Subj(q_0) = (\alpha', H(\alpha:\alpha'), F)$. Therefore q_0 satisfies T1. Recall that $S(p_0, T[\alpha, H, F]) = \{(\alpha, H, f) : f \in F\}$. Hence $S(q_0, T[\alpha, H, F](\alpha:\alpha')) = \{(\alpha', H(\alpha:\alpha'), f) : f \in F\} = S(q_0, T[\alpha', H(\alpha:\alpha'), F])$, by T1. But for each (α, H, f) in $S(p_0, T[\alpha, H, F])$, $T[\alpha, H, f]$ is an evaluation tree of $\mathcal{P}$ and $|T[\alpha, H, f]| \leq n$. So by $Y(n)$, $T[\alpha, H, f](\alpha:\alpha')$ is an evaluation tree of $\mathcal{P}$. Hence $T[\alpha, H, f](\alpha:\alpha') = T[\alpha', H(\alpha:\alpha'), f]$. Thus $T(\alpha:\alpha') = T[\alpha, H, F](\alpha:\alpha') = T[\alpha', H(\alpha:\alpha'), F]$ which is an evaluation tree of $\mathcal{P}$.

Since p_0 has a child, p_0 does not satisfy T2.

If p_0 satisfies T3 then let $Subj(p_0) = (\alpha, H, f)$. So $Subj(q_0) = (\alpha', H(\alpha:\alpha'), f)$. Therefore q_0 satisfies T3. Recall that $S(p_0, T[\alpha, H, f]) = \{(\alpha, H, f, r) : \alpha r \notin H$ and $r \in R_d^s[f]\}$. Since $\alpha r \in H$ iff $\alpha' r \in H(\alpha:\alpha')$ we have $\alpha r \notin H$ iff $\alpha' r \notin H(\alpha:\alpha')$. So $S(q_0, T[\alpha, H, f](\alpha:\alpha'))$
$= \{(\alpha', H(\alpha:\alpha'), f, r) : \alpha r \notin H$ and $r \in R_d^s[f]\}$
$= \{(\alpha', H(\alpha:\alpha'), f, r) : \alpha' r \notin H(\alpha:\alpha')$ and $r \in R_d^s[f]\}$
$= S(q_0, T[\alpha', H(\alpha:\alpha'), f])$, by T3.
But for each (α, H, f, r) in $S(p_0, T[\alpha, H, f])$, $T[\alpha, H, f, r]$ is an evaluation tree of $\mathcal{P}$ and $|T[\alpha, H, f, r]| \leq n$. So by $Y(n)$, $T[\alpha, H, f, r](\alpha:\alpha')$ is an evaluation tree of $\mathcal{P}$. Hence $T[\alpha, H, f, r](\alpha:\alpha') = T[\alpha', H(\alpha:\alpha'), f, r]$. Thus $T(\alpha:\alpha') = T[\alpha, H, f](\alpha:\alpha') = T[\alpha', H(\alpha:\alpha'), f]$ which is an evaluation tree of $\mathcal{P}$.

If p_0 satisfies T4 then let $Subj(p_0) = (\alpha, H, f, r)$. Therefore we have $Subj(q_0) = (\alpha', H(\alpha:\alpha'), f, r)$. Since $\alpha r \in H$ iff $\alpha' r \in H(\alpha:\alpha')$, $\alpha r \notin H$ iff $\alpha' r \notin H(\alpha:\alpha')$. So q_0 satisfies T4. Recall that

6.2 Consistency

$S(p_0, T[\alpha, H, f, r]) = \{(\alpha, H+\alpha r, A(r))\} \cup \{(\alpha, H, f, r, s) : s \in Foe(\alpha, f, r)\}$. Since $Foe(\alpha, f, r) = Foe(\alpha', f, r)$, we have
$S(q_0, T[\alpha, H, f, r](\alpha : \alpha'))$
$= \{(\alpha', H(\alpha : \alpha') + \alpha' r, A(r))\} \cup \{(\alpha', H(\alpha : \alpha'), f, r, s) : s \in Foe(\alpha, f, r)\}$
$= S(q_0, T[\alpha', H(\alpha : \alpha'), f, r])$, by T4.
But $T[\alpha, H+\alpha r, A(r)]$ is an evaluation tree of $\mathcal{P}$ and $|T[\alpha, H+\alpha r, A(r)]| \leq n$. So by $Y(n)$, $T[\alpha, H+\alpha r, A(r)](\alpha : \alpha')$ is an evaluation tree of $\mathcal{P}$. So $T[\alpha, H+\alpha r, A(r)](\alpha : \alpha') = T[\alpha', H(\alpha : \alpha') + \alpha' r, A(r)]$. Also for each (α, H, f, r, s) in $S(p_0, T[\alpha, H, f, r])$, we have that $T[\alpha, H, f, r, s]$ is an evaluation tree of $\mathcal{P}$ and $|T[\alpha, H, f, r, s]| \leq n$. So by $Y(n)$, $T[\alpha, H, f, r, s](\alpha : \alpha')$ is an evaluation tree of $\mathcal{P}$. Hence $T[\alpha, H, f, r, s](\alpha : \alpha') = T[\alpha', H(\alpha : \alpha'), f, r, s]$. Thus $T(\alpha : \alpha') = T[\alpha, H, f, r](\alpha : \alpha') = T[\alpha', H(\alpha : \alpha'), f, r]$ which is an evaluation tree of $\mathcal{P}$.

If p_0 satisfies T5 then let $Subj(p_0) = (\alpha, H, f, r, s)$. Therefore we have $Subj(q_0) = (\alpha', H(\alpha : \alpha'), f, r, s)$. Since $\alpha r \in H$ iff $\alpha' r \in H(\alpha : \alpha')$, $\alpha r \notin H$ iff $\alpha' r \notin H(\alpha : \alpha')$. So q_0 satisfies T5. Recall that $S(p_0, T[\alpha, H, f, r, s]) = \{(\alpha, H+\alpha t, A(t)) : \alpha t \notin H$ and $t \in R_d^s[f; s]\} \cup \{-(\alpha', H+\alpha' s, A(s)) : \alpha' s \notin H\}$. Also since $\alpha' r \in H$ iff $\alpha r \in H(\alpha : \alpha')$ we have $\alpha' r \notin H$ iff $\alpha r \notin H(\alpha : \alpha')$. Therefore
$S(q_0, T[\alpha, H, f, r, s](\alpha : \alpha'))$
$= \{(\alpha', H(\alpha : \alpha') + \alpha' t, A(t)) : \alpha t \notin H$ and $t \in R_d^s[f; s]\} \cup$
$\{-(\alpha, H(\alpha : \alpha') + \alpha s, A(s)) : \alpha' s \notin H\}$
$= \{(\alpha', H(\alpha : \alpha') + \alpha' t, A(t)) : \alpha' t \notin H(\alpha : \alpha')$ and $t \in R_d^s[f; s]\} \cup$
$\{-(\alpha, H(\alpha : \alpha') + \alpha s, A(s)) : \alpha s \notin H(\alpha : \alpha')\}$
$= S(q_0, T[\alpha', H(\alpha : \alpha'), f, r, s])$, by T5.

But for each $(\alpha, H+\alpha t, A(t))$ in $S(p_0, T[\alpha, H, f, r, s])$, $T[\alpha, H+\alpha t, A(t)]$ is an evaluation tree of $\mathcal{P}$ and $|T[\alpha, H+\alpha t, A(t)]| \leq n$. So by $Y(n)$, $T[\alpha, H+\alpha t, A(t)](\alpha : \alpha')$ is an evaluation tree of $\mathcal{P}$. Hence $T[\alpha, H+\alpha t, A(t)](\alpha : \alpha') = T[\alpha', H(\alpha : \alpha') + \alpha t, A(t)]$. Also if $-(\alpha', H+\alpha' s, A(s)) \in S(p_0, T[\alpha, H, f, r, s])$, then $T[-(\alpha', H+\alpha' s, A(s))]$ is an evaluation tree of $\mathcal{P}$ and $|T[-(\alpha', H+\alpha' s, A(s))]| \leq n$. Hence by $Y(n)$, we have that $T[-(\alpha', H+\alpha' s, A(s))](\alpha : \alpha')$ is an evaluation tree of $\mathcal{P}$. Therefore we have $T[-(\alpha', H+\alpha' s, A(s))](\alpha : \alpha') = T[-(\alpha, H(\alpha : \alpha') + \alpha s, A(s))]$.

Thus $T(\alpha : \alpha') = T[\alpha, H, f, r, s](\alpha : \alpha') = T[\alpha', H(\alpha : \alpha'), f, r, s]$ which is an evaluation tree of $\mathcal{P}$.

If p_0 satisfies T6 then let $Subj(p_0) = -(\alpha', H, F)$. Therefore we have $Subj(q_0) = -(\alpha, H(\alpha : \alpha'), F)$. Hence q_0 satisfies T6. Recall that $S(p_0, T[-(\alpha', H, F)]) = \{(\alpha', H, F)\}$. Therefore
$S(q_0, T[-(\alpha', H, F)](\alpha : \alpha'))$
$= \{(\alpha, H(\alpha : \alpha'), F)\}$
$= S(q_0, T[-(\alpha, H(\alpha : \alpha'), F)])$, by T6.
But $T[\alpha', H, F]$ is an evaluation tree of $\mathcal{P}$ and $|T[\alpha', H, F]| \leq n$. Therefore by $Y(n)$, $T[\alpha', H, F](\alpha : \alpha')$ is an evaluation tree of $\mathcal{P}$. Therefore we have $T[\alpha', H, F](\alpha : \alpha') = T[\alpha, H(\alpha : \alpha'), F]$. Thus $T(\alpha : \alpha') = T[-(\alpha', H, F)](\alpha : \alpha') = T[-(\alpha, H(\alpha : \alpha'), F)]$ which is an evaluation tree of $\mathcal{P}$.

Therefore $Y(n+1)$, and hence the lemma, is proved by induction.
EndProofLemma6.2.7

Our next result shows that ψ is smaller than ψ'.

Lemma 6.2.8. Suppose $(R,>)$ is a plausible theory, H is a ψ-history, and x is either a formula or a finite set of formulas. If $(\psi,H) \vdash x$ then $(\psi',H(\psi:\psi')) \vdash x$. Hence $\psi \lesssim \psi'$.

Proof

Suppose $(R,>)$ is a plausible theory. Let $Y(n)$ denote the following conditional statement. "If H is a ψ-history, F is a finite set of formulas, f is a formula, $x \in \{F,f\}$, $T(\psi,H,x) = +1$, and $|T[\psi,H,x]| \leq n$ then $T(\psi',H(\psi:\psi'),x) = +1$."

By Theorem 5.3.17(1,2), and Theorem 5.3.15(1), it suffices to prove $Y(n)$ by induction on n.

Suppose $n = 1$. Let the antecedent of $Y(1)$ hold. Let the only node of $T[\psi,H,x]$ be p_0. Let the root of $T[\psi',H(\psi:\psi'),x]$ be q_0.

If p_0 satisfies T1 then $x = \{\}$ and so q_0 satisfies T1. So by T1, $T[\psi',H(\psi:\psi'),\{\}]$ has only one node and $T(\psi',H(\psi:\psi'),\{\}) = +1$.

If p_0 satisfies T2 then $x = f$ and $Ax \models f$. So q_0 satisfies T2 and therefore $T(\psi',H(\psi:\psi'),f) = +1$.

Since p_0 has no children and the proof value of p_0 is $+1$, p_0 does not satisfy T3.

Since the subject of p_0 is (ψ,H,x), p_0 does not satisfy T4, or T5, or T6. Thus the base case holds.

Take any positive integer n. Suppose $Y(n)$ is true. We shall prove $Y(n+1)$.

Suppose the antecedent of $Y(n+1)$ holds and that $|T[\psi,H,x]| = n+1$. Then $T(\psi,H,x) = +1$. Suppose p_0 denotes the root of $T[\psi,H,x]$ and q_0 denotes the root of $T[\psi',H(\psi:\psi'),x]$.

If p_0 satisfies T1 then $x = F$. We see that q_0 also satisfies T1. Hence we have $t(p_0) = ((\psi,H,F), \min, +1)$ and $t(q_0) = ((\psi',H(\psi:\psi'),F), \min, w_0)$, where $w_0 = T(\psi',H(\psi:\psi'),F) \in \{+1,-1\}$. Let $\{p_f : f \in F\}$ be the set of children of p_0 and $\{q_f : f \in F\}$ be the set of children of q_0. Let f be any formula in F. Then the subject of p_f is (ψ,H,f), and the subject of q_f is $(\psi',H(\psi:\psi'),f)$. Also $|T[\psi,H,f]| \leq n$ and the proof value of p_f is $+1$ because p_0 is a min node with proof value $+1$. So by $Y(n)$ the proof value of q_f is $+1$. But this is true for each f, so $w_0 = +1$, as required.

Since p_0 has a child, p_0 does not satisfy T2.

If p_0 satisfies T3 then $x = f$. We see that q_0 also satisfies T3. Hence we have $t(p_0) = ((\psi,H,f), \max, +1)$ and $t(q_0) = ((\psi',H(\psi:\psi'),f), \max, w_0)$, where $w_0 = T(\psi',H(\psi:\psi'),f) \in \{+1,-1\}$.

We shall adopt the following naming conventions. Each non-root node of T is denoted by $p_l(\#,y)$ where l is the level of the node, $\#$ is the number in $[1..6]$ such that the node satisfies T$\#$, and y is a rule, or a formula, or a set, which distinguishes siblings. The proof value of $p_l(\#,y)$ will be denoted by $v_l(\#,y)$. For non-root nodes in $T[\psi',H(\psi:\psi'),f]$ we shall use $q_l(\#,y)$, and its proof value will be denoted by $w_l(\#,y)$.

Let the set of children of p_0 be $\{p_1(4,r) : \psi r \notin H$ and $r \in R_d^s[f]\}$, where the tags of these children are: $t(p_1(4,r)) = ((\psi,H,f,r), \min, v_1(4,r))$. Therefore we have that $+1 = \max\{v_1(4,r) : \psi r \notin H$ and $r \in R_d^s[f]\}$.

6.2 Consistency

Let the set of children of q_0 be $\{q_1(4,r) : \psi'r \notin H(\psi:\psi')$ and $r \in R_d^s[f]\}$, where the tags of these children are: $t(q_1(4,r)) = ((\psi', H(\psi:\psi'), f, r), \min, w_1(4,r))$. So $w_0 = \max\{w_1(4,r) : \psi'r \notin H(\psi:\psi')$ and $r \in R_d^s[f]\}$.

From above there exists r_0 in $R_d^s[f]$ such that $\psi r_0 \notin H$ and $v_1(4,r_0) = +1$. By using Definition 6.2.5, if $\psi' r_0 \in H(\psi:\psi')$ then $\psi r_0 \in H$. Therefore, if $\psi r_0 \notin H$ then $\psi' r_0 \notin H(\psi:\psi')$. Hence $q_1(4,r_0)$ exists. We shall show that $w_1(4,r_0) = +1$ and hence that $w_0 = +1$, as required.

Let the set of children of $p_1(4,r_0)$ be $\{p_2(1,r_0)\} \cup \{p_2(5,s) : s \in Foe(\psi,f,r_0)\}$, where the tags of these children are: $t(p_2(1,r_0)) = ((\psi, H+\psi r_0, A(r_0)), \min, v_2(1,r_0))$; and $t(p_2(5,s)) = ((\psi, H, f, r_0, s), \max, v_2(5,s))$. So $+1 = v_1(4,r_0) = \min[\{v_2(1,r_0)\} \cup \{v_2(5,s) : s \in Foe(\psi,f,r_0)\}]$. Hence $v_2(1,r_0) = +1$; and for each s in $Foe(\psi,f,r_0)$, $v_2(5,s) = +1$.

Let the set of children of $q_1(4,r_0)$ be $\{q_2(1,r_0)\} \cup \{q_2(5,s) : s \in R[\neg f;r_0]\}$, where the tags of these children are:
$t(q_2(1,r_0)) = ((\psi', H(\psi:\psi')+\psi' r_0, A(r_0)), \min, w_2(1,r_0))$; and
$t(q_2(5,s)) = ((\psi', H(\psi:\psi'), f, r_0, s), \max, w_2(5,s))$.
So $w_1(4,r_0) = \min[\{w_2(1,r_0)\} \cup \{w_2(5,s) : s \in R[\neg f;r_0]\}]$.

Since $v_2(1,r_0) = +1$, $T(\psi, H+\psi r_0, A(r_0)) = +1$ and $|T[\psi, H+\psi r_0, A(r_0)]| \leq n$. So by $Y(n)$, $w_2(1,r_0) = T(\psi', H(\psi:\psi')+\psi' r_0, A(r_0)) = +1$. Therefore $w_1(4,r_0) = \min\{w_2(5,s) : s \in R[\neg f;r_0]\}$. If $R[\neg f;r_0] = \{\}$ then $w_1(4,r_0) = \min\{\} = +1$ as desired. So suppose $R[\neg f;r_0] \neq \{\}$.

Since $R[\neg f;r_0] \subseteq Foe(\psi,f,r_0)$, if $q_2(5,s)$ exists then $p_2(5,s)$ exists.

For each node, $p_2(5,s)$, let the set of children of $p_2(5,s)$ be $\{p_3(1,t) : \psi t \notin H$ and $t \in R_d^s[f;s]\} \cup \{p_3(6,s) : \psi's \notin H\}$, where the tags of these children are:
$t(p_3(1,t)) = ((\psi, H+\psi t, A(t)), \min, v_3(1,t))$; and
$t(p_3(6,s)) = (-(\psi', H+\psi's, A(s)), -, v_3(6,s))$.
So $+1 = v_2(5,s) = \max[\{v_3(1,t) : \psi t \notin H$ and $t \in R_d^s[f;s]\} \cup \{v_3(6,s) : \psi's \notin H\}]$.

For each node, $q_2(5,s)$, let the set of children of $q_2(5,s)$ be $\{q_3(1,t) : \psi't \notin H(\psi:\psi')$ and $t \in R_d^s[f;s]\} \cup \{q_3(6,s) : \psi s \notin H(\psi:\psi')\}$, where the tags of these children are:
$t(q_3(1,t)) = ((\psi', H(\psi:\psi')+\psi't, A(t)), \min, w_3(1,t))$; and
$t(q_3(6,s)) = (-(\psi, H(\psi:\psi')+\psi s, A(s)), -, w_3(6,s))$.
So $w_2(5,s) = \max[\{w_3(1,t) : \psi't \notin H(\psi:\psi')$ and $t \in R_d^s[f;s]\} \cup \{w_3(6,s) : \psi s \notin H(\psi:\psi')\}]$.

By Definition 6.2.5, $\psi t \in H$ iff $\psi't \in H(\psi:\psi')$. So $\psi t \notin H$ iff $\psi't \notin H(\psi:\psi')$. Therefore $p_3(1,t)$ exists iff $q_3(1,t)$ exists. If there exists $t_0 \in R_d^s[f;s]$ such that $v_3(1,t_0) = +1$ then $T(\psi, H+\psi t_0, A(t_0)) = +1$ and $|T[\psi, H+\psi t_0, A(t_0)]| \leq n$ so by $Y(n)$, $w_3(1,t_0) = T(\psi', H(\psi:\psi')+\psi' t_0, A(t_0)) = +1$. Hence $w_2(5,s) = +1$.

So suppose no such $t_0 \in R_d^s[f;s]$ exists. Then $p_3(6,s)$ exists such that $\psi's \notin H$ and $v_3(6,s) = +1$. By Definition 6.2.5, $\psi's \in H$ iff $\psi s \in H(\psi:\psi')$. So $\psi's \notin H$ iff $\psi s \notin H(\psi:\psi')$. Hence $q_3(6,s)$ exists.

Let the child of $p_3(6,s)$ be $p_4(1,s)$ where
$t(p_4(1,s)) = ((\psi', H+\psi's, A(s)), \min, v_4(1,s))$, and $v_3(6,s) = -v_4(1,s)$. Therefore $v_4(1,s) = -1$.

Let the child of $q_3(6,s)$ be $q_4(1,s)$ where
$t(q_4(1,s)) = ((\psi, H(\psi:\psi')+\psi s, A(s)), \min, w_4(1,s))$, and $w_3(6,s) = -w_4(1,s)$.

Let us assume that $w_4(1,s) = +1$. Then $T(\psi, H(\psi:\psi')+\psi s, A(s)) = +1$ and $|T[\psi, H(\psi:\psi')+\psi s, A(s)]| \leq n$. So by $Y(n)$, $T(\psi', H(\psi:\psi')(\psi:\psi')+\psi's, A(s)) = +1$. But $H(\psi:\psi')(\psi:\psi') = H$. So $T(\psi', H+\psi's, A(s)) = +1$. From above $-1 = v_4(1,s) = T(\psi', H+\psi's, A(s)) = +1$. This contradiction shows that $w_4(1,s) = -1$. Therefore $w_3(6,s) = +1$ and so $w_2(5,s) = +1$.

So in both cases, for all s in $R[\neg f; r_0]$, $w_2(5,s) = +1$ and so $w_1(4, r_0) = +1$. Hence $w_0 = +1$, as required.

Since the subject of p_0 is (ψ, H, x), p_0 does not satisfy T4, or T5, or T6.

Thus $Y(n)$, and hence the lemma, is proved by induction.
EndProofLemma6.2.8

We shall conclude this section by showing that Propositional Plausible Logic is strongly 2-consistent. Some other minor consistency results are also proved.

Theorem 6.2.9 (Consistency). Suppose $(R, >)$ is a plausible theory, $Ax = Ax(R)$, $\alpha \in \{\varphi, \pi, \psi, \beta, \beta'\}$, and both f and g are any formulas.
1) If $\alpha \vdash f$ and $\alpha \vdash g$ then $Ax \cup \{f, g\}$ is satisfiable.
   That is, α is strongly 2-consistent.
2) If $(\psi, H) \vdash f$ then $(\psi', H) \not\vdash \neg f$.
3) Suppose that whenever $s \in R_d^s[\neg f]$ and $(\pi', H+\pi's) \vdash A(s)$ then $R_d^s[f;s] = \{\}$.
   If $(\pi, H) \vdash f$ then $(\pi', H) \not\vdash \neg f$.

Proof

Suppose $(R, >)$ is a plausible theory, $Ax = Ax(R)$, $\alpha \in \{\varphi, \pi, \psi, \beta, \beta'\}$, and both f and g are any formulas.

(1) Suppose $\alpha \vdash f$ and $\alpha \vdash g$. So by Theorem 5.3.17(2), $T(\alpha, (), f) = +1$ and $T(\alpha, (), g) = +1$.

Let p_0 be the root of $T[\alpha, (), f]$ and q_0 be the root of $T[\alpha, (), g]$. Since the subject of p_0 is $(\alpha, (), f)$, p_0 does not satisfy T1, or T4, or T5, or T6. Since the subject of q_0 is $(\alpha, (), g)$, q_0 does not satisfy T1, or T4, or T5, or T6. Therefore p_0 satisfies T2 or T3, and q_0 satisfies T2 or T3. So there are four cases to consider.

Case 1: p_0 satisfies T2 and q_0 satisfies T2.
Then $Ax \models f$ and $Ax \models g$. By Lemma 5.1.6(1), Ax is satisfiable. Therefore $Ax \cup \{f, g\}$ is satisfiable.

Case 2: p_0 satisfies T2 and q_0 satisfies T3.
Then $Ax \models f$, $r_{ax} \in R[f]$, $Ax \not\models g$, and $\alpha \neq \varphi$. So $\alpha \in \{\pi, \psi, \beta, \beta'\}$. Assume $Ax \cup \{f, g\}$ is unsatisfiable. By Lemma 5.2.14(6), $R[f] \subseteq R[\neg g]$. So $r_{ax} \in R[\neg g]$. By T3, q_0 has a child, q_1, in $T[\alpha, (), g]$ such that $t(q_1) = ((\alpha, (), g, r_g), \min, +1)$ and $r_g \in R_d^s[g]$. By T4, q_1 has a child, q_2, in $T[\alpha, (), g]$ such that $t(q_2) = ((\alpha, (), g, r_g, r_{ax}), \max, +1)$. By T5, q_2 has a child, q_3, in $T[\alpha, (), g]$ such that $pv(q_3) = +1$. By T5, $Subj(q_3) \in S(q_2)$. By Definition 5.2.4, $R_d^s[g; r_{ax}] = \{\}$ and so $S(q_2) = S(q_2, \alpha)$. However, $A(r_{ax}) = \{\}$.

6.2 Consistency

So $t(q_3) = (-(\alpha', (\alpha' r_{ax}), \{\}), -, +1)$. By T6, q_3 has a child, q_4 in $T[\alpha, (), g]$ such that $Subj(q_4) = (\alpha', (\alpha' r_{ax}), \{\})$. So by T1, $pv(q_4) = +1$. But by T7, $+1 = pv(q_3) = -pv(q_4) = -1$. This contradiction shows that $Ax \cup \{f, g\}$ is satisfiable.

Case 3: p_0 satisfies T3 and q_0 satisfies T2.
This case is the same as Case 2 but with p and q interchanged and with f and g interchanged. So by doing the indicated interchanges the proof for Case 2 becomes a proof for Case 3.

Case 4: p_0 satisfies T3 and q_0 satisfies T3.
Then $Ax \not\models f$, $Ax \not\models g$, and $\alpha \neq \varphi$. So $\alpha \in \{\pi, \psi, \beta, \beta'\}$. By T3, p_0 has a child, p_1, in $T[\alpha, (), f]$ such that $t(p_1) = ((\alpha, (), f, r_f), \min, +1)$ and $r_f \in R_d^s[f]$. By T3, q_0 has a child, q_1, in $T[\alpha, (), g]$ such that $t(q_1) = ((\alpha, (), g, r_g), \min, +1)$ and $r_g \in R_d^s[g]$.

Assume $Ax \cup \{f, g\}$ is unsatisfiable. By Lemma 5.2.14(6), $R_d^s[f] \subseteq R[f] \subseteq R[\neg g]$ and $R_d^s[g] \subseteq R[g] \subseteq R[\neg f]$. So $r_f \in R[\neg g]$ and $r_g \in R[\neg f]$.

By T4, either $r_f > r_g$; or p_1 has a child, $p_2(r_g)$, in $T[\alpha, (), f]$ such that $Subj(p_2(r_g)) = (\alpha, (), f, r_f, r_g)$ and $pv(p_2(r_g)) = +1$. By T5, $p_2(r_g)$ has a child, $p_3(r_g)$ in $T[\alpha, (), f]$ such that $pv(p_3(r_g)) = +1$ and $Subj(p_3(r_g)) \in S(p_2(r_g))$.

Similarly by T4, either $r_g > r_f$; or q_1 has a child, $q_2(r_f)$, in $T[\alpha, (), g]$ such that $Subj(q_2(r_f)) = (\alpha, (), g, r_g, r_f)$ and $pv(q_2(r_f)) = +1$. By T5, $q_2(r_f)$ has a child, $q_3(r_f)$ in $T[\alpha, (), g]$ such that $pv(q_3(r_f)) = +1$ and $Subj(q_3(r_f)) \in S(q_2(r_f))$.

Case 4.1: $Subj(p_3(r_g)) = -(\alpha', (\alpha' r_g), A(r_g))$.
Since $pv(p_3(r_g)) = +1$, $T(\alpha', (\alpha' r_g), A(r_g)) = -1$. Therefore by Theorem 5.3.17(1), $\text{not}[(\alpha', (\alpha' r_g)) \vdash A(r_g)]$. But $Subj(q_2(r_g)) = (\alpha, (\alpha r_g), A(r_g))$ and $pv(q_2(r_g)) = +1$. Therefore $T(\alpha, (\alpha r_g), A(r_g)) = +1$. By Theorem 5.3.17(1), $(\alpha, (\alpha r_g)) \vdash A(r_g)$. By Lemmas 6.2.4, 6.2.8, and 6.2.7, $(\alpha', (\alpha' r_g)) \vdash A(r_g)$. This contradiction shows that Case 4.1 cannot occur. Thus $Subj(p_3(r_g)) = (\alpha, (\alpha t), A(t))$ where $t \in R_d^s[f; r_g]$.

Case 4.2: $Subj(q_3(r_f)) = -(\alpha', (\alpha' r_f), A(r_f))$.
This case is the same as Case 4.1 but with p and q interchanged and with f and g interchanged. So by doing the indicated interchanges the proof that Case 4.1 cannot occur becomes a proof that Case 4.2 cannot occur. Thus $Subj(q_3(r_f)) = (\alpha, (\alpha t), A(t))$ where $t \in R_d^s[g; r_f]$.

In summary Cases 4.1 and 4.2 have shown that we have
either $r_f > r_g$ or there is a t in $R_d^s[f; r_g]$; and also
either $r_g > r_f$ or there is a t in $R_d^s[g; r_f]$.

So there exists $t_f(1)$ in $R_d^s[f; r_g] \subseteq R_d^s[f] \subseteq R[f] \subseteq R[\neg g]$. Hence $t_f(1) > r_g$ and $t_f(1) \in R[\neg g]$. So $q_2(r_f)$ can be replaced by $q_2(t_f(1))$, and $q_3(r_f)$ can be replaced by $q_3(t_f(1))$.

Also there exists $t_g(1)$ in $R_d^s[g; r_f] \subseteq R_d^s[g] \subseteq R[g] \subseteq R[\neg f]$. Hence $t_g(1) > r_f$ and $t_g(1) \in R[\neg f]$. So $p_2(r_g)$ can be replaced by $p_2(t_g(1))$, and $p_3(r_g)$ can be replaced by $p_3(t_g(1))$.

Similarly, the arguments in Cases 4.1 and 4.2 for these new nodes yield rules $t_f(2)$ and $t_g(2)$ with the following properties: $t_f(2) \in R_d^s[f; t_g(1)] \subseteq R_d^s[f] \subseteq R[f] \subseteq R[\neg g]$; and $t_g(2) \in R_d^s[g; t_f(1)] \subseteq R_d^s[g] \subseteq R[g] \subseteq R[\neg f]$. So $t_f(2) > t_g(1)$ and $t_f(2) \in R[\neg g]$ and $t_g(2) > t_f(1)$ and $t_g(2) \in R[\neg f]$. Hence $t_f(2) > t_g(1) > r_f$ and $t_g(2) > t_f(1) > r_g$.

This process can be continued indefinitely to yield the following sequences of rules. $r_f < t_g(1) < t_f(2) < t_g(3) < t_f(4) < \ldots$ and $r_g < t_f(1) < t_g(2) < t_f(3) < t_g(4) < \ldots$ Now each $t_f(i) \in R_d^s[f]$ and each $t_g(i) \in R_d^s[g]$. Since $\alpha \vdash f$ and $\alpha \vdash g$, by Lemma 5.1.6(2), both f and g are satisfiable. But $Ax \not\models f$ and $Ax \not\models g$, so by Lemma 5.3.11(1), both $R_d^s[f]$ and $R_d^s[g]$ are finite. So for some i and some $j > i$, $t_f(2i) = t_f(2j)$. Hence $>$ is cyclic, which contradicts the definition of $>$ as being acyclic. This contradiction shows that $Ax \cup \{f, g\}$ is satisfiable.

(2) If H is a ψ-history then H is also a ψ'-history. Suppose $(\psi, H) \vdash f$. We shall use Definition 5.2.10. Assume $(\psi', H) \vdash \neg f$.

By I3.1 for $\neg f$, (*1) $\exists s_1 \in R_d^s[\neg f]$ such that $(\psi', H + \psi' s_1) \vdash A(s_1)$. By I3.2 for f either $(\psi', H + \psi' s_1) \not\vdash A(s_1)$, which contradicts (*1), or (*2) $\exists r_2 \in R_d^s[f; s_1]$ such that $(\psi, H + \psi r_2) \vdash A(r_2)$. By I3.2 for $\neg f$ either $(\psi, H + \psi r_2) \not\vdash A(r_2)$, which contradicts (*2), or (*3) $\exists s_3 \in R_d^s[\neg f; r_2]$ such that $(\psi', H + \psi' s_3) \vdash A(s_3)$. By I3.2 for f either $(\psi', H + \psi' s_3) \not\vdash A(s_3)$, which contradicts (*3), or (*4) $\exists r_4 \in R_d^s[f; s_3]$ such that $(\psi, H + \psi r_4) \vdash A(r_4)$. So we have $r_4 > s_3 > r_2 > s_1$.

We can continue the reasoning in the above paragraph to create two arbitrarily long sequences $s_1, s_3, \ldots, s_{2i-1}, \ldots$ and $r_2, r_4, \ldots, r_{2i}, \ldots$ such that each $s_{2i-1} \in R_d^s[\neg f]$ and each $r_{2i} \in R_d^s[f]$. Moreover for each odd i, $s_{i+2} > r_{i+1} > s_i$. Since $(\psi, H) \vdash f$, by Lemma 5.1.6(2), f is satisfiable and $Ax \not\models \neg f$. Since $(\psi', H) \vdash \neg f$, by Lemma 5.1.6(2), $\neg f$ is satisfiable and $Ax \not\models f$. So by Lemma 5.3.11(1), both $R_d^s[f]$ and $R_d^s[\neg f]$ are finite. So there is an even j and an even k such that $j < k$ and $r_j = r_k$. Hence $>$ is cyclic, contradicting its acyclicity. Thus (2) is proved.

(3) Suppose that (*) whenever $s \in R_d^s[\neg f]$ and $(\pi', H + \pi' s) \vdash A(s)$ then we have $R_d^s[f; s] = \{\}$.

If H is a π-history then H is also a π'-history. Suppose $(\pi, H) \vdash f$. We shall use Definition 5.2.10. Assume $(\pi', H) \vdash \neg f$.

By I3.1 for $\neg f$, (**) $\exists s_1 \in R_d^s[\neg f]$ such that $(\pi', H + \pi' s_1) \vdash A(s_1)$. By I3.2 for f either $(\pi', H + \pi' s_1) \not\vdash A(s_1)$, which contradicts (**), or $\exists r_2 \in R_d^s[f; s_1]$ such that $(\pi, H + \pi r_2) \vdash A(r_2)$, which contradicts (*). Thus (3) is proved.
EndProofTheorem6.2.9

6.3 Truth Values

In Section 5.4 A Truth Theory, we defined four plausible truth values for Propositional Plausible Logic (PPL) as follows.

Suppose $\mathcal{P} = (R, >)$ is a plausible theory, $\alpha \in Alg$, and f is any formula. The truth function for $\mathcal{P}$, V, from $Alg \times Fml$ to the set of plausible truth values $\{\mathbf{a}, \mathbf{t}, \mathbf{f}, \mathbf{u}\}$ is defined by V1 to V4.
V1) $V(\alpha, f) = \mathbf{a}$ iff $\alpha \vdash f$ and $\alpha \vdash \neg f$.
V2) $V(\alpha, f) = \mathbf{t}$ iff $\alpha \vdash f$ and $\alpha \not\vdash \neg f$.
V3) $V(\alpha, f) = \mathbf{f}$ iff $\alpha \not\vdash f$ and $\alpha \vdash \neg f$.
V4) $V(\alpha, f) = \mathbf{u}$ iff $\alpha \not\vdash f$ and $\alpha \not\vdash \neg f$.

6.3 Truth Values

We say **a** is the ambiguous truth value, **t** is the usually true truth value, **f** is the usually false truth value, and **u** is the undetermined truth value.

The Included Middle Principle (Principle 4.11.2) is the following. "A logic for plausible reasoning must have a truth theory with at least 3 truth values." So PPL satisfies this principle.

Also in Section 4.11 we noted that the closest plausible reasoning can get to the usual relationship between conjunction and **t**, and between disjunction and **t**, is the following.
▷ If $V(\alpha, \wedge\{f,g\}) = \mathbf{t}$ then $V(\alpha, f) = \mathbf{t} = V(\alpha, g)$.
▷ If $V(\alpha, f) = \mathbf{T}$ or $V(\alpha, g) = \mathbf{t}$ then $V(\alpha, \vee\{f,g\}) = \mathbf{t}$.
These two results are a special case of Theorem 6.3.1(6,7) below.

Given the definition of the plausible truth values, it is not surprising that both soundness and completeness hold, as stated in Theorem 6.3.1(10,11) below. Although not surprising, we regard soundness as necessary, and so it is important that it holds.

Theorem 6.3.1 (Truth values). Suppose $(R, >)$ is a plausible theory, $\alpha \in Alg$, F is a finite set of formulas, and f is a formula.
1) $V(\alpha, \neg\neg f) = V(\alpha, f)$.
2) $V(\alpha, f) = \mathbf{t}$ iff $V(\alpha, \neg f) = \mathbf{f}$.
3) $V(\alpha, f) = \mathbf{f}$ iff $V(\alpha, \neg f) = \mathbf{t}$.
4) $V(\alpha, f) = \mathbf{a}$ iff $V(\alpha, \neg f) = \mathbf{a}$.
5) $V(\alpha, f) = \mathbf{u}$ iff $V(\alpha, \neg f) = \mathbf{u}$.
6) If $V(\alpha, \wedge F) = \mathbf{t}$ then for each f in F, $V(\alpha, f) = \mathbf{t}$.
7) If $f \in F$ and $V(\alpha, f) = \mathbf{t}$ then $V(\alpha, \vee F) = \mathbf{t}$.
8) If $\alpha \in \{\varphi, \pi, \psi, \beta, \beta'\}$ then $V(\alpha, f) \in \{\mathbf{t}, \mathbf{f}, \mathbf{u}\}$.
9) If $V(\alpha, f) = \mathbf{a}$ then $\alpha \in \{\psi', \pi'\}$.
10) If $V(\alpha, f) = \mathbf{t}$ then $\alpha \vdash f$. (completeness)
11) If $\alpha \in \{\varphi, \pi, \psi, \beta, \beta'\}$ and $\alpha \vdash f$ then $V(\alpha, f) = \mathbf{t}$. (soundness)

Proof

Suppose $(R, >)$ is a plausible theory, $\alpha \in Alg$, F is a finite set of formulas, and f is a formula. By Theorem 6.1.2(2), $\alpha \vdash f$ iff $\alpha \vdash \neg\neg f$; and so $\alpha \nvdash f$ iff $\alpha \nvdash \neg\neg f$.

(1) This follows from Definition 5.4.1 and the equivalences noted above.

(2) $V(\alpha, f) = \mathbf{t}$ iff $\alpha \vdash f$ and $\alpha \nvdash \neg f$. $V(\alpha, \neg f) = \mathbf{f}$ iff $\alpha \nvdash \neg f$ and $\alpha \vdash \neg\neg f$. So (2) holds.

(3) $V(\alpha, f) = \mathbf{f}$ iff $\alpha \nvdash f$ and $\alpha \vdash \neg f$. $V(\alpha, \neg f) = \mathbf{t}$ iff $\alpha \vdash \neg f$ and $\alpha \nvdash \neg\neg f$. So (3) holds.

(4) $V(\alpha, f) = \mathbf{a}$ iff $\alpha \vdash f$ and $\alpha \vdash \neg f$. $V(\alpha, \neg f) = \mathbf{a}$ iff $\alpha \vdash \neg f$ and $\alpha \vdash \neg\neg f$. So (4) holds.

(5) $V(\alpha, f) = \mathbf{u}$ iff $\alpha \nvdash f$ and $\alpha \nvdash \neg f$. $V(\alpha, \neg f) = \mathbf{u}$ iff $\alpha \nvdash \neg f$ and $\alpha \nvdash \neg\neg f$. So (5) holds.

(6) Suppose $V(\alpha, \wedge F) = \mathbf{t}$. Then $\alpha \vdash \wedge F$ and $\alpha \nvdash \neg \wedge F$. By Theorem 6.1.2(2), for each f in F, $\alpha \vdash f$. Take any f in F and assume $\alpha \vdash \neg f$. By Theorem 6.1.2(2), $\alpha \vdash \vee \neg F$, where $\neg F = \{\neg f : f \in F\}$. So by Theorem 6.1.2(2), $\alpha \vdash \neg \wedge F$. This contradiction shows that for each f in F, $\alpha \nvdash \neg f$. Thus for each f in F, $V(\alpha, f) = \mathbf{t}$.

(7) Suppose $f \in F$ and $V(\alpha, f) = \mathbf{t}$. Then $\alpha \vdash f$ and $\alpha \not\vdash \neg f$. By Theorem 6.1.2(2), $\alpha \vdash \vee F$. Assume $\alpha \vdash \neg \vee F$. By Theorem 6.1.2(2), $\alpha \vdash \wedge \neg F$ and so $\alpha \vdash \neg f$. This contradiction shows that $\alpha \not\vdash \neg \vee F$. Thus $V(\alpha, \vee F) = \mathbf{t}$.

(8) Suppose $\alpha \in \{\varphi, \pi, \psi, \beta, \beta'\}$. Recall $V(\alpha, f) = \mathbf{a}$ iff $\alpha \vdash f$ and $\alpha \vdash \neg f$. So by Theorem 6.2.9(1), $V(\alpha, f) \neq \mathbf{a}$.

(9) This is just the contrapositive of part (8).

(10) Recall $V(\alpha, f) = \mathbf{t}$ iff $\alpha \vdash f$ and $\alpha \not\vdash \neg f$.

(11) Suppose $\alpha \in \{\varphi, \pi, \psi, \beta, \beta'\}$ and $\alpha \vdash f$. By Definition 5.4.1 and $\alpha \vdash f$ we have $V(\alpha, f) \in \{\mathbf{a}, \mathbf{t}\}$. So by part (8), $V(\alpha, f) = \mathbf{t}$.

EndProofTheorem6.3.1

Theorem 6.3.1(1-5) show that the truth values behave as expected with respect to negation. The conditions under which the ambiguous truth value **a** can occur are given in Theorem 6.3.1(8,9).

6.4 Proof Algorithm Hierarchy

In Section 6.2 Consistency, we compared the 'sizes' of some proof algorithms. However, not all proof algorithms were compared. In this section we shall do the missing comparisons, which will culminate in a linear hierarchy.

We start by converting a π-history into a ψ-history.

Definition 6.4.1. If H is a π-history then define $H(\pi := \psi)$ to be the sequence formed from H by just replacing each π by ψ, and each π' by ψ'.

The following lemma shows that π is smaller than ψ.

Lemma 6.4.2. Suppose $(R, >)$ is a plausible theory, H is a π-history, and x is either a formula or a finite set of formulas. If $(\pi, H) \vdash x$ then $(\psi, H(\pi := \psi)) \vdash x$. Hence $\pi \lesssim \psi$.

Proof

Suppose $(R, >)$ is a plausible theory and Ax is its set of axioms. Let $Y(n)$ denote the following conditional statement. "If H is a π-history, x is either a formula or a finite set of formulas, $T(\pi, H, x) = +1$, and $|T[\pi, H, x]| \leq n$ then $T(\psi, H(\pi := \psi), x) = +1$." By Theorem 5.3.17(1,2), and Theorem 5.3.15(1), it suffices to prove $Y(n)$ by induction on n.

Suppose $n = 1$. Let the antecedent of $Y(1)$ hold. Let p_0 be the root of $T[\pi, H, x]$ and q_0 be the root of $T[\psi, H(\pi := \psi), x]$. Then p_0 has no children.

If p_0 satisfies T1 then $x = \{\}$ and so q_0 satisfies T1. So by T1, $T[\psi, H(\pi := \psi), \{\}]$ has only one node and $T(\psi, H(\pi := \psi), \{\}) = +1$.

If p_0 satisfies T2 then $x = f$ and $Ax \models f$. Therefore q_0 satisfies T2 and hence $T(\psi, H(\pi := \psi), f) = +1$.

Since p_0 has no children and the proof value of p_0 is $+1$, p_0 does not satisfy T3.

6.4 Proof Algorithm Hierarchy

Since the subject of p_0 is (π,H,x), p_0 does not satisfy T4, or T5, or T6. Thus the base case holds.

Take any positive integer n. Suppose that $Y(n)$ is true. We shall prove $Y(n+1)$.

Suppose the antecedent of $Y(n+1)$ holds and that $|T[\pi,H,x]| = n+1$. Let p_0 be the root of $T[\pi,H,x]$ and q_0 be the root of $T[\psi,H(\pi:=\psi),x]$.

If p_0 satisfies T1 then let x be F. We see that q_0 also satisfies T1. Therefore $t(p_0) = ((\pi,H,F),\min,+1)$ and $t(q_0) = ((\psi,H(\pi:=\psi),F),\min,w_0)$, where $w_0 = T(\psi,H(\pi:=\psi),F) \in \{+1,-1\}$. Let $\{p_f : f \in F\}$ be the set of children of p_0 and $\{q_f : f \in F\}$ be the set of children of q_0. Let f be any formula in F. Then the subject of p_f is (π,H,f), and the subject of q_f is $(\psi,H(\pi:=\psi),f)$. Also $|T[\pi,H,f]| \leq n$ and the proof value of p_f is $+1$ because p_0 is a min node with proof value $+1$. So by $Y(n)$ the proof value of q_f is $+1$. But this is true for each f, so $w_0 = +1$, as required.

Since p_0 has a child, p_0 does not satisfy T2.

If p_0 satisfies T3 then let x be f. We see that q_0 also satisfies T3. Therefore $t(p_0) = ((\pi,H,f),\max,+1)$ and $t(q_0) = ((\psi,H(\pi:=\psi),f),\max,w_0)$, where $w_0 = T(\psi,H(\pi:=\psi),f) \in \{+1,-1\}$.

We shall adopt the following naming conventions. Each non-root node of $T[\pi,H,f]$ is denoted by $p_l(\#,y)$ where l is the level of the node, $\#$ is the number in $[1..6]$ such that the node satisfies T#, and y is a rule, or a formula, or a set, which distinguishes siblings. The proof value of $p_l(\#,y)$ will be denoted by $v_l(\#,y)$. For non-root nodes in $T[\psi,H(\pi:=\psi),f]$ we shall use $q_l(\#,y)$, and its proof value will be denoted by $w_l(\#,y)$.

Let the set of children of p_0 be $\{p_1(4,r) : \pi r \notin H \text{ and } r \in R_d^s[f]\}$, where the tags of these children are: $t(p_1(4,r)) = ((\pi,H,f,r),\min,v_1(4,r))$. So $+1 = \max\{v_1(4,r) : \pi r \notin H \text{ and } r \in R_d^s[f]\}$.

Let the set of children of q_0 be $\{q_1(4,r) : \psi r \notin H(\pi:=\psi) \text{ and } r \in R_d^s[f]\}$, where the tags of these children are: $t(q_1(4,r)) = ((\psi,H(\pi:=\psi),f,r),\min,w_1(4,r))$. So $w_0 = \max\{w_1(4,r) : \psi r \notin H(\pi:=\psi) \text{ and } r \in R_d^s[f]\}$.

From above there exists r_0 in $R_d^s[f]$ such that $\pi r_0 \notin H$ and $v_1(4,r_0) = +1$. So $t(p_1(4,r_0)) = ((\pi,H,f,r_0),\min,+1)$.

Let the set of children of $p_1(4,r_0)$ be $\{p_2(1,r_0)\} \cup \{p_2(5,s) : s \in Foe(\pi,f,r_0)\}$, where the tags of these children are: $t(p_2(1,r_0)) = ((\pi,H+\pi r_0,A(r_0)),\min,v_2(1,r_0))$; and $t(p_2(5,s)) = ((\pi,H,f,r_0,s),\max,v_2(5,s))$. So $+1 = v_1(4,r_0) = \min[\{v_2(1,r_0)\} \cup \{v_2(5,s) : s \in Foe(\pi,f,r_0)\}]$. Hence $v_2(1,r_0) = +1$; and for each s in $Foe(\pi,f,r_0)$, $v_2(5,s) = +1$.

Let the set of children of $q_1(4,r_0)$ be $\{q_2(1,r_0)\} \cup \{q_2(5,s) : s \in Foe(\psi,f,r_0)\}$, where the tags of these children are:
$t(q_2(1,r_0)) = ((\psi,H(\pi:=\psi)+\psi r_0,A(r_0)),\min,w_2(1,r_0))$; and
$t(q_2(5,s)) = ((\psi,H(\pi:=\psi),f,r_0,s),\max,w_2(5,s))$.
So $w_1(4,r_0) = \min[\{w_2(1,r_0)\} \cup \{w_2(5,s) : s \in Foe(\psi,f,r_0)\}]$.

Since $|T[\pi,H+\pi r_0,A(r_0)]| < n$ and $T(\pi,H+\pi r_0,A(r_0)) = v_2(1,r_0) = +1$, by $Y(n)$ we have $T(\psi,H(\pi:=\psi)+\psi r_0,A(r_0)) = w_2(1,r_0) = +1$. Therefore
(*) $w_1(4,r_0) = \min\{w_2(5,s) : s \in Foe(\psi,f,r_0)\}$.

For each s in $Foe(\pi,f,r_0)$ let the set of children of $p_2(5,s)$ be $\{p_3(1,t) : \pi t \notin H$ and $t \in R_d^s[f;s]\} \cup \{p_3(6,s) : \pi's \notin H\}$, where the tags of these children are: $t(p_3(1,t)) = ((\pi,H+\pi t,A(t)),\min,v_3(1,t))$; and $t(p_3(6,s)) = (-(\pi',H+\pi's,A(s)),-,v_3(6,s))$. So $+1 = v_2(5,s) = \max[\{v_3(1,t) : \pi t \notin H$ and $t \in R_d^s[f;s]\} \cup \{v_3(6,s) : \pi's \notin H\}$.

For each s in $Foe(\psi,f,r_0)$ let the set of children of $q_2(5,s)$ be $\{q_3(1,t) : \psi t \notin H(\pi:=\psi)$ and $t \in R_d^s[f;s]\} \cup \{q_3(6,s) : \psi's \notin H(\pi:=\psi)\}$, where the tags of these children are: $t(q_3(1,t)) = ((\psi,H(\pi:=\psi)+\psi t,A(t)),\min,w_3(1,t))$; and $t(q_3(6,s)) = (-(\psi',H(\pi:=\psi)+\psi's,A(s)),-,w_3(6,s))$. So $w_2(5,s) = \max[\{w_3(1,t) : \psi t \notin H(\pi:=\psi)$ and $t \in R_d^s[f;s]\} \cup \{w_3(6,s) : \psi's \notin H(\pi:=\psi)\}]$.

Take any s in $Foe(\psi,f,r_0)$. We shall show that $w_2(5,s) = +1$.

Suppose there exists t_0 such that $\pi t_0 \notin H$ and $t_0 \in R_d^s[f;s]$ and $+1 = v_3(1,t_0) = T(\pi,H+\pi t_0,A(t_0))$. Then $p_3(1,t_0)$ exists, and $\psi t_0 \notin H(\pi:=\psi)$ and so $q_3(1,t_0)$ exists. Since $|T[\pi,H+\pi t_0,A(t_0)]| < n$, by $Y(n)$ we have $T(\psi,H(\pi:=\psi)+\psi t_0,A(t_0)) = w_3(1,t_0) = +1$. Hence $w_2(5,s) = +1$.

So suppose that such a t_0 does not exist. Then $v_3(6,s) = +1$ and so $p_3(6,s)$ exists and $\pi's \notin H$. Hence $\psi's \notin H(\pi:=\psi)$, and so $q_3(6,s)$ exists. Let the child of $p_3(6,s)$ be $p_4(1,s)$ where $t(p_4(1,s)) = ((\pi',H+\pi's,A(s)),\min,v_4(1,s))$. Then $+1 = v_3(6,s) = -v_4(1,s)$. So $v_4(1,s) = -1$ and hence $T(\pi',H+\pi's,A(s)) = -1$. By Theorem 5.3.17(1), $(\pi',H+\pi's) \not\vdash A(s)$. By Definition 6.2.3 with $\alpha = \psi'$ and Definition 6.4.1, $H(\pi:=\psi)(\psi':=\pi') = H$ and $(\psi's)(\psi':=\pi') = (\pi's)$.

Let the child of $q_3(6,s)$ be $q_4(1,s)$ where $t(q_4(1,s)) = ((\psi',H(\pi:=\psi)+\psi's,A(s)),\min,w_4(1,s))$. Then $w_3(6,s) = -w_4(1,s)$. Let us assume $w_4(1,s) = +1$. Then $T(\psi',H(\pi:=\psi)+\psi's,A(s)) = +1$ and so by Theorem 5.3.17(1), $(\psi',H(\pi:=\psi)+\psi's) \vdash A(s)$. By Lemma 6.2.4 with $\alpha = \psi'$, $(\pi',H(\pi:=\psi)(\psi':=\pi')++(\psi's)(\psi':=\pi')) \vdash A(s)$. But from the previous paragraph, this simplifies to $(\pi',H+\pi's) \vdash A(s)$. This contradiction shows that $w_4(1,s) = -1$. Hence $w_3(6,s) = +1$. Therefore $w_2(5,s) = +1$.

Thus for all s in $Foe(\psi,f,r_0)$, $w_2(5,s) = +1$. So by (*), $w_1(4,r_0) = +1$ and hence $w_0 = +1$, as required.

Since the subject of p_0 is (π,H,x), p_0 does not satisfy T4, or T5, or T6.

Thus $Y(n)$, and hence the lemma, is proved by induction.
EndProofLemma6.4.2

To compare ψ and β we shall need to convert a ψ-history into a β-history. This is done by the next definition.

Definition 6.4.3. If H is a ψ-history then define $H(\psi:=\beta)$ to be the sequence formed from H by just replacing each ψ by β, and each ψ' by β'.

The following result not only shows that ψ is smaller than β, but also that β' is smaller than ψ'.

Lemma 6.4.4. Suppose $(R,>)$ is a plausible theory, H is a ψ-history, and x is either a formula or a finite set of formulas.
1) If $(\psi,H) \vdash x$ then $(\beta,H(\psi:=\beta)) \vdash x$.

6.4 Proof Algorithm Hierarchy

2) If $(\psi', H) \not\vdash x$ then $(\beta', H(\psi := \beta)) \not\vdash x$.
Hence $\psi \lesssim \beta$ and $\beta' \lesssim \psi'$.

Proof

Suppose $(R, >)$ is a plausible theory, Ax is its set of axioms, H is a ψ-history, and x is either a formula or a finite set of formulas. Note that H is a ψ-history iff H is a ψ'-history.

Let $Y(k)$ and $Z(k)$ denote the following conditional statements.
$Y(k)$: If H is a ψ-history, x is either a formula or a finite set of formulas,
$T(\psi, H, x) = +1$, and $|T[\psi, H, x]| \leq k$ then $T(\beta, H(\psi := \beta), x) = +1$.
$Z(k)$: If H is a ψ'-history, x is either a formula or a finite set of formulas,
$T(\psi', H, x) = -1$, and $|T[\psi', H, x]| \leq k$ then $T(\beta', H(\psi := \beta), x) = -1$.
By Theorem 5.3.17(1,2) and Theorem 5.3.15(2) it suffices to prove $Y(k)$ and $Z(k)$ by induction on k.

Suppose $k = 1$.

Let the antecedent of $Y(1)$ hold. Let p_0 be the root of $T[\psi, H, x]$ and q_0 be the root of $T[\beta, H(\psi := \beta), x]$. Then p_0 has no children.

If p_0 satisfies T1 then $x = \{\}$ and so q_0 satisfies T1. So by T1, $T[\beta, H(\psi := \beta), \{\}]$ has only one node and $T(\beta, H(\psi := \beta), \{\}) = +1$.

If p_0 satisfies T2 then $x = f$ and $Ax \models f$. Hence q_0 satisfies T2 and therefore $T(\beta, H(\psi := \beta), f) = +1$.

Since p_0 has no children and the proof value of p_0 is $+1$, p_0 does not satisfy T3.
Since the subject of p_0 is (ψ, H, x), p_0 does not satisfy T4 or T5 or T6.
Thus $Y(1)$ holds.

Let the antecedent of $Z(1)$ hold. Let m_0 be the root of $T[\psi', H, x]$ and n_0 be the root of $T[\beta', H(\psi := \beta), x]$. Then m_0 has no children.

Since $pv(m_0) = T(\psi', H, x) = -1$, m_0 does not satisfy T1 or T2.

If m_0 satisfies T3 then $x = f$ and for each $r \in R_d^s[f]$, $\psi' r \in H$. So n_0 satisfies T3 and for each $r \in R_d^s[f]$, $\beta' r \in H(\psi := \beta)$. Therefore by T3, n_0 has no children and so $-1 = pv(n_0) = T(\beta', H(\psi := \beta), f)$.

Since the subject of m_0 is (ψ', H, x), m_0 does not satisfy T4 or T5 or T6.
Thus $Z(1)$ holds.

Take any positive integer k. Suppose that both $Y(k)$ and $Z(k)$ are true. We shall prove both $Y(k+1)$ and $Z(k+1)$.

Suppose the antecedent of $Y(k+1)$ holds and $|T[\psi, H, x]| = k+1$. We must show $T(\beta, H(\psi := \beta), x) = +1$. Let p_0 be the root of $T[\psi, H, x]$ and q_0 be the root of $T[\beta, H(\psi := \beta), x]$.

If p_0 satisfies T1 then let x be F. We see that q_0 also satisfies T1. Therefore $t(p_0) = ((\psi, H, F), \min, +1)$ and $t(q_0) = ((\beta, H(\psi := \beta), F), \min, w_0)$, where $w_0 = T(\beta, H(\psi := \beta), F) \in \{+1, -1\}$. Let $\{p_f : f \in F\}$ be the set of children of p_0 and $\{q_f : f \in F\}$ be the set of children of q_0. Let f be any formula in F. Then the subject of p_f is (ψ, H, f), and the subject of q_f is $(\beta, H(\psi := \beta), f)$. Also $|T[\psi, H, f]| \leq k$ and the proof value of p_f is $+1$ because p_0 is a min node with proof value $+1$. So by $Y(k)$ the proof value of q_f is $+1$. But this is true for each f, so by T1, $w_0 = +1$, as required.

Since p_0 has a child, p_0 does not satisfy T2.

If p_0 satisfies T3 then let x be f. We see that q_0 also satisfies T3. Therefore $t(p_0) = ((\psi, H, f), \max, +1)$ and $t(q_0) = ((\beta, H(\psi := \beta), f), \max, w_0)$, where $w_0 = T(\beta, H(\psi := \beta), f) \in \{+1, -1\}$.

We shall adopt the following naming conventions. Each non-root node of $T[\psi, H, f]$ is denoted by $p_l(\#, y)$ where l is the level of the node, $\#$ is the number in $[1..6]$ such that the node satisfies T$\#$, and y is a rule, or a formula, or a set, which distinguishes siblings. The proof value of $p_l(\#, y)$ will be denoted by $v_l(\#, y)$. For non-root nodes in $T[\beta, H(\psi := \beta), f]$ we shall use $q_l(\#, y)$, and its proof value will be denoted by $w_l(\#, y)$.

Let the set of children of p_0 be $\{p_1(4, r) : \psi r \notin H$ and $r \in R_d^s[f]\}$, where the tags of these children are: $t(p_1(4, r)) = ((\psi, H, f, r), \min, v_1(4, r))$. So $+1 = \max\{v_1(4, r) : \psi r \notin H$ and $r \in R_d^s[f]\}$. Hence there exists r_0 in $R_d^s[f]$ such that $\psi r_0 \notin H$ and $v_1(4, r_0) = +1$. So $t(p_1(4, r_0)) = ((\psi, H, f, r_0), \min, +1)$.

Let the set of children of q_0 be $\{q_1(4, r) : \beta r \notin H(\psi := \beta)$ and $r \in R_d^s[f]\}$, where the tags of these children are: $t(q_1(4, r)) = ((\beta, H(\psi := \beta), f, r), \min, w_1(4, r))$. So $w_0 = \max\{w_1(4, r) : \beta r \notin H(\psi := \beta)$ and $r \in R_d^s[f]\}$.

Let the set of children of $p_1(4, r_0)$ be $\{p_2(1, r_0)\} \cup \{p_2(5, s) : s \in Foe(\psi, f, r_0)\}$, where the tags of these children are: $t(p_2(1, r_0)) = ((\psi, H + \psi r_0, A(r_0)), \min, v_2(1, r_0))$; and $t(p_2(5, s)) = ((\psi, H, f, r_0, s), \max, v_2(5, s))$. So $+1 = v_1(4, r_0) = \min[\{v_2(1, r_0)\} \cup \{v_2(5, s) : s \in Foe(\psi, f, r_0)\}]$. Hence $v_2(1, r_0) = +1$; and for each s in $Foe(\psi, f, r_0)$, $v_2(5, s) = +1$.

Let the set of children of $q_1(4, r_0)$ be $\{q_2(1, r_0)\} \cup \{q_2(5, s) : s \in Foe(\beta, f, r_0)\}$, where the tags of these children are:
$t(q_2(1, r_0)) = ((\beta, H(\psi := \beta) + \beta r_0, A(r_0)), \min, w_2(1, r_0))$; and
$t(q_2(5, s)) = ((\beta, H(\psi := \beta), f, r_0, s), \max, w_2(5, s))$.
So $w_1(4, r_0) = \min[\{w_2(1, r_0)\} \cup \{w_2(5, s) : s \in Foe(\beta, f, r_0)\}]$.

Since $|T[\psi, H + \psi r_0, A(r_0)]| \leq k$ and $T(\psi, H + \psi r_0, A(r_0)) = v_2(1, r_0) = +1$, by $Y(k)$ we have $T(\beta, H(\psi := \beta) + \beta r_0, A(r_0)) = w_2(1, r_0) = +1$. Hence
(*) $w_1(4, r_0) = \min\{w_2(5, s) : s \in Foe(\beta, f, r_0)\}$.

For each s in $Foe(\psi, f, r_0)$ let the set of children of $p_2(5, s)$ be $\{p_3(1, t) : \psi t \notin H$ and $t \in R_d^s[f; s]\} \cup \{p_3(6, s) : \psi' s \notin H\}$, where the tags of these children are: $t(p_3(1, t)) = ((\psi, H + \psi t, A(t)), \min, v_3(1, t))$; and $t(p_3(6, s)) = (-(\psi', H + \psi' s, A(s)), -, v_3(6, s))$. So $+1 = v_2(5, s) = \max[\{v_3(1, t) : \psi t \notin H$ and $t \in R_d^s[f; s]\} \cup \{v_3(6, s) : \psi' s \notin H\}]$.

For each s in $Foe(\beta, f, r_0)$ let the set of children of $q_2(5, s)$ be $\{q_3(1, t) : \beta t \notin H(\psi := \beta)$ and $t \in R_d^s[f; s]\} \cup \{q_3(6, s) : \beta' s \notin H(\psi := \beta)\}$, where the tags of these children are: $t(q_3(1, t)) = ((\beta, H(\psi := \beta) + \beta t, A(t)), \min, w_3(1, t))$; and $t(q_3(6, s)) = (-(\beta', H(\psi := \beta) + \beta' s, A(s)), -, w_3(6, s))$. So $w_2(5, s) = \max[\{w_3(1, t) : \beta t \notin H(\psi := \beta)$ and $t \in R_d^s[f; s]\} \cup \{w_3(6, s) : \beta' s \notin H(\psi := \beta)\}]$.

Take any s in $Foe(\beta, f, r_0)$. We shall show that $w_2(5, s) = +1$. Observe that $Foe(\psi, f, r_0) = Foe(\beta, f, r_0)$.

Suppose there exists t_0 such that $\psi t_0 \notin H$ and $t_0 \in R_d^s[f; s]$ and $+1 = v_3(1, t_0) = T(\psi, H + \psi t_0, A(t_0))$. Then $p_3(1, t_0)$ exists. Since $\psi t_0 \in H$ iff $\beta t_0 \in H(\psi := \beta)$, we have

6.4 Proof Algorithm Hierarchy

$\beta t_0 \notin H(\psi := \beta)$ and so $q_3(1, t_0)$ exists. Since $|T[\psi, H+\psi t_0, A(t_0)]| \leq k$, by $Y(k)$ we have $T(\beta, H(\psi := \beta) + \beta t_0, A(t_0)) = w_3(1, t_0) = +1$. Hence $w_2(5, s) = +1$.

So suppose that such a t_0 does not exist. Then $v_3(6, s) = +1$ and hence $p_3(6, s)$ exists and $\psi' s \notin H$. Since $\psi' s \in H$ iff $\beta' s \in H(\psi := \beta)$, we have $\beta' s \notin H(\psi := \beta)$, and so $q_3(6, s)$ exists. Let the child of $p_3(6, s)$ be $p_4(1, s)$ where $t(p_4(1, s)) = ((\psi', H+\psi' s, A(s)), \min, v_4(1, s))$. Then $+1 = v_3(6, s) = -v_4(1, s)$. So $v_4(1, s) = -1$ and hence $T(\psi', H+\psi' s, A(s)) = -1$. Since $|T[\psi', H+\psi' s, A(s)]| \leq k$, by $Z(k)$ we have $T(\beta', H(\psi := \beta) + \beta' s, A(s)) = -1$.

Let the child of $q_3(6, s)$ be $q_4(1, s)$ where $t(q_4(1, s)) = ((\beta', H(\psi := \beta) + \beta' s, A(s)), \min, w_4(1, s))$. Then $w_3(6, s) = -w_4(1, s) = -T(\beta', H(\psi := \beta) + \beta' s, A(s)) = +1$. Therefore $w_2(5, s) = +1$.

Thus for all s in $Foe(\beta, f, r_0)$, $w_2(5, s) = +1$. So by (*), $w_1(4, r_0) = +1$ and hence $w_0 = +1$, as required.

Since the subject of p_0 is (ψ, H, x), p_0 does not satisfy T4, or T5, or T6.

Thus $Y(k+1)$ is proved.

To prove $Z(k+1)$ we suppose that the antecedent of $Z(k+1)$ holds and that $|T[\psi', H, x]| = k+1$. We must show $T(\beta', H(\psi := \beta), x) = -1$. Let m_0 be the root of $T[\psi', H, x]$ and n_0 be the root of $T[\beta', H(\psi := \beta), x]$. Then m_0 has a child and $pv(m_0) = T(\psi', H, x) = -1$.

If m_0 satisfies T1 then let x be F. We see that n_0 also satisfies T1. So $t(m_0) = ((\psi', H, F), \min, -1)$ and $t(n_0) = ((\beta', H(\psi := \beta), F), \min, pv(n_0))$, where $pv(n_0) = T(\beta', H(\psi := \beta), F) \in \{+1, -1\}$. Let $\{m_f : f \in F\}$ be the set of children of m_0 and $\{n_f : f \in F\}$ be the set of children of n_0. Let f be any formula in F. Then the subject of m_f is (ψ', H, f), and the subject of n_f is $(\beta', H(\psi := \beta), f)$. Also $|T[\psi', H, f]| \leq k$. There exists $f_0 \in F$ such at $pv(m_{f_0}) = -1$ because m_0 is a min node with proof value -1. So by $Z(k)$, $pv(n_{f_0}) = T(\beta', H(\psi := \beta), f_0) = -1$. But n_0 is a min node, so $pv(n_0) = -1$, as required.

Since $pv(m_0) = T(\psi', H, x) = -1$, m_0 does not satisfy T2.

If m_0 satisfies T3 then let x be f. We see that n_0 also satisfies T3. Therefore we have $t(m_0) = ((\psi', H, f), \max, -1)$ and $t(n_0) = ((\beta', H(\psi := \beta), f), \max, pv(n_0))$, where $pv(n_0) = T(\beta', H(\psi := \beta), f) \in \{+1, -1\}$.

It is convenient to adopt the following naming conventions. Each non-root node of $T[\psi', H, f]$ is denoted by $m_l(\#, y)$ where l is the level of the node, # is the number in $[1..6]$ such that the node satisfies T#, and y is a rule, or a formula, or a set, which distinguishes siblings. For non-root nodes in $T[\beta', H(\psi := \beta), f]$ we shall use $n_l(\#, y)$.

Let the set of children of m_0 be $\{m_1(4, r) : \psi' r \notin H \text{ and } r \in R_d^s[f]\}$, where the tags of these children are: $t(m_1(4, r)) = ((\psi', H, f, r), \min, pv(m_1(4, r)))$. Recall that m_0 has at least one child. So $-1 = \max\{pv(m_1(4, r)) : \psi' r \notin H \text{ and } r \in R_d^s[f]\}$. Hence if $\psi' r \notin H$ and $r \in R_d^s[f]$ then $pv(m_1(4, r)) = -1$. Therefore $t(m_1(4, r)) = ((\psi', H, f, r), \min, -1)$.

Let the set of children of n_0 be $\{n_1(4, r) : \beta' r \notin H(\psi := \beta) \text{ and } r \in R_d^s[f]\}$, where the tags of these children are: $t(n_1(4, r)) = ((\beta', H(\psi := \beta), f, r), \min, pv(n_1(4, r)))$. So $pv(n_0) = \max\{pv(n_1(4, r)) : \beta' r \notin H(\psi := \beta) \text{ and } r \in R_d^s[f]\}$. If n_0 does not have a child then $pv(n_0) = \max\{\} = -1$, as required. So suppose that n_0 has at least one child.

If $\psi'r \notin H$ and $r \in R_d^s[f]$ then let the set of children of $m_1(4,r)$ be $\{m_2(1,r)\} \cup \{m_2(5,s) : s \in R[\neg f; r]\}$, where the tags of these children are:
$t(m_2(1,r)) = ((\psi', H + \psi'r, A(r)), \min, pv(m_2(1,r)))$; and
$t(m_2(5,s)) = ((\psi', H, f, r, s), \max, pv(m_2(5,s)))$.
Therefore $-1 = pv(m_1(4,r)) = \min[\{pv(m_2(1,r))\} \cup \{pv(m_2(5,s)) : s \in R[\neg f; r]\}]$.
So either $pv(m_2(1,r)) = -1$ or there is an s_0 in $R[\neg f; r]$ such that $pv(m_2(5,s_0)) = -1$.

If $\beta'r \notin H(\psi := \beta)$ and $r \in R_d^s[f]$ then let the set of children of $n_1(4,r)$ be $\{n_2(1,r)\} \cup \{n_2(5,s) : s \in Foe(\beta', f, r)\}$, where the tags of these children are:
$t(n_2(1,r)) = ((\beta', H(\psi := \beta) + \beta'r, A(r)), \min, pv(n_2(1,r)))$; and
$t(n_2(5,s)) = ((\beta', H(\psi := \beta), f, r, s), \max, pv(n_2(5,s)))$.
So $pv(n_1(4,r)) = \min[\{pv(n_2(1,r))\} \cup \{pv(n_2(5,s)) : s \in Foe(\beta', f, r)\}]$.

We show that for each r in $R_d^s[f]$ such that $\beta'r \notin H(\psi := \beta)$, $pv(n_1(4,r)) = -1$.

We have $|T[\psi', H + \psi'r, A(r)]| \le k$. If $-1 = pv(m_2(1,r)) = T(\psi', H + \psi'r, A(r))$ then by $Z(k)$, $pv(n_2(1,r)) = T(\beta', H(\psi := \beta) + \beta'r, A(r)) = -1$. So $pv(n_1(4,r)) = -1$.

So suppose there exists s_0 in $R[\neg f; r]$ such that $T(\psi', H, f, r, s_0) = pv(m_2(5,s_0)) = -1$. Then $s_0 \in Foe(\beta', f, r)$. We show that $T(\beta', H(\psi := \beta), f, r, s_0) = pv(n_2(5,s_0)) = -1$, and hence that $pv(n_1(4,r)) = -1$.

Let the set of children of $m_2(5,s_0)$ be $\{m_3(1,t) : \psi't \notin H$ and $t \in R_d^s[f; s_0]\} \cup \{m_3(6,s_0) : \psi s_0 \notin H\}$, where the tags of these children are:
$t(m_3(1,t)) = ((\psi', H + \psi't, A(t)), \min, pv(m_3(1,t)))$, and
$t(m_3(6,s_0)) = (-(\psi, H + \psi s_0, A(s_0)), -, pv(m_3(6,s_0)))$. So $-1 = pv(m_2(5,s_0)) = \max[\{pv(m_3(1,t)) : \psi't \notin H$ and $t \in R_d^s[f; s_0]\} \cup \{pv(m_3(6,s_0)) : \psi s_0 \notin H\}]$. Hence for all t in $R_d^s[f; s_0]$ such that $\psi't \notin H$ we have $-1 = pv(m_3(1,t)) = T(\psi', H + \psi't, A(t))$. Also if $\psi s_0 \notin H$ then $-1 = pv(m_3(6,s_0)) = T(-(\psi, H + \psi s_0, A(s_0)))$.

Let the set of children of $n_2(5,s_0)$ be $\{n_3(1,t) : \beta't \notin H(\psi := \beta)$ and $t \in R_d^s[f; s_0]\} \cup \{n_3(6,s_0) : \beta s_0 \notin H(\psi := \beta)\}$, where the tags of these children are:
$t(n_3(1,t)) = ((\beta', H(\psi := \beta) + \beta't, A(t)), \min, pv(n_3(1,t)))$, and
$t(n_3(6,s_0)) = (-(\beta, H(\psi := \beta) + \beta s_0, A(s_0)), -, pv(n_3(6,s_0)))$. Therefore
$pv(n_2(5,s_0)) = \max[\{pv(n_3(1,t)) : \beta't \notin H(\psi := \beta)$ and $t \in R_d^s[f; s_0]\} \cup \{pv(n_3(6,s_0)) : \beta s_0 \notin H(\psi := \beta)\}]$. If $n_2(5,s_0)$ has no children then $pv(n_2(5,s_0)) = \max\{\} = -1$, as required. So suppose that $n_2(5,s_0)$ has at least one child.

Case 1: $n_3(1,t)$ is a child of $n_2(5,s_0)$.
Then $\beta't \notin H(\psi := \beta)$ and $t \in R_d^s[f; s_0]$. Therefore $\psi't \notin H$. Hence from above we have, $-1 = pv(m_3(1,t)) = T(\psi', H + \psi't, A(t))$. Also $|T[\psi', H + \psi't, A(t)]| \le k$. So by $Z(k)$, $-1 = T(\beta', H(\psi := \beta) + \beta't, A(t)) = pv(t(n_3(1,t)))$.

Case 2: $n_3(6,s_0)$ is a child of $n_2(5,s_0)$.
Then $\beta s_0 \notin H(\psi := \beta)$. So $\psi s_0 \notin H$. Hence from above, $-1 = pv(m_3(6,s_0)) = T(-(\psi, H + \psi s_0, A(s_0)))$. Therefore $T(\psi, H + \psi s_0, A(s_0)) = +1$. We also have that $|T[\psi, H + \psi s_0, A(s_0)]| \le k$. Therefore by $Y(k)$, $T(\beta, H(\psi := \beta) + \beta s_0, A(s_0)) = +1$. But $pv(n_3(6,s_0)) = -T(\beta, H(\psi := \beta) + \beta s_0, A(s_0)) = -+1 = -1$.

These two cases show that $pv(n_2(5,s_0)) = -1$, as required. So $pv(n_1(4,r)) = -1$. Therefore $pv(n_0) = -1$. Thus $Z(k+1)$ is proved.

6.4 Proof Algorithm Hierarchy

Therefore both $Y(k)$ and $Z(k)$ are proved by induction, and hence the lemma is also proved.
EndProofLemma6.4.4

We generalise Definition 6.2.5, which interchanges proof algorithms, as follows.

Definition 6.4.5. Suppose $\{\alpha, \gamma, \lambda\} \subseteq Alg$, H is a α-history, and T is an evaluation tree of some plausible theory.
1) If $\alpha \notin \{\gamma, \gamma', \lambda, \lambda'\}$ then define $\alpha(\gamma:\lambda) = \alpha$;
   else define $\gamma(\gamma:\lambda) = \lambda$, $\gamma'(\gamma:\lambda) = \lambda'$, $\lambda(\gamma:\lambda) = \gamma$, and $\lambda'(\gamma:\lambda) = \gamma'$.
2) If $H = (\alpha_1 r_1, ..., \alpha_n r_n)$ then define $H(\gamma:\lambda) = (\alpha_1(\gamma:\lambda)r_1, ..., \alpha_n(\gamma:\lambda)r_n)$.
3) Define $T(\gamma:\lambda)$ to be the tree formed from T by only changing the subject of each node as follows. For each node p of T replace $alg(p)$ by $alg(p)(\gamma:\lambda)$, and replace $Hist(p)$ by $Hist(p)(\gamma:\lambda)$.

Lemma 6.2.7 shows that β and β' are the same size. The following lemma shows that, if the priority relation $>$ is empty, then π and ψ are the same size, and also π' and ψ' are the same size.

Lemma 6.4.6. Suppose $\mathcal{P} = (R, >)$ is a plausible theory such that $>$ is empty. If T is an evaluation tree of $\mathcal{P}$ then $T(\pi:\psi)$ is an evaluation tree of $\mathcal{P}$. Hence $\psi \approx \pi$ and $\pi' \approx \psi'$.

Proof

Let $\mathcal{P} = (R, >)$ be a plausible theory such that $>$ is empty. Then $Foe(\psi', f, r) = \{\} = Foe(\pi', f, r)$ and $Foe(\pi, f, r) = R[\neg f] = Foe(\psi, f, r)$. Let $Y(n)$ denote the following conditional statement. "If T is an evaluation tree of $\mathcal{P}$ and $|T| \leq n$ then $T(\pi:\psi)$ is an evaluation tree of $\mathcal{P}$." By Definition 5.3.4, it suffices to prove $Y(n)$ by induction on n.

Suppose $n = 1$ and the antecedent of $Y(1)$ holds. Let the root of T be p_0 and $alg(p_0) = \alpha$. If $\alpha \notin \{\pi, \psi, \psi', \pi'\}$ then $T(\pi:\psi) = T$ and so $Y(1)$ holds. So suppose $\alpha \in \{\pi, \psi, \psi', \pi'\}$. Let the root of $T(\pi:\psi)$ be q_0, and let $alg(q_0) = \lambda$. So $\alpha(\pi:\psi) = \lambda$. Since p_0 has no children, q_0 has no children.

If p_0 satisfies T1 then let $Subj(p_0) = (\alpha, H, \{\})$. Therefore we have $Subj(q_0) = (\lambda, H(\pi:\psi), \{\})$ and so q_0 satisfies T1. Thus $T(\pi:\psi)$ is an evaluation tree of $\mathcal{P}$.

If p_0 satisfies T2 then let $Subj(p_0) = (\alpha, H, f)$. Hence $Subj(q_0) = (\lambda, H(\pi:\psi), f)$ and so q_0 satisfies T2. Thus $T(\pi:\psi)$ is an evaluation tree of $\mathcal{P}$.

If p_0 satisfies T3 then let $Subj(p_0) = (\alpha, H, f)$. Hence $Subj(q_0) = (\lambda, H(\pi:\psi), f)$. Since p_0 has no children, $S(p_0) = \{\}$. So if $r \in R_d^s[f]$ then $\alpha r \in H$. Now $\alpha r \in H$ iff $\lambda r \in H(\pi:\psi)$. Hence $S(q_0) = \{\}$ and so q_0 satisfies T3. Thus $T(\pi:\psi)$ is an evaluation tree of $\mathcal{P}$.

Since p_0 and q_0 have no children, p_0 and q_0 satisfy neither T4 nor T6.

If p_0 satisfies T5 then let $Subj(p_0) = (\alpha, H, f, r, s)$. Therefore we have $Subj(q_0) = (\lambda, H(\pi:\psi), f, r, s)$. Since p_0 has no children, $S(p_0) = \{\}$. Hence if $t \in R_d^s[f; s]$ then $\alpha t \in H$. Now $\alpha t \in H$ iff $\lambda t \in H(\pi:\psi)$. Also $\alpha' s \in H$. Hence $\lambda' s \in H(\pi:\psi)$. So $S(q_0) = \{\}$. Thus q_0 satisfies T5 and so $T(\pi:\psi)$ is an evaluation tree of $\mathcal{P}$.

All cases have been considered and so $Y(1)$ holds.

If T is any tree and p is any node of T for which $Subj(p)$ is defined, then define the set $S(p,T)$ of subjects of the children of p in T by $S(p,T) = \{Subj(c) : c$ is a child of p in $T\}$.

Take any integer n such that $n \geq 1$. Suppose that $Y(n)$ is true. We shall prove $Y(n+1)$. Suppose the antecedent of $Y(n+1)$ holds and that $|T| = n+1$. Let the root of T be p_0 and $alg(p_0) = \alpha$. If $\alpha \notin \{\pi, \psi, \psi', \pi'\}$ then $T(\pi:\psi) = T$ and so $Y(n+1)$ holds. So suppose $\alpha \in \{\pi, \psi, \psi', \pi'\}$. Let the root of $T(\pi:\psi)$ be q_0, and let $alg(q_0) = \lambda$. So $\alpha(\pi:\psi) = \lambda$.

If p_0 satisfies T1 then let $Subj(p_0) = (\alpha,H,F)$. So $Subj(q_0) = (\lambda,H(\pi:\psi),F)$. Hence q_0 satisfies T1. Recall that $S(p_0,T[\alpha,H,F]) = \{(\alpha,H,f) : f \in F\}$. Therefore $S(q_0,T[\alpha,H,F](\pi:\psi)) = \{(\lambda,H(\pi:\psi),f) : f \in F\} = S(q_0,T[\lambda,H(\pi:\psi),F])$, by T1. But for each (α,H,f) in $S(p_0,T[\alpha,H,F])$, $T[\alpha,H,f]$ is an evaluation tree of $\mathcal{P}$ and $|T[\alpha,H,f]| \leq n$. So by $Y(n)$, $T[\alpha,H,f](\pi:\psi)$ is an evaluation tree of $\mathcal{P}$. Hence $T[\alpha,H,f](\pi:\psi) = T[\lambda,H(\pi:\psi),f]$. Therefore $T(\pi:\psi) = T[\alpha,H,F](\pi:\psi) = T[\lambda,H(\pi:\psi),F]$ which is an evaluation tree of $\mathcal{P}$.

Since p_0 has a child, p_0 does not satisfy T2.

If p_0 satisfies T3 then let $Subj(p_0) = (\alpha,H,f)$. Therefore we have $Subj(q_0) = (\lambda,H(\pi:\psi),f)$. Therefore q_0 satisfies T3. Recall the following: $S(p_0,T[\alpha,H,f]) = \{(\alpha,H,f,r) : \alpha r \notin H$ and $r \in R_d^s[f]\}$. Since $\alpha r \in H$ iff $\lambda r \in H(\pi:\psi)$ we have $\alpha r \notin H$ iff $\lambda r \notin H(\pi:\psi)$. Hence
$S(q_0,T[\alpha,H,f](\pi:\psi))$
$= \{(\lambda,H(\pi:\psi),f,r) : \alpha r \notin H$ and $r \in R_d^s[f]\}$
$= \{(\lambda,H(\pi:\psi),f,r) : \lambda r \notin H(\pi:\psi)$ and $r \in R_d^s[f]\}$
$= S(q_0,T[\lambda,H(\pi:\psi),f])$, by T3.
But for each (α,H,f,r) in $S(p_0,T[\alpha,H,f])$, $T[\alpha,H,f,r]$ is an evaluation tree of $\mathcal{P}$ and $|T[\alpha,H,f,r]| \leq n$. So by $Y(n)$, $T[\alpha,H,f,r](\pi:\psi)$ is an evaluation tree of $\mathcal{P}$. Hence $T[\alpha,H,f,r](\pi:\psi) = T[\lambda,H(\pi:\psi),f,r]$. Thus $T(\pi:\psi) = T[\alpha,H,f](\pi:\psi) = T[\lambda,H(\pi:\psi),f]$ which is an evaluation tree of $\mathcal{P}$.

If p_0 satisfies T4 then let $Subj(p_0) = (\alpha,H,f,r)$. Therefore we have $Subj(q_0) = (\lambda,H(\pi:\psi),f,r)$. Since $\alpha r \in H$ iff $\lambda r \in H(\pi:\psi)$, $\alpha r \notin H$ iff $\lambda r \notin H(\pi:\psi)$. So q_0 satisfies T4. Recall that the following holds: $S(p_0,T[\alpha,H,f,r]) = \{(\alpha,H+\alpha r,A(r))\} \cup \{(\alpha,H,f,r,s) : s \in Foe(\alpha,f,r)\}$. Since $Foe(\alpha,f,r) = Foe(\lambda,f,r)$, we have
$S(q_0,T[\alpha,H,f,r](\pi:\psi))$
$= \{(\lambda,H(\pi:\psi)+\lambda r,A(r))\} \cup \{(\lambda,H(\pi:\psi),f,r,s) : s \in Foe(\alpha,f,r)\}$
$= S(q_0,T[\lambda,H(\pi:\psi),f,r])$, by T4.
But $T[\alpha,H+\alpha r,A(r)]$ is an evaluation tree of $\mathcal{P}$ and $|T[\alpha,H+\alpha r,A(r)]| \leq n$. So by $Y(n)$ we have that $T[\alpha,H+\alpha r,A(r)](\pi:\psi)$ is an evaluation tree of $\mathcal{P}$. Therefore $T[\alpha,H+\alpha r,A(r)](\pi:\psi) = T[\lambda,H(\pi:\psi)+\lambda r,A(r)]$. Also for each (α,H,f,r,s) in $S(p_0,T[\alpha,H,f,r])$, $T[\alpha,H,f,r,s]$ is an evaluation tree of $\mathcal{P}$ and $|T[\alpha,H,f,r,s]| \leq n$. So by $Y(n)$ we have that $T[\alpha,H,f,r,s](\pi:\psi)$ is an evaluation tree of $\mathcal{P}$. Therefore $T[\alpha,H,f,r,s](\pi:\psi) = T[\lambda,H(\pi:\psi),f,r,s]$. Thus $T(\pi:\psi) = T[\alpha,H,f,r](\pi:\psi) = T[\lambda,H(\pi:\psi),f,r]$ which is an evaluation tree of $\mathcal{P}$.

6.4 Proof Algorithm Hierarchy

If p_0 satisfies T5 then let $Subj(p_0) = (\alpha, H, f, r, s)$. Therefore $\alpha \neq \pi'$. Since $s \in Foe(\alpha, f, r)$, $Foe(\alpha, f, r) \neq \{\}$ and so $\alpha \neq \psi'$. Therefore $\alpha \in \{\pi, \psi\}$ and so $\lambda \in \{\pi, \psi\}$. Now $Subj(q_0) = (\lambda, H(\pi:\psi), f, r, s)$. Since $\alpha r \in H$ iff $\lambda r \in H(\pi:\psi)$ we have $\alpha r \notin H$ iff $\lambda r \notin H(\pi:\psi)$. So q_0 satisfies T5. Recall that $S(p_0, T[\alpha, H, f, r, s])$
$= \{(\alpha, H+\alpha t, A(t)) : \alpha t \notin H$ and $t \in R_d^s[f;s]\} \cup \{-(\alpha', H+\alpha' s, A(s)) : \alpha' s \notin H\}$.
Also since $\alpha' s \in H$ iff $\lambda' s \in H(\pi:\psi)$ we have $\alpha' s \notin H$ iff $\lambda' s \notin H(\pi:\psi)$. So
$S(q_0, T[\alpha, H, f, r, s](\pi:\psi))$
$= \{(\lambda, H(\pi:\psi)+\lambda t, A(t)) : \alpha t \notin H$ and $t \in R_d^s[f;s]\} \cup$
  $\{-(\lambda', H(\pi:\psi)+\lambda' s, A(s)) : \alpha' s \notin H\}$
$= \{(\lambda, H(\pi:\psi)+\lambda t, A(t)) : \lambda t \notin H(\pi:\psi)$ and $t \in R_d^s[f;s]\} \cup$
  $\{-(\lambda', H(\pi:\psi)+\lambda' s, A(s)) : \lambda' s \notin H(\pi:\psi)\}$
$= S(q_0, T[\lambda, H(\pi:\psi), f, r, s])$, by T5.

But for each $(\alpha, H+\alpha t, A(t))$ in $S(p_0, T[\alpha, H, f, r, s])$, we have $T[\alpha, H+\alpha t, A(t)]$ is an evaluation tree of $\mathcal{P}$ and $|T[\alpha, H+\alpha t, A(t)]| \leq n$. Therefore by $Y(n)$ we have $T[\alpha, H+\alpha t, A(t)](\pi:\psi)$ is an evaluation tree of $\mathcal{P}$. Hence $T[\alpha, H+\alpha t, A(t)](\pi:\psi) = T[\lambda, H(\pi:\psi)+\lambda t, A(t)]$. Also if $-(\alpha', H+\alpha' s, A(s)) \in S(p_0, T[\alpha, H, f, r, s])$, then $T[-(\alpha', H+\alpha' s, A(s))]$ is an evaluation tree of $\mathcal{P}$ and $|T[-(\alpha', H+\alpha' s, A(s))]| \leq n$. Therefore by $Y(n)$, $T[-(\alpha', H+\alpha' s, A(s))](\pi:\psi)$ is an evaluation tree of $\mathcal{P}$. Hence $T[-(\alpha', H+\alpha' s, A(s))](\pi:\psi) = T[-(\lambda', H(\pi:\psi)+\lambda' s, A(s))]$.

Thus $T(\pi:\psi) = T[\alpha, H, f, r, s](\pi:\psi) = T[\lambda, H(\pi:\psi), f, r, s]$ which is an evaluation tree of $\mathcal{P}$.

If p_0 satisfies T6 then let $Subj(p_0) = -(\alpha', H, F)$. Then $\alpha \in \{\pi, \psi\}$ and so $\lambda \in \{\psi, \pi\}$. Hence $Subj(q_0) = -(\lambda', H(\pi:\psi), F)$. So q_0 satisfies T6. Recall that $S(p_0, T[-(\alpha', H, F)]) = \{(\alpha', H, F)\}$. Therefore
$S(q_0, T[-(\alpha', H, F)](\pi:\psi))$
$= \{(\lambda', H(\pi:\psi), F)\}$
$= S(q_0, T[-(\lambda', H(\pi:\psi), F)])$, by T6.
But $T[\alpha', H, F]$ is an evaluation tree of $\mathcal{P}$ and $|T[\alpha', H, F]| \leq n$. Therefore by $Y(n)$, $T[\alpha', H, F](\pi:\psi)$ is an evaluation tree of $\mathcal{P}$. Hence we have $T[\alpha', H, F](\pi:\psi) = T[\lambda', H(\pi:\psi), F]$. Thus $T(\pi:\psi) = T[-(\alpha', H, F)](\pi:\psi) = T[-(\lambda', H(\pi:\psi), F)]$ which is an evaluation tree of $\mathcal{P}$.

Therefore $Y(n+1)$, and hence the lemma, is proved by induction.
EndProofLemma6.4.6

The previous comparison results are summarised in the following theorem, which shows that all the proof algorithms of Propositional Plausible Logic form a linear hierarchy.

Theorem 6.4.7 (Proof Algorithm Hierarchy). Suppose $\mathcal{P} = (R, >)$ is a plausible theory.
1) $\mathcal{P}(\varphi) \subseteq \mathcal{P}(\pi) \subseteq \mathcal{P}(\psi) \subseteq \mathcal{P}(\beta) = \mathcal{P}(\beta') \subseteq \mathcal{P}(\psi') \subseteq \mathcal{P}(\pi')$.
2) If $>$ is empty then $\mathcal{P}(\varphi) \subseteq \mathcal{P}(\pi) = \mathcal{P}(\psi) \subseteq \mathcal{P}(\beta) = \mathcal{P}(\beta') = \mathcal{P}(\psi') = \mathcal{P}(\pi')$.
Proof

Suppose $\mathcal{P} = (R, >)$ is a plausible theory. By Lemma 6.2.2, $\mathcal{P}(\varphi) \subseteq \mathcal{P}(\pi)$. By Lemma 6.4.2, $\mathcal{P}(\pi) \subseteq \mathcal{P}(\psi)$. By Lemma 6.4.4(1), $\mathcal{P}(\psi) \subseteq \mathcal{P}(\beta)$. By Lemma 6.2.7,

$\mathcal{P}(\beta) = \mathcal{P}(\beta')$. By Lemma 6.4.4(2), $\mathcal{P}(\beta') \subseteq \mathcal{P}(\psi')$. By Lemma 6.2.4, $\mathcal{P}(\psi') \subseteq \mathcal{P}(\pi')$. So part (1) holds.

Part (2) holds by part (1) and Lemma 6.4.6.
EndProofTheorem6.4.7

The proof algorithm hierarchy of Theorem 6.4.7 can be visualised by the following partial order diagrams. If two algorithms are joined by a line then every formula proved by the higher algorithm can be proved by the lower algorithm. The diagram on the right shows the relationships when the priority relation is empty. There is a greater distance between the β algorithms and the ψ' algorithm. This is to indicate that the algorithms above this big gap at least plausibly prove formulas, whereas the algorithms below the big gap do not even plausibly prove formulas, they only assist their non-primed co-algorithms to plausibly prove formulas.

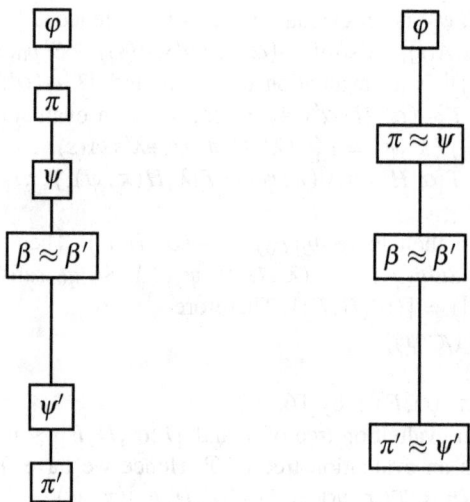

Most logics have only one proof algorithm. But Propositional Plausible Logic has 4 proof algorithms. So which algorithm should be used in a given situation? If only facts are involved then φ is the appropriated choice. If an ambiguity blocking algorithm is needed, for instance in civil law cases, then β is the appropriated choice.

But if an ambiguity propagating algorithm is needed, for instance in criminal law cases, then there are two to choose from, namely π and ψ. If the priority relation is empty then both π and ψ will give the same answer. But if the priority relation is not empty then π and ψ may give different answers. Which should be chosen?

This is a special case of the question asked in the paragraph after the diagrams. If there is no guidance provided by the situation then which proof algorithm is the right one? The linear hierarchy of proof algorithms provides 4 different non-numeric confidence levels or levels of reliability. The φ confidence level, given by the φ algorithm, is the highest confidence level. Indeed since it only reasons with facts its answers are

6.4 Proof Algorithm Hierarchy

always true. The π confidence level, given by the π algorithm, is the highest confidence level when plausible information is used. The second highest confidence level when plausible information is used is the ψ confidence level, given by the ψ algorithm. The β confidence level, given by the β algorithm, is the least reliable level when using plausible information. Nevertheless it is still reliable enough for its answers to be acted on.

So the answer to the question in the previous paragraph is to use the algorithm highest in the hierarchy that proves the formula under consideration. This will give the level of confidence that you may have in the answer. So after a formula f is proved by using the α proof algorithm, one can say that f is usually true with a confidence or reliability level of α.

Chapter 7
Examples

Abstract The sections of Chapter 7 Examples are: 7.1 Introduction; 7.2 The Non-Monotonicity Example; 7.3 The Ambiguity Puzzle; 7.4 The 3-lottery Example; 7.5 The Left Factual Disjunction Example; 7.6 The 4-lottery Example; 7.7 PPL, the Signpost Examples, and the Principles; 7.8 Priority; 7.9 Warning Rules; and 7.10 Self Attack.

Apart from Sections 7.1 and 7.7, this chapter contains detailed demonstrations of how Propositional Plausible Logic (PPL) represents and reasons with various examples. Each of the five sections from 7.2 to 7.6 demonstrates how PPL represents and reasons with one of the signpost examples from Chapter 4 Principles of Plausible Reasoning. Section 7.7 shows that PPL satisfies all the plausible reasoning principles in Chapter 4. An example requiring a non-empty priority relation is considered in Section 7.8. Section 7.9 considers two examples that each require a warning rule. A self-contradictory example is considered in the last section.

7.1 Introduction

The three kinds of rules of Propositional Plausible Logic (PPL), and the priority relation on them, make it straightforward to represent a plausible-reasoning situation. Moreover PPL has been designed so that it is straightforward to use PPL to solve plausible-reasoning problems. We shall demonstrate how to use PPL to represent and reason with problems by considering various examples. Among these examples will be all the signpost examples in Chapter 4 Principles of Plausible Reasoning.

As indicated in the paragraph just before Definition 5.3.16, we shall use the proof function P defined in Definition 5.3.16 and Theorem 5.3.17(1,2) to prove or disprove formulas. The evaluation of $P(.,.,.)$ will be given a number, just like each lemma has a number. Some evaluations are the same for several proof algorithms. So rather than repeat such evaluations we shall use a parameter, α, and give the range of the parameter after the number of the evaluation. Other shortcuts or condensations are used and it is hoped that these will be self-explanatory. If an evaluation proves or disproves a formula then this is stated after the number of the evaluation. An evaluation consists of a sequence of numbered lines. If the evaluation is using a parameter then an appropriate proof algorithm is often attached to the line number.

In this chapter there are many examples, each with its own set of axioms Ax and its own set of rules R. Using subscripts to differentiate between each different set leads to an unnecessarily cluttered notation. So we shall restrict the scope of Ax and R to its subsection, or section if there is no subsection. That is, the Ax and R of one [sub]section may be different to the Ax and R of a different [sub]section.

To save space and effort we shall use some of the theorems in Chapters 5 and 6; this will also illustrate some of the utility of these theorems. In some of the examples we shall use the following results denoted by † and □.

†) If $\alpha \in Alg$, H is an α-history, and f is a formula then, by P1,
$P(\alpha, H, \{f\}) = P(\alpha, H, f)$.

□) If $\alpha \in Alg$ and H is an α-history, then, by P1, $P(\alpha, H, \{\}) = \min\{\} = +1$.

7.2 The Non-Monotonicity Example

We shall show that the plausible proof algorithms, namely π, ψ, and β, are non-monotonic. Recall Example 4.3.2.

Cephalopods are marine animals that have tentacles; octopuses, squids, cuttlefish, and nautiluses are all cephalopods. Consider the following four statements.
1) Nautiluses are cephalopods.
2) Cephalopods usually do not have external shells.
3) Nautiluses have external shells.
4) Nancy is a cephalopod.

From these four statements it is reasonable to conclude that Nancy probably does not have an external shell. But suppose that later we discover the following statement.
5) Nancy is a nautilus.

Then it is not reasonable to conclude that Nancy probably does not have an external shell. Indeed from these five statements it is reasonable to conclude that Nancy has an external shell.

Statement (1) is a fact that can be represented by the clause $\vee\{\neg n, c\}$.
Statement (2) can be represented by the defeasible rule $\{c\} \Rightarrow \neg s$.
Statement (3) is a fact that can be represented by the clause $\vee\{\neg n, s\}$.
Statement (4) is a fact that can be represented by the clause c.
Statement (5) is a fact that can be represented by the clause n.

Our first task is to represent statements (1), (2), (3), and (4) as a plausible theory. We begin by generating the axioms from the given facts.
$Ax = CorRes(\{\vee\{\neg n, c\}, \vee\{\neg n, s\}, c\})$
$= Cor(\{\vee\{\neg n, c\}, \vee\{\neg n, s\}, c\})$
$= \{\vee\{\neg n, s\}, c\}$.
The strict rules can now be generated from Ax.

7.2 The Non-Monotonicity Example

Let r_{Ax} be $\{\} \to \wedge Ax$,
r_{ns} be $\{n\} \to s$,
$r_{\neg s \neg n}$ be $\{\neg s\} \to \neg n$, and
$r_{c \neg s}$ be $\{c\} \Rightarrow \neg s$.
The plausible theory $(R, >)$ which models (1), (2), (3), and (4) is defined as follows: $R = \{r_{Ax}, r_{ns}, r_{\neg s \neg n}, r_{c \neg s}\}$ and the priority relation $>$ is empty. We want to prove $\neg s$ is likely; but with which proof algorithm? The answer to this question is at the end of the previous chapter. Since it cannot be certain that Nancy does not have an external shell, we should try the next best proof algorithm π. To help with the following evaluation we note the following results.

†) If $\alpha \in Alg$, H is an α-history, and f is a formula then, by P1,
$$P(\alpha, H, \{f\}) = P(\alpha, H, f).$$
7.2.e1) $R_d^s[\neg s] = \{r_{c \neg s}\}$.
7.2.e2) $Foe(\pi, \neg s, r_{c \neg s}) = \{r_{ns}\}$.
7.2.e3) $R_d^s[\neg s; r_{ns}] = \{\}$.
7.2.e4) $R_d^s[n] = \{\}$.

Evaluation 7.2.1. $\pi \vdash \neg s$
1) $P(\pi, (), \neg s) = For(\pi, (), \neg s, r_{c \neg s})$ by P3, 7.2.e1,
2) $= \min\{P(\pi, (\pi r_{c \neg s}), \{c\}), Dftd(\pi, (), \neg s, r_{c \neg s}, r_{ns})\}$ by P4, 7.2.e2,
3) $= \min\{P(\pi, (\pi r_{c \neg s}), c), Dftd(\pi, (), \neg s, r_{c \neg s}, r_{ns})\}$ by †,
4) $= \min\{+1, Dftd(\pi, (), \neg s, r_{c \neg s}, r_{ns})\}$ by P2,
5) $= Dftd(\pi, (), \neg s, r_{c \neg s}, r_{ns})$
6) $= -P(\pi', (\pi' r_{ns}), \{n\})$ by P5, 7.2.e3,
7) $= -P(\pi', (\pi' r_{ns}), n)$ by †,
8) $= -\max\{\}$ by P3, 7.2.e4,
9) $= --1$
10) $= +1$.

Therefore $\pi \vdash \neg s$, and by Theorem 6.4.7(Proof Algorithm Hierarchy), $\psi \vdash \neg s$ and $\beta \vdash \neg s$.

To show that π, ψ, and β are non-monotonic we need to add statement (5) to the previous four statements, and then show that we cannot prove $\neg s$. As before we begin by generating the axioms from the given facts.
$Ax' = CorRes(\{\vee\{\neg n, c\}, \vee\{\neg n, s\}, c, n\})$
$= Cor(\{\vee\{\neg n, c\}, \vee\{\neg n, s\}, c, n, s\})$
$= \{c, n, s\}$.
This leads to only one strict rule. Let r_{cns} be $\{\} \to \wedge\{c, n, s\}$, and $r_{c \neg s}$ be $\{c\} \Rightarrow \neg s$. The plausible theory $(R', >)$ which models (1), (2), (3), (4), and (5) is defined as follows: $R' = \{r_{cns}, r_{c \neg s}\}$ and the priority relation $>$ is empty. We shall show that $\neg s$ cannot be proved by doing even more, namely proving s. This time we can use the φ proof algorithm.

Evaluation 7.2.2. $\varphi \vdash s$
1) $P(\varphi, (), s) = +1$ by P2.

So $\varphi \vdash s$ and by Theorem 6.4.7(Proof Algorithm Hierarchy), $\pi \vdash s$ and $\psi \vdash s$ and $\beta \vdash s$. However, by Theorem 6.2.9(Consistency)(1), we have $\pi \not\vdash \neg s$ and $\psi \not\vdash \neg s$ and $\beta \not\vdash \neg s$. Thus π, ψ, and β are non-monotonic and so the Non-Monotonicity Principle (Principle 4.3.4) is satisfied.

7.3 The Ambiguity Puzzle

We show that the π and ψ proof algorithms are ambiguity propagating and that the β proof algorithm is ambiguity blocking.

Recall from Example 4.9.1 that the Ambiguity Puzzle is the following.
1) a is a fact. b is a fact. c is a fact.
2) If a is true then e is likely.
3) If d is true then $\neg e$ is likely.
4) If b is true then d is likely.
5) If c is true then $\neg d$ is likely.

From (1) we generate the set of axioms $Ax = CorRes(\{a,b,c\}) = \{a,b,c\}$.
From (2) we get the rule r_{ae} where r_{ae} is $\{a\} \Rightarrow e$.
From (3) we get the rule $r_{d\neg e}$ where $r_{d\neg e}$ is $\{d\} \Rightarrow \neg e$.
From (4) we get the rule r_{bd} where r_{bd} is $\{b\} \Rightarrow d$.
From (5) we get the rule $r_{c\neg d}$ where $r_{c\neg d}$ is $\{c\} \Rightarrow \neg d$.
This can be represented by the following diagram.

Formally the plausible theory $(R,>)$ which models the Ambiguity Puzzle is defined as follows. The set of rules, R, is given by $R = \{r_{abc}, r_{ae}, r_{d\neg e}, r_{bd}, r_{c\neg d}\}$, where r_{abc} is $\{\} \rightarrow \wedge \{a,b,c\}$. The priority relation $>$ is empty.

So $R_d^s = R_d = \{r_{ae}, r_{d\neg e}, r_{bd}, r_{c\neg d}\}$, $R_d^s[d] = \{r_{bd}\}$, $R_d^s[\neg d] = \{r_{c\neg d}\}$, $R_d^s[e] = \{r_{ae}\}$, and $R_d^s[\neg e] = \{r_{d\neg e}\}$. If $l \in \{a, \neg a, b, \neg b, c, \neg c, d, \neg d, e, \neg e\}$ and $s \in R$ then $R[l;s] = \{\}$.

We need to evaluate $P(\alpha, (), e)$ for each α in $\{\pi, \psi, \beta\}$. It happens that the evaluations of $P(\pi, (), e)$ and $P(\psi, (), e)$ are the same except for the proof algorithm. Also the first 7 lines of the evaluation of $P(\beta, (), e)$ is the same as the evaluation of $P(\pi, (), e)$ except for the proof algorithm. Hence we can condense these three evaluations by using a proof algorithm parameter α.

The following results are useful.

7.4 The 3-lottery Example

†) If $\alpha \in Alg$, H is an α-history, and f is a formula then, by P1,
$P(\alpha,H,\{f\}) = P(\alpha,H,f)$.
7.3.e1) $Foe(\alpha,e,r_{ae}) = R[\neg e] = \{r_{d \neg e}\}$.
7.3.e2) $Foe(\alpha',d,r_{bd}) = \{\}$.
7.3.e3) $Foe(\beta',d,r_{bd}) = R[\neg d] = \{r_{c \neg d}\}$.

Evaluation 7.3.1. $\alpha \in \{\pi, \psi, \beta\}$
1α) $P(\alpha,(),e) = For(\alpha,(),e,r_{ae})$ by P3,
2α) $= \min\{P(\alpha,(\alpha r_{ae}),\{a\}), Dftd(\alpha,(),e,r_{ae},r_{d \neg e})\}$ by P4, 7.3.e1,
3α) $= \min\{P(\alpha,(\alpha r_{ae}),a), Dftd(\alpha,(),e,r_{ae},r_{d \neg e})\}$ by †,
4α) $= Dftd(\alpha,(),e,r_{ae},r_{d \neg e})\}$ by P2,
5α) $= -P(\alpha',(\alpha' r_{d \neg e}),\{d\})$ by P5,
6α) $= -P(\alpha',(\alpha' r_{d \neg e}),d)$ by †,
7α) $= -For(\alpha',(\alpha' r_{d \neg e}),d,r_{bd})$ by P3,

Evaluation 7.3.2. $\alpha \in \{\pi, \psi\}$ and $\alpha \not\vdash e$
7α) $P(\alpha,(),e) = -For(\alpha',(\alpha' r_{d \neg e}),d,r_{bd})$ by Evaluation 7.3.1,
8α) $= -P(\alpha',(\alpha' r_{d \neg e}, \alpha' r_{bd}),\{b\})$ by P4, 7.3.e2,
9α) $= -P(\alpha',(\alpha' r_{d \neg e}, \alpha' r_{bd}),b)$ by †,
10α) $= -1$ by P2.

Evaluation 7.3.3. $\beta \vdash e$
7β) $P(\beta,(),e) = -For(\beta',(\beta' r_{d \neg e}),d,r_{bd})$ by Evaluation 7.3.1,
8β) $= -\min\{P(\beta',(\beta' r_{d \neg e}, \beta' r_{bd}),\{b\}), Dftd(\beta',(\beta' r_{d \neg e}),d,r_{bd},r_{c \neg d})\}$
 by P4, 7.3.e3,
9β) $= -\min\{P(\beta',(\beta' r_{d \neg e}, \beta' r_{bd}),b), Dftd(\beta',(\beta' r_{d \neg e}),d,r_{bd},r_{c \neg d})\}$ by †,
10α) $= -Dftd(\beta',(\beta' r_{d \neg e}),d,r_{bd},r_{c \neg d})$ by P2,
11α) $= --P(\beta,(\beta' r_{d \neg e}, \beta r_{c \neg d}),\{c\})$ by P5,
12α) $= P(\beta,(\beta' r_{d \neg e}, \beta r_{c \neg d}),c)$ by †,
13α) $= +1$ by P2.

By Evaluation 7.3.2 and Theorem 5.3.15(Decisiveness), π and ψ cannot prove e and so they are ambiguity propagating. By Evaluation 7.3.3, β proves e and so is ambiguity blocking. Thus the Many Proof Algorithms Principle (Principle 4.9.2) is satisfied.

7.4 The 3-lottery Example

Recall the following from the 3-lottery example (Example 4.1.1).
1) Exactly one element of $\{s_1, s_2, s_3\}$ is true.
2) Each element of $\{\neg s_1, \neg s_2, \neg s_3\}$ is likely.
3) The disjunction of any pair of elements of $\{s_1, s_2, s_3\}$ is likely.

From (1) we get the following facts: $\vee\{s_1,s_2,s_3\}$, $\neg\wedge\{s_1,s_2\}$, $\neg\wedge\{s_1,s_3\}$, and $\neg\wedge\{s_2,s_3\}$. Converting these facts to clauses gives the following set C of clauses. $C = \{\vee\{s_1,s_2,s_3\}, \vee\{\neg s_1,\neg s_2\}, \vee\{\neg s_1,\neg s_3\}, \vee\{\neg s_2,\neg s_3\}\}$. From C we generate the set of axioms Ax as follows.

$Ax = CorRes(C)$
$= CorRes(\{\vee\{s_1,s_2,s_3\}, \vee\{\neg s_1,\neg s_2\}, \vee\{\neg s_1,\neg s_3\}, \vee\{\neg s_2,\neg s_3\}\})$
$= Cor(\{\vee\{s_1,s_2,s_3\}, \vee\{\neg s_1,\neg s_2\}, \vee\{\neg s_1,\neg s_3\}, \vee\{\neg s_2,\neg s_3\}, T\})$, where

every element of T is either a tautology or a superclause of a clause in C
$= \{\vee\{s_1,s_2,s_3\}, \vee\{\neg s_1,\neg s_2\}, \vee\{\neg s_1,\neg s_3\}, \vee\{\neg s_2,\neg s_3\}\}$
$= C$.

The strict rules generated from Ax are r_1 to r_{10} below. From (2) we get the defeasible rules r_{11} to r_{13} below. From (3) we get the defeasible rules r_{14} to r_{16} below.

The plausible theory $(R,>)$ which models this situation is defined as follows. The priority relation $>$ is empty, and $R = \{r_1, r_2, ..., r_{16}\}$, where $r_1: \{\} \to \wedge Ax$,

$r_2: \{\neg s_1\} \to \vee\{s_2,s_3\}$, $r_5: \{\wedge\{\neg s_2,\neg s_3\}\} \to s_1$, $r_8: \{s_1\} \to \wedge\{\neg s_2,\neg s_3\}$,
$r_3: \{\neg s_2\} \to \vee\{s_1,s_3\}$, $r_6: \{\wedge\{\neg s_1,\neg s_3\}\} \to s_2$, $r_9: \{s_2\} \to \wedge\{\neg s_1,\neg s_3\}$,
$r_4: \{\neg s_3\} \to \vee\{s_1,s_2\}$, $r_7: \{\wedge\{\neg s_1,\neg s_2\}\} \to s_3$, $r_{10}: \{s_3\} \to \wedge\{\neg s_1,\neg s_2\}$,

$r_{11}: \{\} \Rightarrow \neg s_1$, $r_{14}: \{\} \Rightarrow \vee\{s_1,s_2\}$,
$r_{12}: \{\} \Rightarrow \neg s_2$, $r_{15}: \{\} \Rightarrow \vee\{s_1,s_3\}$,
$r_{13}: \{\} \Rightarrow \neg s_3$, $r_{16}: \{\} \Rightarrow \vee\{s_2,s_3\}$.

Let $U = \{\neg s_1, \neg s_2, \vee\{s_1,s_2\}\}$. If $\alpha \in \{\pi, \psi, \beta\}$ then we show
▷ α proves each element of U,
▷ α cannot prove the negation of each element of U, and
▷ α cannot prove $\wedge\{\neg s_1, \neg s_2\}$.

The following results are used in the next evaluation.
†) If $\alpha \in Alg$, H is an α-history, and f is a formula then, by P1,
$P(\alpha, H, \{f\}) = P(\alpha, H, f)$.
□) If $\alpha \in Alg$ and H is an α-history, then, by P1, $P(\alpha, H, \{\}) = \min\{\} = +1$.
7.4.e1) $R_d^s[\neg s_1] = \{r_2, r_6, r_7, r_9, r_{10}, r_{11}, r_{16}\}$.
7.4.e2) $Foe(\pi, \neg s_1, r_{11}) = R[s_1] = \{r_5, r_8\}$.
7.4.e3) If f is a formula and r is a rule then $R_d^s[f;r] = \{\}$.
7.4.e4) $R_d^s[\wedge\{\neg s_2, \neg s_3\}] = \{r_5, r_8\}$.
7.4.e5) If f is a formula and r is a rule then $Foe(\pi', f, r) = \{\}$.
7.4.e6) $R_d^s[s_1] = \{r_5, r_8\}$.
We start by showing that $\pi \vdash \neg s_1$.

Evaluation 7.4.1. $\pi \vdash \neg s_1$
1) $P(\pi, (), \neg s_1) = \max\{For(\pi, (), \neg s_1, r_i) : i \in \{2, 6, 7, 9, 10, 11, 16\}\}$ by P3, 7.4.e1.
2) $For(\pi, (), \neg s_1, r_{11})$
$= \min\{P(\pi, (\pi r_{11}), \{\}), Dftd(\pi, (), \neg s_1, r_{11}, r_5), Dftd(\pi, (), \neg s_1, r_{11}, r_8)\}$
by P4, 7.4.e2,
3) $= \min\{-P(\pi', (\pi' r_5), \wedge\{\neg s_2, \neg s_3\}), -P(\pi', (\pi' r_8), s_1)\}$ by □, P5, 7.4.e3, †.
4) $P(\pi', (\pi' r_5), \wedge\{\neg s_2, \neg s_3\}) = For(\pi', (\pi' r_5), \wedge\{\neg s_2, \neg s_3\}, r_8)$ by P3, 7.4.e4,
5) $= P(\pi', (\pi' r_5, \pi' r_8), s_1)$ by P4, †, 7.4.e5,

6) $= \max\{\}$ by P3, 7.4.e6,
7) $= -1$.
8) $\therefore For(\pi,(),\neg s_1,r_{11}) = -P(\pi',(\pi'r_8),s_1)$ by (7) to (2),
9) $= -For(\pi',(\pi'r_8),s_1,r_5)$ by P3, 7.4.e6,
10) $= -P(\pi',(\pi'r_8,\pi'r_5),\wedge\{\neg s_2,\neg s_3\})$ by P4, †, 7.4.e5,
11) $= -\max\{\}$ by P3, 7.4.e4,
12) $= +1$.
13) $\therefore P(\pi,(),\neg s_1) = +1$ by (12) to (8), and (1).

Because the 3-lottery example is symmetric in s_1, s_2, and s_3, a very similar evaluation gives $P(\pi,(),\neg s_2) = +1$ and $P(\pi,(),\neg s_3) = +1$. Hence by †, $P(\pi,(),\{\neg s_3\}) = +1$. By Theorem 5.3.17, $\pi \vdash \neg s_1$, $\pi \vdash \neg s_2$, and $\pi \vdash \{\neg s_3\}$. By using r_4 and Theorem 6.1.2(3)(Modus Ponens for strict rules), we get $\pi \vdash \vee\{s_1,s_2\}$. Thus π proves each element in $U = \{\neg s_1, \neg s_2, \vee\{s_1,s_2\}\}$.

Suppose $\alpha \in \{\pi, \psi, \beta\}$. Then by Theorem 6.4.7(Proof Algorithm Hierarchy), α proves each element of U. By Theorem 6.2.9(Consistency)(1), the negation of each element of U cannot be proved by α. Hence by Theorem 6.1.2(2)(Right Weakening), α cannot prove $\wedge\{\neg s_1, \neg s_2\}$.

Thus π, ψ, and β are not conjunctive and so the Non-Conjunctive Principle (Principle 4.4.2) is satisfied. Moreover π, ψ, and β are not 3-consistent and so the Non-3-Consistency Principle (Principle 4.8.4) is satisfied.

7.5 The Left Factual Disjunction Example

Recall the following from the Left Factual Disjunction example (Example 4.5.4).
Let S be a plausible-structure that models the 7-lottery based on $[1..7]$.
Let g be $\wedge\{\neg s_1, \neg s_2\}$ and $S+g = (Fact(S) \cup \{g\}, Plaus(S))$.
Let h be $\wedge\{\neg s_3, \neg s_4\}$ and $S+h = (Fact(S) \cup \{g\}, Plaus(S))$.
Let $S+\vee\{g,h\} = (Fact(S) \cup \{\vee\{g,h\}\}, Plaus(S))$.
Let f be $\vee\{s_5, s_6, s_7\}$.
Then we have the following.
1) In S exactly one element of $\{s_1,s_2,s_3,s_4,s_5,s_6,s_7\}$ is true. Hence in S, f is unlikely.
2) In $S+g$ exactly one element of $\{s_3,s_4,s_5,s_6,s_7\}$ is true. Hence in $S+g$, f is likely.
3) In $S+h$ exactly one element of $\{s_1,s_2,s_5,s_6,s_7\}$ is true. Hence in $S+h$, f is likely. But $Fact(S) \equiv Fact(S) \cup \{\vee\{g,h\}\}$. Therefore in $S+\vee\{g,h\}$ exactly one element of $\{s_1,s_2,s_3,s_4,s_5,s_6,s_7\}$ is true. Hence in $S+\vee\{g,h\}$, f is unlikely.

We shall translate this example into Propositional Plausible Logic (PPL) and then prove that the plausible proof algorithms, namely π, ψ, and β, are not left factually disjunctive. This will show that PPL satisfies Principle 4.5.5 the Not Left Factually Disjunctive Principle.

7.5.0.1 The 7-Lottery

The set of atoms of the alphabet used to describe the 7-lottery based on $[1..7]$ is $\{s_1, s_2, s_3, s_4, s_5, s_6, s_7\}$. So let $A = \{s_1, s_2, s_3, s_4, s_5, s_6, s_7\}$. The characteristic property of the 7-lottery is that exactly one element of A is true. A formula that has this property is the conjunction of the following 22 formulas: $\vee A$, and $\neg \wedge \{s_i, s_j\}$, where $1 \leq i < j \leq 7$. First we convert the 21 non-clauses, that characterise the "at most one" part of the 7-lottery, into the following set, C_{1-7}, of 21 clauses.
$C_{1-7} = \{$
$\vee\{\neg s_1, \neg s_2\}, \vee\{\neg s_1, \neg s_3\}, \vee\{\neg s_1, \neg s_4\}, \vee\{\neg s_1, \neg s_5\}, \vee\{\neg s_1, \neg s_6\}, \vee\{\neg s_1, \neg s_7\},$
$\vee\{\neg s_2, \neg s_3\}, \vee\{\neg s_2, \neg s_4\}, \vee\{\neg s_2, \neg s_5\}, \vee\{\neg s_2, \neg s_6\}, \vee\{\neg s_2, \neg s_7\},$
$\vee\{\neg s_3, \neg s_4\}, \vee\{\neg s_3, \neg s_5\}, \vee\{\neg s_3, \neg s_6\}, \vee\{\neg s_3, \neg s_7\},$
$\vee\{\neg s_4, \neg s_5\}, \vee\{\neg s_4, \neg s_6\}, \vee\{\neg s_4, \neg s_7\},$
$\vee\{\neg s_5, \neg s_6\}, \vee\{\neg s_5, \neg s_7\},$
$\vee\{\neg s_6, \neg s_7\}\}.$

The set of axioms Ax is generated as follows.
$Ax = CorRes(\{\vee A\} \cup C_{1-7})$
$= Cor(\{\vee A\} \cup C_{1-7} \cup T)$, where every element of T is either a tautology
$\phantom{= Cor(\{\vee A\} \cup C_{1-7} \cup T),}$ or a superclause of a clause in C_{1-7}
$= \{\vee A\} \cup C_{1-7}.$

Let r_{Ax} be $\{\} \to \wedge Ax$.
For each i in $[1..7]$ let r_i be $\{s_i\} \to \wedge(\neg A - \{\neg s_i\})$.
If $\{\} \subset K \subset A$ let $r_{ns(\vee K)}$ be $\{ns(\wedge \neg(A - K))\} \to ns(\vee K)$.
Let r_{12} be $\{\neg s_1, \neg s_2\} \Rightarrow \vee\{s_5, s_6, s_7\}$.
Let r_{34} be $\{\neg s_3, \neg s_4\} \Rightarrow \vee\{s_5, s_6, s_7\}$.

The defeasible rule r_{12} says that if we know that 1 was not selected and we know that 2 was not selected then it is likely that one of 5 or 6 or 7 was selected. Similarly, r_{34} says that if we know that 3 was not selected and we know that 4 was not selected then it is likely that one of 5 or 6 or 7 was selected.

The plausible theory $(R, >)$ which models S is defined as follows.
The priority relation $>$ is empty, and
$R = \{r_{Ax}\} \cup \{r_i : i \in [1..7]\} \cup \{r_{ns(\vee K)} : \{\} \subset K \subset A\} \cup \{r_{12}, r_{34}\}.$

We shall show (after Evaluation 7.5.6) that $\vee\{s_5, s_6, s_7\}$, that is f, is not provable by π or ψ or β. However, before that it will useful to have the following preparatory results.

†) If $\alpha \in Alg$, H is an α-history, and f' is a formula then, by P1,
$P(\alpha, H, \{f'\}) = P(\alpha, H, f').$

7.5.e1) $R_d^s[s_i] = \{r_i, r_{s_i}\}.$
7.5.e2) $R_d^s[\wedge(\neg A - \{\neg s_i\})] = R_d^s[s_i] = \{r_i, r_{s_i}\}$ by Lemma 5.2.14(3).

Rather than state a lemma and then claim that the following evaluation proves the lemma, we shall combine the two by stating the lemma after the number of the evaluation. Recall that the concatenation of a finite sequence H onto the left of another sequence H' is denoted by H++H'. Adding or appending an element h onto the right end of a finite sequence H is denoted by $H+h$ and is defined by $H+h = H$++(h).

7.5 The Left Factual Disjunction Example

Evaluation 7.5.1. If H is a β-history and $i \in [1..7]$ then $P(\beta, H, s_i) = -1$.
1) $P(\beta, H, s_i) = \max\{For(\beta, H, s_i, r_j) : j \in \{i, s_i\} \text{ and } \beta r_j \notin H\}$ by P3, 7.5.e1.
2) $For(\beta, H, s_i, r_i) = \min[\{P(\beta, H+\beta r_i, \{s_i\})\} \cup \{...\}]$ by P4,
3) $= \min[\{P(\beta, H+\beta r_i, s_i)\} \cup \{...\}]$ by †.
4) $P(\beta, H+\beta r_i, s_i) = \max\{For(\beta, H+\beta r_i, s_i, r_{s_i}) : \beta r_{s_i} \notin H+\beta r_i\}$ by P3, 7.5.e1.
5) $For(\beta, H+\beta r_i, s_i, r_{s_i}) = \min[\{P(\beta, H++(\beta r_i, \beta r_{s_i}), A(r_{s_i}))\} \cup \{...\}]$ by P4,
6) $= \min[\{P(\beta, H++(\beta r_i, \beta r_{s_i}), \wedge(\neg A - \{\neg s_i\}))\} \cup \{...\}]$ by †.
7) $P(\beta, H++(\beta r_i, \beta r_{s_i}), \wedge(\neg A - \{\neg s_i\})) = \max\{\}$ by P3, 7.5.e2,
8) $= -1$.
9) $\therefore For(\beta, H+\beta r_i, s_i, r_{s_i}) = -1$ by (7) to (5).
10) $\therefore P(\beta, H+\beta r_i, s_i) = -1$ by (9) and (4).
11) $\therefore P(\beta, H, s_i) = \max\{For(\beta, H, s_i, r_{s_i}) : \beta r_{s_i} \notin H\}$ by (10) and (3) to (1).
12) $For(\beta, H, s_i, r_{s_i}) = \min[\{P(\beta, H+\beta r_{s_i}, A(r_{s_i}))\} \cup \{...\}]$ by P4,
13) $= \min[\{P(\beta, H+\beta r_{s_i}, \wedge(\neg A - \{\neg s_i\}))\} \cup \{...\}]$ by †.
14) $P(\beta, H+\beta r_{s_i}, \wedge(\neg A - \{\neg s_i\}))$
    $= \max\{For(\beta, H+\beta r_{s_i}, \wedge(\neg A - \{\neg s_i\}), r_i) : \beta r_i \notin H+\beta r_{s_i}\}$ by P3, 7.5.e2.
15) $For(\beta, H+\beta r_{s_i}, \wedge(\neg A - \{\neg s_i\}), r_i)$
    $= \min[\{P(\beta, H++(\beta r_{s_i}, \beta r_i), \{s_i\})\} \cup \{...\}]$ by P4,
16) $= \min[\{P(\beta, H++(\beta r_{s_i}, \beta r_i), s_i)\} \cup \{...\}]$ by †.
17) $P(\beta, H++(\beta r_{s_i}, \beta r_i), s_i) = \max\{\}$ by P3, 7.5.e1,
18) $= -1$.
19) $\therefore For(\beta, H+\beta r_{s_i}, \wedge(\neg A - \{\neg s_i\}), r_i) = -1$ by (18) to (15).
20) $\therefore P(\beta, H+\beta r_{s_i}, \wedge(\neg A - \{\neg s_i\})) = -1$ by (19) and (14).
21) $\therefore For(\beta, H, s_i, r_{s_i}) = -1$ by (20), (13), and (12).
22) $\therefore P(\beta, H, s_i) = -1$ by (21) and (11).

Evaluation 7.5.2. If H is a β-history, f' is a formula, and $j \in [1..7]$ then we have $For(\beta, H, f', r_j) = -1$.
1) $For(\beta, H, f', r_j) = \min[\{P(\beta, H+\beta r_j, \{s_j\})\} \cup \{...\}]$ by P4,
2) $= \min[\{P(\beta, H+\beta r_j, s_j)\} \cup \{...\}]$ by †,
3) $= -1$ by Evaluation 7.5.1.

The following results are used in the next evaluation.
†) If $\alpha \in Alg$, H is an α-history, and f' is a formula then, by P1,
$P(\alpha, H, \{f'\}) = P(\alpha, H, f')$.
7.5.e3) $R_d^s[\neg s_i] = \{r_1, r_2, r_3, r_4, r_5, r_6, r_7, r_{ns(\vee K_i)}\} - \{r_i\}$, where $K_i = A - \{s_i\}$.

Evaluation 7.5.3. If H is a β-history and $i \in [1..7]$ then $P(\beta, H, \neg s_i) = -1$.
1) $P(\beta, H, \neg s_i) = \max\{For(\beta, H, \neg s_i, r_j) : r_j \in R_d^s[\neg s_i] \text{ and } \beta r_j \notin H\}$ by P3, 7.5.e3.
2) $= \max\{For(\beta, H, \neg s_i, r_{ns(\vee K_i)}) : \beta r_{ns(\vee K_i)} \notin H\}$ by Evaluation 7.5.2.
3) $For(\beta, H, \neg s_i, r_{ns(\vee K_i)}) = \min[\{P(\beta, H+\beta r_{ns(\vee K_i)}, \{\neg s_i\})\} \cup \{...\}]$ by P4,
4) $= \min[\{P(\beta, H+\beta r_{ns(\vee K_i)}, \neg s_i)\} \cup \{...\}]$ by †.
5) $P(\beta, H+\beta r_{ns(\vee K_i)}, \neg s_i)$
    $= \max\{For(\beta, H+\beta r_{ns(\vee K_i)}, \neg s_i, r_j) : r_j \in R_d^s[\neg s_i] \text{ and } \beta r_j \notin H+\beta r_{ns(\vee K_i)}\}$
    by P3, 7.5.e3.

6) $= \max\{For(\beta, H+\beta r_{ns(\vee K_i)}, \neg s_i, r_j) : j \in [1..7] - \{i\}$ and $\beta r_j \notin H+\beta r_{ns(\vee K_i)}\}$,
7) $= -1$ by Evaluation 7.5.2.
8) $\therefore P(\beta, H, \neg s_i) = -1$ by (7) to (1).

Lemma 7.5.4. If $\{\} \subset B \subseteq A$ and H is a β-history then $P(\beta, H, ns(\wedge \neg B)) = -1$.
Proof
By Theorem 5.3.17(2) and the Decisiveness theorem, Theorem 5.3.15(2), we have the following.
▷ If $\alpha \in Alg$, H_0 is an α-history, and f' is a formula then
$P(\alpha, H_0, f') = +1$ iff $(\alpha, H_0) \vdash f'$ iff $T(\alpha, H_0, f') = +1$.
▷ Either $T(\alpha, H_0, f') = +1$ or $T(\alpha, H_0, f') = -1$ but not both.
We shall use these results without explicit reference.
Suppose $\{\} \subset B \subseteq A$ and H is any β-history. Take any s_i in B. Then we have $ns(\wedge \neg B) \models \neg s_i$. By Evaluation 7.5.3, $P(\beta, H, \neg s_i) = -1$. However, by Right Weakening (Theorem 6.1.2(2)) we have, if $P(\beta, H, ns(\wedge \neg B)) = +1$ and $\wedge \neg B \models \neg s_i$ then $P(\beta, H, \neg s_i) = +1$. Hence $P(\beta, H, ns(\wedge \neg B)) \neq +1$ and so $P(\beta, H, ns(\wedge \neg B)) = -1$.
EndProofLemma7.5.4

Evaluation 7.5.5. If H is a β-history, f' is a formula, and $j \in \{12, 34\}$ then we have $For(\beta, H, f', r_j) = -1$.
1) $For(\beta, H, f', r_j) = \min[\{P(\beta, H+\beta r_j, \{\neg s_i, \neg s_k\})\} \cup \{...\}]$ by P4,
2) $= \min[\{P(\beta, H+\beta r_j, \neg s_i), P(\beta, H+\beta r_j, \neg s_k)\} \cup \{...\}]$ by P1,
3) $= -1$ by Evaluation 7.5.3.

Recall that $f = \vee\{s_5, s_6, s_7\}$. We now have all the preparatory results to enable us to show that f is not provable by π or ψ or β.
The following results are used in the next evaluation.
†) If $\alpha \in Alg$, H is an α-history, and f' is a formula then, by P1,
$P(\alpha, H, \{f'\}) = P(\alpha, H, f')$.
7.5.e4) $R_d^s[f] = \{r_5, r_6, r_7, r_{12}, r_{34}\} \cup \{r_{ns(\vee K)} : \{\} \subset K \subseteq \{s_5, s_6, s_7\}\}$.

Evaluation 7.5.6. $\beta \nvdash f$
1) $P(\beta, (), f) = \max\{For(\beta, (), f, r_i) : r_i \in R_d^s[f]\}$ by P3, 7.5.e4,
2) $= \max\{For(\beta, (), f, r_{ns(\vee K)}) : \{\} \subset K \subseteq \{s_5, s_6, s_7\}\}$
$\qquad$ by Evaluation 7.5.2, Evaluation 7.5.5.
3) $For(\beta, (), f, r_{ns(\vee K)}) = \min[\{P(\beta, (\beta r_{ns(\vee K)}), \{ns(\wedge \neg(A-K))\}\} \cup \{...\}]$ by P4,
4) $= \min[\{P(\beta, (\beta r_{ns(\vee K)}), ns(\wedge \neg(A-K)))\} \cup \{...\}]$ by †,
5) $= -1$ by Lemma 7.5.4.
6) $\therefore P(\beta, (), f) = -1$ by (5) to (1).

Hence by Evaluation 7.5.6 and the Proof Algorithm Hierarchy theorem (Theorem 6.4.7), $\beta \nvdash f$ and $\psi \nvdash f$ and $\pi \nvdash f$.

7.5 The Left Factual Disjunction Example

7.5.0.2 The 7-Lottery and $\wedge\{\neg s_1, \neg s_2\}$

Recall that $f = \vee\{s_5, s_6, s_7\}$ and $g = \wedge\{\neg s_1, \neg s_2\}$. In Subsubsection 7.5.0.1 we defined $A = \{s_1, s_2, s_3, s_4, s_5, s_6, s_7\}$ and $Ax = \{\vee A\} \cup C_{1-7}$ to be the 22 clauses that characterise the 7-lottery based on [1..7]. Before adding g to Ax we must convert g into its set of clauses, namely $\{\neg s_1, \neg s_2\}$. The new set of axioms will need the following sets.

$A_{3-7} = \{s_3, s_4, s_5, s_6, s_7\}$.
$C_{3-7} = \{\vee\{\neg s_3, \neg s_4\}, \vee\{\neg s_3, \neg s_5\}, \vee\{\neg s_3, \neg s_6\}, \vee\{\neg s_3, \neg s_7\},$
$\qquad \vee\{\neg s_4, \neg s_5\}, \vee\{\neg s_4, \neg s_6\}, \vee\{\neg s_4, \neg s_7\},$
$\qquad \vee\{\neg s_5, \neg s_6\}, \vee\{\neg s_5, \neg s_7\},$
$\qquad \vee\{\neg s_6, \neg s_7\}\}$.

From $Ax \cup \{\neg s_1, \neg s_2\}$ we generate the set of axioms Axg as follows.
$Axg = CorRes(\{\vee A, \neg s_1, \neg s_2\} \cup C_{1-7})$
$= Cor(\{\vee A_{3-7}, \neg s_1, \neg s_2\} \cup C_{1-7} \cup T_g)$, where every element of T_g is either
$\qquad$ a tautology, or a superclause of a clause in C_{1-7}, or a superclause of $\vee A_{3-7}$
$= \{\vee A_{3-7}, \neg s_1, \neg s_2\} \cup C_{3-7}$.

Not surprisingly $\{\vee A_{3-7}\} \cup C_{3-7}$ is the set of clauses that characterise the 5-lottery based on [3..7].

We shall now name the strict rules generated from Axg. But to avoid arbitrary or clumsy notation we shall use the following convention.

Notational Convention

The definition of each of r_i, $r_{ns(\vee K)}$, and R in this subsubsection only applies to this subsubsection. They must not be confused with different definitions of the same names in places other than this subsubsection.

$\quad$ Let r_{Axg} be $\{\} \to \wedge Axg$.
For each i in [3..7] let r_i be $\{s_i\} \to \wedge(\neg A_{3-7} - \{\neg s_i\})$.
If $\{\} \subset K \subset A_{3-7}$ let $r_{ns(\vee K)}$ be $\{ns(\wedge \neg(A_{3-7}-K))\} \to ns(\vee K)$.
Let r_{12} be $\{\neg s_1, \neg s_2\} \Rightarrow \vee\{s_5, s_6, s_7\}$.
Let r_{34} be $\{\neg s_3, \neg s_4\} \Rightarrow \vee\{s_5, s_6, s_7\}$.

$\quad$ The plausible theory $(R, >)$ which models $\mathcal{S}+g$ is defined as follows. The priority relation $>$ is empty, and
$R = \{r_{Axg}\} \cup \{r_i : i \in [3..7]\} \cup \{r_{ns(\vee K)} : \{\} \subset K \subset A_{3-7}\} \cup \{r_{12}, r_{34}\}$.

We shall show (after Evaluation 7.5.11) that $\vee\{s_5, s_6, s_7\}$, that is f, is provable by π and ψ and β. But before that it will help to have the following preparatory results.
†) If $\alpha \in Alg$, H is an α-history, and f' is a formula then, by P1,
$\qquad P(\alpha, H, \{f'\}) = P(\alpha, H, f')$.
7.5.e5) If $i \in [3..7]$ then $R^s_d[s_i] = \{r_i, r_{s_i}\}$.
7.5.e6) If f' is a formula and $r \in R - \{r_{Axg}\}$ then $Foe(\pi', f', r) = \{\}$.
7.5.e7) If $i \in [3..7]$ then $R^s_d[\wedge(\neg A_{3-7} - \{\neg s_i\})] = R^s_d[s_i] = \{r_i, r_{s_i}\}$ by
$\qquad$ Lemma 5.2.14(3).

Evaluation 7.5.7. If H is a π'-history and $i \in [3..7]$ then $P(\pi',H,s_i) = -1$.
1) $P(\pi',H,s_i) = \max\{For(\pi',H,s_i,r_j) : j \in \{i,s_i\} \text{ and } \pi'r_j \notin H\}$ by P3, 7.5.e5.
2) $For(\pi',H,s_i,r_i) = P(\pi',H+\pi'r_i,s_i)$ by P4, 7.5.e6, †,
3) $= \max\{For(\pi',H+\pi'r_i,s_i,r_{s_i}) : \pi'r_{s_i} \notin H+\pi'r_i\}$ by P3, 7.5.e5.
4) $For(\pi',H+\pi'r_i,s_i,r_{s_i}) = P(\pi',H++(\pi'r_i,\pi'r_{s_i}),\wedge(\neg A_{3-7} - \{\neg s_i\}))$
   by P4, 7.5.e6, †,
5) $= \max\{\}$ by P3, 7.5.e7,
6) $= -1$.
7) $\therefore P(\pi',H,s_i) = \max\{For(\pi',H,s_i,r_{s_i}) : \pi'r_{s_i} \notin H\}$ by (6) to (1).
8) $For(\pi',H,s_i,r_{s_i}) = P(\pi',H+\pi'r_{s_i},\wedge(\neg A_{3-7} - \{\neg s_i\}))$ by P4, 7.5.e6, †,
9) $= \max\{For(\pi',H+\pi'r_{s_i},\wedge(\neg A_{3-7} - \{\neg s_i\}),r_i) : \pi'r_i \notin H+\pi'r_{s_i}\}$ by P3, 7.5.e7.
10) $For(\pi',H+\pi'r_{s_i},\wedge(\neg A_{3-7} - \{\neg s_i\}),r_i)$
    $= P(\pi',H++(\pi'r_{s_i},\pi'r_i),s_i)$ by P4, 7.5.e6, †,
11) $= \max\{\}$ by P3, 7.5.e5,
12) $= -1$.
13) $\therefore P(\pi',H,s_i) = -1$ by (12) to (7).

Evaluation 7.5.8. If H is a π'-history, f' is a formula, and $i \in [3..7]$ then we have $For(\pi',H,f',r_i) = -1$.
1) $For(\pi',H,f',r_i) = P(\pi',H+\pi'r_i,s_i)$ by P4, 7.5.e6, †,
2) $= -1$ by Evaluation 7.5.7.

Let $A_{3-6} = \{s_3,s_4,s_5,s_6\}$.
7.5.e8) $R_d^s[\neg s_7] = \{r_3,r_4,r_5,r_6,r_{\vee A_{3-6}}\}$.

Evaluation 7.5.9. If H is a π'-history then $P(\pi',H,\neg s_7) = -1$.
1) $P(\pi',H,\neg s_7)$
   $= \max\{For(\pi',H,\neg s_7,r) : r \in \{r_3,r_4,r_5,r_6,r_{\vee A_{3-6}}\} \text{ and } \pi'r \notin H\}$ by P3, 7.5.e8,
2) $= \max\{For(\pi',H,\neg s_7,r_{\vee A_{3-6}}) : \pi'r_{\vee A_{3-6}} \notin H\}$ by Evaluation 7.5.8.
3) $For(\pi',H,\neg s_7,r_{\vee A_{3-6}}) = P(\pi',H+\pi'r_{\vee A_{3-6}},\neg s_7)$ by P4, 7.5.e6, †,
4) $= \max\{For(\pi',H+\pi'r_{\vee A_{3-6}},\neg s_7,r) : r \in \{r_3,r_4,r_5,r_6\} \text{ and } \pi'r \notin H+\pi'r_{\vee A_{3-6}}\}$
   by P3, 7.5.e8,
5) $= -1$ by Evaluation 7.5.8.
6) $\therefore P(\pi',H,\neg s_7) = -1$ by (5) to (1).

Lemma 7.5.10. If H is a π'-history and $B \subseteq A_{3-7}$ then $P(\pi',H,\wedge(\{\neg s_7\} \cup \neg B)) = -1$.

Proof
   By Theorem 5.3.17(2) and the Decisiveness theorem, Theorem 5.3.15(2), we have the following.
▷ If $\alpha \in Alg$, H_0 is an α-history, and f' is a formula then
   $P(\alpha,H_0,f') = +1$ iff $(\alpha,H_0) \vdash f'$ iff $T(\alpha,H_0,f') = +1$.
▷ Either $T(\alpha,H_0,f') = +1$ or $T(\alpha,H_0,f') = -1$ but not both.
We shall use these results without explicit reference.
   Suppose H is a π'-history and $B \subseteq A_{3-7}$. Then $Axg \cup \{\wedge(\{\neg s_7\} \cup \neg B)\} \models \neg s_7$. So by the Strong Right Weakening, Theorem 6.1.2(1), if $(\pi',H) \vdash \wedge(\{\neg s_7\} \cup \neg B)$

7.5 The Left Factual Disjunction Example

then $(\pi', H) \vdash \neg s_7$. But by Evaluation 7.5.9, $P(\pi', H, \neg s_7) = -1$. Therefore we have $P(\pi', H, \wedge(\{\neg s_7\} \cup \neg B)) = -1$.
EndProofLemma7.5.10

Recall that $f = \vee\{s_5, s_6, s_7\}$.
7.5.e9) $R_d^s[f] = \{r_5, r_6, r_7, r_{12}, r_{34}\} \cup \{r_{ns(\vee K)} : \{\} \subset K \subseteq \{s_5, s_6, s_7\}\}$.
7.5.e10) $Foe(\pi, f, r_{12}) = R[\neg f] = R[\wedge\{\neg s_5, \neg s_6, \neg s_7\}] = R[\vee\{s_3, s_4\}]$
$= \{r_3, r_4\} \cup \{r_{ns(\vee K)} : \{\} \subset K \subseteq \{s_3, s_4\}\}$
$= \{r_3, r_4, r_{s_3}, r_{s_4}, r_{\vee\{s_3, s_4\}}\}$ by Lemma 5.2.14(3).
7.5.e11) If r is any rule then $R_d^s[f; r] = \{\}$.

Evaluation 7.5.11. $\pi \vdash f$
1) $P(\pi, (), f) = \max\{For(\pi, (), f, r) : r \in R_d^s[f]\}$ by P3, 7.5.e9.
2) $For(\pi, (), f, r_{12}) = \min[\{P(\pi, (\pi r_{12}), \{\neg s_1, \neg s_2\})\} \cup$
$\{Dftd(\pi, (), f, r_{12}, r) : r \in \{r_3, r_4, r_{s_3}, r_{s_4}, r_{\vee\{s_3, s_4\}}\}\}]$
by P4, 7.5.e10.
3) $P(\pi, (\pi r_{12}), \{\neg s_1, \neg s_2\}) = \min\{P(\pi, (\pi r_{12}), \neg s_1), P(\pi, (\pi r_{12}), \neg s_2)\}$ by P1.
4) $P(\pi, (\pi r_{12}), \neg s_1) = +1$ by P2.
5) $P(\pi, (\pi r_{12}), \neg s_2) = +1$ by P2.
6) $\therefore For(\pi, (), f, r_{12}) = \min\{Dftd(\pi, (), f, r_{12}, r) : r \in \{r_3, r_4, r_{s_3}, r_{s_4}, r_{\vee\{s_3, s_4\}}\}\}$
by (5) to (2).
7) $Dftd(\pi, (), f, r_{12}, r_3) = -P(\pi', (\pi' r_3), s_3)$ by P5, 7.5.e11, †,
8) $= +1$ by Evaluation 7.5.7.
9) $Dftd(\pi, (), f, r_{12}, r_4) = -P(\pi', (\pi' r_4), s_4)$ by P5, 7.5.e11, †,
10) $= +1$ by Evaluation 7.5.7.
11) $Dftd(\pi, (), f, r_{12}, r_{s_3}) = -P(\pi', (\pi' r_{s_3}), \wedge(\neg A_{3-7} - \{\neg s_3\}))$ by P5, 7.5.e11, †,
12) $= +1$ by Lemma 7.5.10.
13) $Dftd(\pi, (), f, r_{12}, r_{s_4}) = -P(\pi', (\pi' r_{s_4}), \wedge(\neg A_{3-7} - \{\neg s_4\}))$ by P5, 7.5.e11, †,
14) $= +1$ by Lemma 7.5.10.
15) $\therefore For(\pi, (), f, r_{12}) = Dftd(\pi, (), f, r_{12}, r_{\vee\{s_3, s_4\}})$ by (14) to (6),
16) $= -P(\pi', (\pi' r_{\vee\{s_3, s_4\}}), \wedge\neg(A_{3-7} - \{s_3, s_4\}))$ by P5, 7.5.e11, †,
17) $= -P(\pi', (\pi' r_{\vee\{s_3, s_4\}}), \wedge\neg\{s_5, s_6, s_7\})$
18) $= +1$ Lemma 7.5.10.
19) $\therefore P(\pi, (), f) = +1$ by (18) to (15), and (1).

Hence by Evaluation 7.5.11 and the Proof Algorithm Hierarchy theorem (Theorem 6.4.7), $\pi \vdash f$ and $\psi \vdash f$ and $\beta \vdash f$.

7.5.0.3 The 7-Lottery and $\wedge\{\neg s_3, \neg s_4\}$

Recall that $f = \vee\{s_5, s_6, s_7\}$ and $h = \wedge\{\neg s_3, \neg s_4\}$. In Subsubsection 7.5.0.1 we defined $A = \{s_1, s_2, s_3, s_4, s_5, s_6, s_7\}$ and $Ax = \{\vee A\} \cup C_{1-7}$ to be the 22 clauses that characterise the 7-lottery based on $[1..7]$. Before adding h to Ax we must convert h

into its set of clauses, namely $\{\neg s_3, \neg s_4\}$. The new set of axioms will need the following sets.

$A_{12567} = \{s_1, s_2, s_5, s_6, s_7\}$.
$C_{12567} = \{\vee\{\neg s_1, \neg s_2\}, \vee\{\neg s_1, \neg s_5\}, \vee\{\neg s_1, \neg s_6\}, \vee\{\neg s_1, \neg s_7\},$
$\qquad \vee\{\neg s_2, \neg s_5\}, \vee\{\neg s_2, \neg s_6\}, \vee\{\neg s_2, \neg s_7\},$
$\qquad \vee\{\neg s_5, \neg s_6\}, \vee\{\neg s_5, \neg s_7\},$
$\qquad \vee\{\neg s_6, \neg s_7\}\}$.

From $Ax \cup \{\neg s_3, \neg s_3\}$ we generate the set of axioms Axh as follows.
$Axh = CorRes(\{\vee A, \neg s_3, \neg s_4\} \cup C_{1-7})$
$= Cor(\{\vee A_{12567}, \neg s_3, \neg s_4\} \cup C_{1-7} \cup T_h)$, where every element of T_h is either
  a tautology, or a superclause of a clause in C_{1-7}, or a superclause of $\vee A_{12567}$
$= \{\vee A_{12567}, \neg s_3, \neg s_4\} \cup C_{12567}$.

Not surprisingly $\{\vee A_{12567}\} \cup C_{12567}$ is the set of clauses that characterise the 5-lottery based on $\{1, 2, 5, 6, 7\}$.

We shall now name the strict rules generated from Axh. But to avoid arbitrary or clumsy notation we shall use the following convention.

Notational Convention

The definition of each of r_i, $r_{ns(\vee K)}$, and R in this subsubsection only applies to this subsubsection. They must not be confused with different definitions of the same names in places other than this subsubsection.

Let r_{Axh} be $\{\} \to \wedge Axh$.
For each i in $\{1, 2, 5, 6, 7\}$ let r_i be $\{s_i\} \to \wedge(\neg A_{12567} - \{\neg s_i\})$.
If $\{\} \subset K \subset A_{12567}$ let $r_{ns(\vee K)}$ be $\{ns(\wedge \neg (A_{12567} - K))\} \to ns(\vee K)$.
Let r_{12} be $\{\neg s_1, \neg s_2\} \Rightarrow \vee\{s_5, s_6, s_7\}$.
Let r_{34} be $\{\neg s_3, \neg s_4\} \Rightarrow \vee\{s_5, s_6, s_7\}$.

The plausible theory $(R, >)$ which models $S+h$ is defined as follows. The priority relation $>$ is empty, and
$R = \{r_{Axh}\} \cup \{r_i : i \in \{1, 2, 5, 6, 7\}\} \cup \{r_{ns(\vee K)} : \{\} \subset K \subset A_{12567}\} \cup \{r_{12}, r_{34}\}$.

We shall show (after Evaluation 7.5.16) that $\vee\{s_5, s_6, s_7\}$, that is f, is provable by π and ψ and β. But before that it will help to have the following preparatory results.
†) If $\alpha \in Alg$, H is an α-history, and f' is a formula then, by P1,
$\qquad P(\alpha, H, \{f'\}) = P(\alpha, H, f')$.
7.5.e12) If $i \in \{1, 2, 5, 6, 7\}$ then $R_d^s[s_i] = \{r_i, r_{s_i}\}$.
7.5.e13) If f' is a formula and $r \in R - \{r_{Axh}\}$ then $Foe(\pi', f', r) = \{\}$.
7.5.e14) If $i \in \{1, 2, 5, 6, 7\}$ then $R_d^s[\wedge(\neg A_{12567} - \{\neg s_i\})] = R_d^s[s_i] = \{r_i, r_{s_i}\}$ by
  Lemma 5.2.14(3).

Evaluation 7.5.12. If H is a π'-history and $i \in \{1, 2, 5, 6, 7\}$ then $P(\pi', H, s_i) = -1$.
1) $P(\pi', H, s_i) = \max\{For(\pi', H, s_i, r_j) : j \in \{i, s_i\}$ and $\pi' r_j \notin H\}$ by P3, 7.5.e12.
2) $For(\pi', H, s_i, r_i) = P(\pi', H + \pi' r_i, s_i)$ by P4, 7.5.e13, †,
3) $= \max\{For(\pi', H + \pi' r_i, s_i, r_{s_i}) : \pi' r_{s_i} \notin H + \pi' r_i\}$ by P3, 7.5.e12.

7.5 The Left Factual Disjunction Example

4) $For(\pi',H+\pi'r_i,s_i,r_{s_i})$
   $= P(\pi',H++(\pi'r_i,\pi'r_{s_i}),\wedge(\neg A_{12567}-\{\neg s_i\}))$ by P4, 7.5.e13, †,
5) $= \max\{\}$ by P3, 7.5.e14,
6) $= -1$.
7) $\therefore P(\pi',H,s_i) = \max\{For(\pi',H,s_i,r_{s_i}) : \pi'r_{s_i} \notin H\}$ by (6) to (1).
8) $For(\pi',H,s_i,r_{s_i}) = P(\pi',H+\pi'r_{s_i},\wedge(\neg A_{12567}-\{\neg s_i\}))$ by P4, 7.5.e13, †,
9) $= \max\{For(\pi',H+\pi'r_{s_i},\wedge(\neg A_{12567}-\{\neg s_i\}),r_i) : \pi'r_i \notin H+\pi'r_{s_i}\}$ by P3, 7.5.e14.
10) $For(\pi',H+\pi'r_{s_i},\wedge(\neg A_{12567}-\{\neg s_i\}),r_i)$
    $= P(\pi',H++(\pi'r_{s_i},\pi'r_i),s_i)$ by P4, 7.5.e13, †,
11) $= \max\{\}$ by P3, 7.5.e12,
12) $= -1$.
13) $\therefore P(\pi',H,s_i) = -1$ by (12) to (7).

Evaluation 7.5.13. If H is a π'-history, f' is a formula, and $i \in \{1,2,5,6,7\}$ then $For(\pi',H,f',r_i) = -1$.
1) $For(\pi',H,f',r_i) = P(\pi',H+\pi'r_i,s_i)$ by P4, 7.5.e13, †,
2) $= -1$ by Evaluation 7.5.12.

Let $A_{1256} = \{s_1,s_2,s_5,s_6\}$.
7.5.e15) $R_d^s[\neg s_7] = \{r_1,r_2,r_5,r_6,r_{\vee A_{1256}}\}$.

Evaluation 7.5.14. If H is a π'-history then $P(\pi',H,\neg s_7) = -1$.
1) $P(\pi',H,\neg s_7)$
   $= \max\{For(\pi',H,\neg s_7,r) : r \in \{r_1,r_2,r_5,r_6,r_{\vee A_{1256}}\}$ and $\pi'r \notin H\}$ by P3, 7.5.e15,
2) $= \max\{For(\pi',H,\neg s_7,r_{\vee A_{1256}}) : \pi'r_{\vee A_{1256}} \notin H\}$ by Evaluation 7.5.13.
3) $For(\pi',H,\neg s_7,r_{\vee A_{1256}}) = P(\pi',H+\pi'r_{\vee A_{1256}},\neg s_7)$ by P4, 7.5.e13, †,
4) $= \max\{For(\pi',H+\pi'r_{\vee A_{1256}},\neg s_7,r) : r \in \{r_1,r_2,r_5,r_6\}$ and $\pi'r \notin H+\pi'r_{\vee A_{1256}}\}$
   by P3, 7.5.e15,
5) $= -1$ by Evaluation 7.5.13.
6) $\therefore P(\pi',H,\neg s_7) = -1$ by (5) to (1).

Lemma 7.5.15. If H is a π'-history and $B \subseteq A_{12567}$ then $P(\pi',H,\wedge(\{\neg s_7\} \cup \neg B)) = -1$.

Proof

By Theorem 5.3.17(2) and the Decisiveness theorem, Theorem 5.3.15(2), we have the following.
▷ If $\alpha \in Alg$, H_0 is an α-history, and f' is a formula then
   $P(\alpha,H_0,f') = +1$ iff $(\alpha,H_0) \vdash f'$ iff $T(\alpha,H_0,f') = +1$.
▷ Either $T(\alpha,H_0,f') = +1$ or $T(\alpha,H_0,f') = -1$ but not both.
We shall use these results without explicit reference.

Suppose H is a π'-history and $B \subseteq A_{12567}$. Then $Axh \cup \{\wedge(\{\neg s_7\} \cup \neg B)\} \models \neg s_7$. So by the Strong Right Weakening, Theorem 6.1.2(1), if $(\pi',H) \vdash \wedge(\{\neg s_7\} \cup \neg B)$ then $(\pi',H) \vdash \neg s_7$. But by Evaluation 7.5.14, $P(\pi',H,\neg s_7) = -1$. Therefore we have $P(\pi',H,\wedge(\{\neg s_7\} \cup \neg B)) = -1$.
EndProofLemma7.5.15

Recall that $f = \vee\{s_5, s_6, s_7\}$.

7.5.e16) $R_d^s[f] = \{r_5, r_6, r_7, r_{12}, r_{34}\} \cup \{r_{ns(\vee K)} : \{\} \subset K \subseteq \{s_5, s_6, s_7\}\}$.

7.5.e17) $Foe(\pi, f, r_{34}) = R[\neg f] = R[\wedge\{\neg s_5, \neg s_6, \neg s_7\}] = R[\vee\{s_1, s_2\}]$
$= \{r_1, r_2\} \cup \{r_{ns(\vee K)} : \{\} \subset K \subseteq \{s_1, s_2\}\}$
$= \{r_1, r_2, r_{s_1}, r_{s_2}, r_{\vee\{s_1, s_2\}}\}$ by Lemma 5.2.14(3).

7.5.e18) If r is any rule then $R_d^s[f;r] = \{\}$.

Evaluation 7.5.16. $\pi \vdash f$

1) $P(\pi, (), f) = \max\{For(\pi, (), f, r) : r \in R_d^s[f]\}$ by P3, 7.5.e16.
2) $For(\pi, (), f, r_{34}) = \min[\{P(\pi, (\pi r_{34}), \{\neg s_3, \neg s_4\})\} \cup$
$\{Dftd(\pi, (), f, r_{34}, r) : r \in \{r_1, r_2, r_{s_1}, r_{s_2}, r_{\vee\{s_1, s_2\}}\}\}]$
by P4, 7.5.e17.
3) $P(\pi, (\pi r_{34}), \{\neg s_3, \neg s_4\}) = \min\{P(\pi, (\pi r_{34}), \neg s_3), P(\pi, (\pi r_{34}), \neg s_4)\}$ by P1.
4) $P(\pi, (\pi r_{34}), \neg s_3) = +1$ by P2.
5) $P(\pi, (\pi r_{34}), \neg s_4) = +1$ by P2.
6) $\therefore For(\pi, (), f, r_{34}) = \min\{Dftd(\pi, (), f, r_{34}, r) : r \in \{r_1, r_2, r_{s_1}, r_{s_2}, r_{\vee\{s_1, s_2\}}\}\}$
by (5) to (2).
7) $Dftd(\pi, (), f, r_{34}, r_1) = -P(\pi', (\pi' r_1), s_1)$ by P5, 7.5.e18, †,
8) $= +1$ by Evaluation 7.5.12.
9) $Dftd(\pi, (), f, r_{34}, r_2) = -P(\pi', (\pi' r_2), s_2)$ by P5, 7.5.e18, †,
10) $= +1$ by Evaluation 7.5.12.
11) $Dftd(\pi, (), f, r_{34}, r_{s_1}) = -P(\pi', (\pi' r_{s_3}), \wedge(\neg A_{12567} - \{\neg s_1\}))$ by P5, 7.5.e18, †,
12) $= +1$ by Lemma 7.5.15.
13) $Dftd(\pi, (), f, r_{34}, r_{s_2}) = -P(\pi', (\pi' r_{s_2}), \wedge(\neg A_{12567} - \{\neg s_2\}))$ by P5, 7.5.e18, †,
14) $= +1$ by Lemma 7.5.15.
15) $\therefore For(\pi, (), f, r_{34}) = Dftd(\pi, (), f, r_{34}, r_{\vee\{s_1, s_2\}})$ by (14) to (6),
16) $= -P(\pi', (\pi' r_{\vee\{s_1, s_2\}}), \wedge\neg(A_{12567} - \{s_1, s_2\}))$ by P5, 7.5.e18, †,
17) $= -P(\pi', (\pi' r_{\vee\{s_1, s_2\}}), \wedge\neg\{s_5, s_6, s_7\})$
18) $= +1$ Lemma 7.5.15.
19) $\therefore P(\pi, (), f) = +1$ by (18) to (15), and (1).

Hence by Evaluation 7.5.16 and the Proof Algorithm Hierarchy theorem (Theorem 6.4.7), $\pi \vdash f$ and $\psi \vdash f$ and $\beta \vdash f$.

7.5.0.4 The 7-Lottery and $\vee\{\wedge\{\neg s_1, \neg s_2\}, \wedge\{\neg s_3, \neg s_4\}\}$

Recall that $f = \vee\{s_5, s_6, s_7\}$, $g = \wedge\{\neg s_1, \neg s_2\}$, and $h = \wedge\{\neg s_3, \neg s_4\}$. In Subsubsection 7.5.0.1 we defined $A = \{s_1, s_2, s_3, s_4, s_5, s_6, s_7\}$ and $Ax = \{\vee A\} \cup C_{1-7}$ to be the 22 clauses that characterise the 7-lottery based on [1..7]. Before adding $\vee\{g, h\}$ to Ax, we must convert $\vee\{g, h\}$ into its set of clauses, namely $Cl(\vee\{g, h\})$. It is straightforward to see that $Cl(\vee\{g, h\}) = \{\vee\{\neg s_1, \neg s_3\}, \vee\{\neg s_1, \neg s_4\}, \vee\{\neg s_2, \neg s_3\}, \vee\{\neg s_2, \neg s_4\}\}$. So $Cl(\vee\{g, h\}) \subseteq C_{1-7}$. Hence $Ax \cup Cl(\vee\{g, h\}) = Ax$.

7.6 The 4-lottery Example

Therefore the plausible theory that models $S+\vee\{g,h\}$ is that same as the plausible theory in Subsubsection 7.5.0.1 that modelled S. Hence by Evaluation 7.5.6 and the Proof Algorithm Hierarchy theorem (Theorem 6.4.7), $\beta \not\vdash f$ and $\psi \not\vdash f$ and $\pi \not\vdash f$.

7.5.0.5 Left Factual Disjunction

From Subsubsections 7.5.0.1, 7.5.0.2, 7.5.0.3, and 7.5.0.4, we see that the plausible proof algorithms of Propositional Plausible Logic (PPL), namely π, ψ, and β, are not left factually disjunctive. Thus PPL satisfies Principle 4.5.5 the Not Left Factually Disjunctive Principle.

7.6 The 4-lottery Example

We have considered four of the five signpost examples in Chapter 4. The fifth signpost example (Example 4.11.1) is based on a 4-lottery.

The set of atoms of the alphabet used to describe the 4-lottery based on $[1..4]$ is $\{s_1, s_2, s_3, s_4\}$. So let $A = \{s_1, s_2, s_3, s_4\}$. Recall $\neg A = \{\neg s_1, \neg s_2, \neg s_3, \neg s_4\}$. Then we have the following.
1) Exactly one element of A is true.
2) Each element of $\neg A$ is likely.
3) The disjunction of any 3 elements of A is likely.
4) Suppose $\{i,j\} \subseteq [1..4]$ and $i \neq j$. If $\neg s_i$ is true then $\vee(A - \{s_i, s_j\})$ is likely.

The purpose of this example is to show that there needs to be at least 3 truth values. In particular, it shows that the disjunction of any 2 different elements of A is as likely to be true as false.

In Section 5.4 a 4-valued truth theory for Propositional Plausible Logic (PPL) was defined as follows. Suppose $\mathcal{P} = (R, >)$ is a plausible theory, $\alpha \in Alg$, and f is any formula. The truth function for $\mathcal{P}$, V, from $Alg \times Fml$ to the set of plausible truth values $\{\mathbf{a}, \mathbf{t}, \mathbf{f}, \mathbf{u}\}$ is defined by V1 to V4.
V1) $V(\alpha, f) = \mathbf{a}$ iff $\alpha \vdash f$ and $\alpha \vdash \neg f$.
V2) $V(\alpha, f) = \mathbf{t}$ iff $\alpha \vdash f$ and $\alpha \not\vdash \neg f$.
V3) $V(\alpha, f) = \mathbf{f}$ iff $\alpha \not\vdash f$ and $\alpha \vdash \neg f$.
V4) $V(\alpha, f) = \mathbf{u}$ iff $\alpha \not\vdash f$ and $\alpha \not\vdash \neg f$.
Each truth value has a name: $\mathbf{a}$ is the ambiguous truth value, $\mathbf{t}$ is the usually true truth value, $\mathbf{f}$ is the usually false truth value, and $\mathbf{u}$ is the undetermined truth value.

We shall show that if $i \neq j$ then the truth value of $\vee\{s_i, s_j\}$ is undetermined. That means showing $V(\beta, \vee\{s_i, s_j\}) = \mathbf{u}$. To do this we must show that $\beta \not\vdash \vee\{s_i, s_j\}$ and that $\beta \not\vdash \neg\vee\{s_i, s_j\}$. That is, we must demonstrate that $P(\beta, (), \vee\{s_i, s_j\}) = -1$ and that $P(\beta, (), \neg\vee\{s_i, s_j\}) = -1$.

The characteristic property of the 4-lottery is that exactly one element of A is true. A formula that has this property is the conjunction of the following 7 formulas: $\vee A$, and

$\neg \wedge \{s_i, s_j\}$, where $1 \leq i < j \leq 4$. First we convert the 6 non-clauses, that characterise the "at most one" part of the 4-lottery, into the following set, C_6, of 6 clauses.
$C_6 = \{\vee\{\neg s_1, \neg s_2\}, \vee\{\neg s_1, \neg s_3\}, \vee\{\neg s_1, \neg s_4\},$
$\vee\{\neg s_2, \neg s_3\}, \vee\{\neg s_2, \neg s_4\}, \vee\{\neg s_3, \neg s_4\}\}.$

The set of axioms Ax is generated as follows.
$Ax = CorRes(\{\vee A\} \cup C_6)$
$= Cor(\{\vee A\} \cup C_6 \cup T)$, where every element of T is either a tautology
or a superclause of a clause in C_6
$= \{\vee A\} \cup C_6.$

The strict rules generated from Ax are as follows.
Let r_{Ax} be $\{\} \to \wedge Ax$.
For each i in $[1..4]$ let r_i be $\{s_i\} \to \wedge(\neg A - \{\neg s_i\})$.
If $\{\} \subset K \subset A$ let $r_{ns(\vee K)}$ be $\{ns(\wedge\neg(A - K))\} \to ns(\vee K)$.

The defeasible rules are the following.
For each i in $[1..4]$ let $r_{\neg s_i}$ be $\{\} \Rightarrow \neg s_i$.
For each i in $[1..4]$ let $r_{\neg i}$ be $\{\} \Rightarrow \vee(A - \{s_i\})$.
If $\{i, j\} \subseteq [1..4]$ and $i \neq j$ then let r_{ij} be $\{\neg s_i\} \Rightarrow \vee(A - \{s_i, s_j\})$.

The plausible theory $(R, >)$ which models this 4-lottery is defined as follows. The priority relation $>$ is empty, and
$R = \{r_{Ax}\} \cup$
$\{r_i : i \in [1..4]\} \cup$
$\{r_{ns(\vee K)} : \{\} \subset K \subset A\} \cup$
$\{r_{\neg s_i} : i \in [1..4]\} \cup$
$\{r_{\neg i} : i \in [1..4]\} \cup$
$\{r_{ij} : \{i, j\} \subseteq [1..4] \text{ and } i \neq j\}.$

Before we show $P(\beta, (), \vee\{s_i, s_j\}) = -1$, in Evaluation 7.6.4, it will help to have the following preparatory results.

†) If $\alpha \in Alg$, H is an α-history, and f' is a formula then, by P1,
$P(\alpha, H, \{f'\}) = P(\alpha, H, f').$
7.6.e1) If $i \in [1..4]$ then $R_d^s[s_i] = R[s_i] = \{r_i, r_{s_i}\}$.
7.6.e2) $R_d^s[\wedge\neg(A - \{s_i\})] = R_d^s[s_i] = \{r_i, r_{s_i}\}$ by Lemma 5.2.14(3).

Evaluation 7.6.1. If $i \in [1..4]$ and H is an β-history then $P(\beta, H, s_i) = -1$.
1) $P(\beta, H, s_i) = \max\{For(\beta, H, s_i, r_j) : j \in \{i, s_i\} \text{ and } \beta r_j \notin H\}$ by P3, 7.6.e1.
2) $For(\beta, H, s_i, r_i) = \min[\{P(\beta, H + \beta r_i, s_i)\} \cup \{...\}]$ by P4, †.
3) $P(\beta, H + \beta r_i, s_i) = \max\{For(\beta, H + \beta r_i, s_i, r_{s_i}) : \beta r_{s_i} \notin H + \beta r_i\}$ by P3, 7.6.e1.
4) $For(\beta, H + \beta r_i, s_i, r_{s_i}) = \min[\{P(\beta, H + +(\beta r_i, \beta r_{s_i}), A(r_{s_i}))\} \cup \{...\}]$ by P4,
5) $= \min[\{P(\beta, H + +(\beta r_i, \beta r_{s_i}), \wedge\neg(A - \{s_i\}))\} \cup \{...\}]$ by †.
6) $P(\beta, H + +(\beta r_i, \beta r_{s_i}), \wedge\neg(A - \{s_i\})) = \max\{\}$ by P3, 7.6.e2,
7) $= -1$.
8) $\therefore P(\beta, H, s_i) = \max\{For(\beta, H, s_i, r_{s_i}) : \beta r_{s_i} \notin H\}$ by (7) to (1).
9) $For(\beta, H, s_i, r_{s_i}) = \min[\{P(\beta, H + \beta r_{s_i}, A(r_{s_i}))\} \cup \{...\}]$ by P4,
10) $= \min[\{P(\beta, H + \beta r_{s_i}, \wedge\neg(A - \{s_i\}))\} \cup \{...\}]$ by †.
11) $P(\beta, H + \beta r_{s_i}, \wedge\neg(A - \{s_i\}))$
$= \max\{For(\beta, H + \beta r_{s_i}, \wedge\neg(A - \{s_i\}), r_i) : \beta r_i \notin H + \beta r_{s_i}\}$ by P3, 7.6.e2.

7.6 The 4-lottery Example

12) $For(\beta, H+\beta r_{s_i}, \wedge\neg(A-\{s_i\}), r_i)$
   $= \min[\{P(\beta, H++(\beta r_{s_i}, \beta r_i), s_i)\} \cup \{...\}]$ by P4, †.
13) $P(\beta, H++(\beta r_{s_i}, \beta r_i), s_i) = \max\{\}$ by P3, 7.6.e1,
14) $= -1$.
15) $\therefore P(\beta, H, s_i) = -1$ by (14) to (8).

Lemma 7.6.2. If $i \in [1..4]$ and H is an β-history then $P(\beta, H, \wedge\neg(A-\{s_i\})) = -1$.
Proof
  By Theorem 5.3.17(2) and the Decisiveness theorem, Theorem 5.3.15(2), we have the following.
▷ If $\alpha \in Alg$, H_0 is an α-history, and f' is a formula then
   $P(\alpha, H_0, f') = +1$ iff $(\alpha, H_0) \vdash f'$ iff $T(\alpha, H_0, f') = +1$.
▷ Either $T(\alpha, H_0, f') = +1$ or $T(\alpha, H_0, f') = -1$ but not both.
We shall use these results without explicit reference.
  Suppose $i \in [1..4]$ and H is an β-history. Then $Ax \cup \{\wedge\neg(A-\{s_i\})\} \models s_i$. Therefore by the Strong Right Weakening, Theorem 6.1.2(1), if $(\beta, H) \vdash \wedge\neg(A-\{s_i\})$ then $(\beta, H) \vdash s_i$. But by Evaluation 7.6.1, $P(\beta, H, s_i) = -1$. Thus $P(\beta, H, \wedge\neg(A-\{s_i\})) \neq +1$ and so $P(\beta, H, \wedge\neg(A-\{s_i\})) = -1$.
EndProofLemma7.6.2

□) If $\alpha \in Alg$ and H is an α-history, then, by P1, $P(\alpha, H, \{\}) = \min\{\} = +1$.
7.6.e3) If $i \in [1..4]$ then $R^s_d[\neg s_i] = \{r_{\neg s_i}\}$.
7.6.e4) If $i \in [1..4]$ then $Foe(\beta', \neg s_i, r_{\neg s_i}) = R[s_i] = \{r_i, r_{s_i}\}$ by 7.6.e1.
7.6.e5) If f is any formula and r is any rule then $R^s_d[f; r] = \{\}$.

Evaluation 7.6.3. If $i \in [1..4]$ and H_i is a β'-history such that $H_i \cap \{\beta' r_{\neg s_i}, \beta r_i, \beta r_{s_i}\} = \{\}$ then $P(\beta', H_i, \neg s_i) = +1$.
1) $P(\beta', H_i, \neg s_i) = For(\beta', H_i, \neg s_i, r_{\neg s_i})$ by P3, 7.6.e3,
2) $= \min[\{P(\beta', H_i + \beta' r_{\neg s_i}, \{\})\} \cup$
   $\{Dftd(\beta', H_i, \neg s_i, r_{\neg s_i}, r_i), Dftd(\beta', H_i, \neg s_i, r_{\neg s_i}, r_{s_i})\}]$ by P4, 7.6.e4,
3) $= \min\{Dftd(\beta', H_i, \neg s_i, r_{\neg s_i}, r_i), Dftd(\beta', H_i, \neg s_i, r_{\neg s_i}, r_{s_i})\}$ by □.
4) $Dftd(\beta', H_i, \neg s_i, r_{\neg s_i}, r_i) = -P(\beta, H_i + \beta r_i, s_i)$ by P5, 7.6.e5, †,
5) $= +1$ by Evaluation 7.6.1.
6) $Dftd(\beta', H_i, \neg s_i, r_{\neg s_i}, r_{s_i}) = -P(\beta, H_i + \beta r_{s_i}, \wedge\neg(A-\{s_i\}))$ by P5, 7.6.e5, †,
7) $= +1$ by Lemma 7.6.2.
8) $\therefore P(\beta', H_i, \neg s_i) = +1$ by (7) to (1).

7.6.e6) If $[1..4] = \{i, j, k, l\}$ then
   $R^s_d[\vee\{s_i, s_j\}] = R[\vee\{s_i, s_j\}] = \{r_i, r_j, r_{s_i}, r_{s_j}, r_{\vee\{s_i, s_j\}}, r_{kl}, r_{lk}\}$.
7.6.e7) If $[1..4] = \{i, j, k, l\}$ and r is any rule except r_{Ax} then
   $Foe(\beta, \vee\{s_i, s_j\}, r) = R[\neg(\vee\{s_i, s_j\})] = R[\vee\{s_k, s_l\}]$
   $= \{r_k, r_l, r_{s_k}, r_{s_l}, r_{\vee\{s_k, s_l\}}, r_{ij}, r_{ji}\}$ by Lemma 5.2.14(3) and 7.6.e6.
7.6.e8) If $[1..4] = \{i, j, k, l\}$ and r is any rule except r_{Ax} then
   $Foe(\beta, \wedge\neg(A-\{s_i, s_j\}), r) = R[\neg\wedge\neg(A-\{s_i, s_j\})] = R[\vee\{s_k, s_l\}]$
   $= \{r_k, r_l, r_{s_k}, r_{s_l}, r_{\vee\{s_k, s_l\}}, r_{ij}, r_{ji}\}$ by Lemma 5.2.14(3) and 7.6.e7.

218　　7　Examples

7.6.e9) If $[1..4] = \{i,j,k,l\}$ then
$R_d^s[\wedge\neg(A - \{s_i,s_j\})] = R[\wedge\neg\{s_k,s_l\}] = R[\vee\{s_i,s_j\}]$
$= \{r_i, r_j, r_{s_i}, r_{s_j}, r_{\vee\{s_i,s_j\}}, r_{kl}, r_{lk}\}$ by Lemma 5.2.14(3) and 7.6.e6.

Evaluation 7.6.4. If $i \neq j$ then $\beta \not\vdash \vee\{s_i,s_j\}$.

1) $P(\beta,(),\vee\{s_i,s_j\}) = \max\{For(\beta,(),\vee\{s_i,s_j\},r) : r \in R[\vee\{s_i,s_j\}]\}$
　　　　　　　　　　　　　　　　　　　　　　　　　　by P3, 7.6.e6.
2) $For(\beta,(),\vee\{s_i,s_j\},r_i) = \min[\{P(\beta,(\beta r_i),s_i)\} \cup \{...\}]$ by P4, †,
3) $= -1$ by Evaluation 7.6.1.
4) $For(\beta,(),\vee\{s_i,s_j\},r_j) = \min[\{P(\beta,(\beta r_j),s_i)\} \cup \{...\}]$ by P4, †,
5) $= -1$ by Evaluation 7.6.1.
6) $For(\beta,(),\vee\{s_i,s_j\},r_{s_i}) = \min[\{P(\beta,(\beta r_{s_i}),A(r_{s_i}))\} \cup \{...\}]$ by P4,
7) $= \min[\{P(\beta,(\beta r_{s_i}),\wedge\neg(A - \{s_i\}))\} \cup \{...\}]$ by †,
8) $= -1$ by Lemma 7.6.2.
9) $For(\beta,(),\vee\{s_i,s_j\},r_{s_j}) = \min[\{P(\beta,(\beta r_{s_j}),A(r_{s_j}))\} \cup \{...\}]$ by P4,
10) $= \min[\{P(\beta,(\beta r_{s_j}),\wedge\neg(A - \{s_j\}))\} \cup \{...\}]$ by †,
11) $= -1$ by Lemma 7.6.2.
12) $For(\beta,(),\vee\{s_i,s_j\},r_{kl}) = \min[\{...\} \cup \{Dftd(\beta,(),\vee\{s_i,s_j\},r_{kl},r) :$
　　　　　　　　$r \in \{r_k, r_l, r_{s_k}, r_{s_l}, r_{\vee\{s_k,s_l\}}, r_{ij}, r_{ji}\}\}]$ by P4, 7.6.e7.
13) $Dftd(\beta,(),\vee\{s_i,s_j\},r_{kl},r_{ij}) = -P(\beta',(\beta' r_{ij}),\neg s_i)$ by P5, 7.6.e5, †,
14) $= -1$ by Evaluation 7.6.3.
15) $For(\beta,(),\vee\{s_i,s_j\},r_{lk}) = \min[\{...\} \cup \{Dftd(\beta,(),\vee\{s_i,s_j\},r_{lk},r) :$
　　　　　　　　$r \in \{r_k, r_l, r_{s_k}, r_{s_l}, r_{\vee\{s_k,s_l\}}, r_{ij}, r_{ji}\}\}]$ by P4, 7.6.e7.
16) $Dftd(\beta,(),\vee\{s_i,s_j\},r_{lk},r_{ij}) = -P(\beta',(\beta' r_{ij}),\neg s_i)$ by P5, 7.6.e5, †,
17) $= -1$ by Evaluation 7.6.3.
18) $\therefore P(\beta,(),\vee\{s_i,s_j\}) = For(\beta,(),\vee\{s_i,s_j\},r_{\vee\{s_i,s_j\}})$ by (17) to (1),
19) $= \min[\{P(\beta,(\beta r_{\vee\{s_i,s_j\}}),\wedge\neg(A - \{s_i,s_j\}))\} \cup \{...\}]$ by P4, †.
20) $P(\beta,(\beta r_{\vee\{s_i,s_j\}}),\wedge\neg(A - \{s_i,s_j\}))$
　　$= \max\{For(\beta,(\beta r_{\vee\{s_i,s_j\}}),\wedge\neg(A - \{s_i,s_j\}),r) :$
　　　　$\beta r \notin (\beta r_{\vee\{s_i,s_j\}})$ and $r \in R_d^s[\wedge\neg(A - \{s_i,s_j\})]\}$ by P3,
21) $= \max\{For(\beta,(\beta r_{\vee\{s_i,s_j\}}),\wedge\neg(A - \{s_i,s_j\}),r) : r \in \{r_i, r_j, r_{s_i}, r_{s_j}, r_{kl}, r_{lk}\}\}$
　　　　　　　　　　　　　　　　　　　　　　　　　　by 7.6.e9.
22) $For(\beta,(\beta r_{\vee\{s_i,s_j\}}),\wedge\neg(A - \{s_i,s_j\}),r_i)$
　　$= \min[\{P(\beta,(\beta r_{\vee\{s_i,s_j\}},\beta r_i),s_i)\} \cup \{...\}]$ by P4, †,
23) $= -1$ by Evaluation 7.6.1.
24) $For(\beta,(\beta r_{\vee\{s_i,s_j\}}),\wedge\neg(A - \{s_i,s_j\}),r_j)$
　　$= \min[\{P(\beta,(\beta r_{\vee\{s_i,s_j\}},\beta r_j),s_j)\} \cup \{...\}]$ by P4, †,
25) $= -1$ by Evaluation 7.6.1.
26) $For(\beta,(\beta r_{\vee\{s_i,s_j\}}),\wedge\neg(A - \{s_i,s_j\}),r_{s_i})$
　　$= \min[\{P(\beta,(\beta r_{\vee\{s_i,s_j\}},\beta r_{s_i}),\wedge\neg(A - \{s_i\}))\} \cup \{...\}]$ by P4, †,
27) $= -1$ by Lemma 7.6.2.
28) $For(\beta,(\beta r_{\vee\{s_i,s_j\}}),\wedge\neg(A - \{s_i,s_j\}),r_{s_j})$
　　$= \min[\{P(\beta,(\beta r_{\vee\{s_i,s_j\}},\beta r_{s_j}),\wedge\neg(A - \{s_j\}))\} \cup \{...\}]$ by P4, †,
29) $= -1$ by Lemma 7.6.2.

30) $For(\beta,(\beta r_{\vee\{s_i,s_j\}}),\wedge\neg(A-\{s_i,s_j\}),r_{kl})$
   $= \min[\{...\} \cup \{Dftd(\beta,(\beta r_{\vee\{s_i,s_j\}}),\wedge\neg(A-\{s_i,s_j\}),r_{kl},r):$
   $r \in \{r_k,r_l,r_{s_k},r_{s_l},r_{\vee\{s_k,s_l\}},r_{ij},r_{ji}\}\}]$ by P4, 7.6.e8.
31) $Dftd(\beta,(\beta r_{\vee\{s_i,s_j\}}),\wedge\neg(A-\{s_i,s_j\}),r_{kl},r_{ij})$
   $= -P(\beta',(\beta r_{\vee\{s_i,s_j\}},\beta' r_{ij}),\neg s_i)$ by P5, 7.6.e5, †,
32) $= -1$ by Evaluation 7.6.3.
33) $For(\beta,(\beta r_{\vee\{s_i,s_j\}}),\wedge\neg(A-\{s_i,s_j\}),r_{lk})$
   $= \min[\{...\} \cup \{Dftd(\beta,(\beta r_{\vee\{s_i,s_j\}}),\wedge\neg(A-\{s_i,s_j\}),r_{lk},r):$
   $r \in \{r_k,r_l,r_{s_k},r_{s_l},r_{\vee\{s_k,s_l\}},r_{ij},r_{ji}\}\}]$ by P4, 7.6.e8.
34) $Dftd(\beta,(\beta r_{\vee\{s_i,s_j\}}),\wedge\neg(A-\{s_i,s_j\}),r_{lk},r_{ij})$
   $= -P(\beta',(\beta r_{\vee\{s_i,s_j\}},\beta' r_{ij}),\neg s_i)$ by P5, 7.6.e5, †,
35) $= -1$ by Evaluation 7.6.3.
36) $\therefore P(\beta,(),\vee\{s_i,s_j\}) = -1$ by (35) to (18).

Lemma 7.6.5. If $\{i,j\} \subseteq [1..4]$ and $i \neq j$ then $P(\beta,(),\neg\vee\{s_i,s_j\}) = -1$.
Proof

By Theorem 5.3.17(2) and the Decisiveness theorem, Theorem 5.3.15(2), we have the following.
▷ If $\alpha \in Alg$, H_0 is an α-history, and f' is a formula then
  $P(\alpha,H_0,f') = +1$ iff $(\alpha,H_0) \vdash f'$ iff $T(\alpha,H_0,f') = +1$.
▷ Either $T(\alpha,H_0,f') = +1$ or $T(\alpha,H_0,f') = -1$ but not both.
We shall use these results without explicit reference.

Suppose $[1..4] = \{i,j,k,l\}$. Then $Ax \cup \{\neg\vee\{s_i,s_j\}\} \models \vee\{s_k,s_l\}$. So by the Strong Right Weakening, Theorem 6.1.2(1), if $(\beta,()) \vdash \neg\vee\{s_i,s_j\}$ then $(\beta,()) \vdash \vee\{s_k,s_l\}$. But by Evaluation 7.6.4, $P(\beta,(),\vee\{s_k,s_l\}) = -1$. Thus $P(\beta,(),\neg\vee\{s_i,s_j\}) \neq +1$ and so $P(\beta,(),\neg\vee\{s_i,s_j\}) = -1$.
EndProofLemma7.6.5

By Evaluation 7.6.4 and Lemma 7.6.5, if $i \neq j$ then $P(\beta,(),\vee\{s_i,s_j\}) = -1$ and $P(\beta,(),\neg\vee\{s_i,s_j\}) = -1$. By the Proof Algorithm Hierarchy theorem (Theorem 6.4.7), if $\alpha \in \{\varphi,\pi,\psi\}$ then $P(\alpha,(),\vee\{s_i,s_j\}) = -1$ and $P(\alpha,(),\neg\vee\{s_i,s_j\}) = -1$. Therefore $V(\alpha,\vee\{s_i,s_j\}) = \mathbf{u}$; that is, the truth value of $\vee\{s_i,s_j\}$ is undetermined.

7.7 PPL, the Signpost Examples, and the Principles

We shall begin this section by showing how Propositional Plausible Logic (PPL) represents and reasons with all five signpost examples in Chapter 4 Principles of Plausible Reasoning. Then we shall show that PPL satisfies all the principles in Chapter 4.

The first signpost example, the 3-lottery example (Example 4.1.1), is considered in the 3-lottery example section (Section 7.4). The second signpost example, the Non-Monotonicity example (Example 4.3.2), is considered in the Non-Monotonicity example section (Section 7.2). The third signpost example, the Left Factual Disjunction example (Example 4.5.4), is considered in the Left Factual Disjunction example section

(Section 7.5). The fourth signpost example, the Ambiguity Puzzle (Example 4.9.1), is considered in the Ambiguity Puzzle section (Section 7.3). The fifth signpost example, the 4-lottery example (Example 4.11.1), is considered in the 4-lottery example section (Section 7.6).

Thus PPL represents and reasons with all five signpost examples in Chapter 4.

We shall now show that PPL satisfies all the principles in Chapter 4.

PPL has strict and defeasible rules, and so can represent and distinguish between factual and plausible statements. Hence the Representation Principle (Principle 4.2.1) is satisfied.

Moreover PPL does not use numbers, like probabilities, that could lead to a proved formula being more precise than the information used to derive it. So the Precision Principle (Principle 4.2.2) is satisfied.

The correspondence between the general 'plausible-structure' notation of Sections 4.1 and 4.3 and the particular notation of PPL is as follows. The plausible-structure S corresponds to the plausible description $\mathcal{P} = (R, >)$. If we let $Ax = Ax(R)$, then $Fact(S)$ corresponds to Ax, $Thm(Fact(S))$ corresponds to $\{f : Ax \models f\}$, and $Thm(\mathcal{L}, \alpha, S)$ corresponds to $\mathcal{P}(\alpha)$. For the rest of this section suppose α is in $\{\pi, \psi, \beta\}$.

The discussion in Section 5.2 up to Definition 5.2.10 shows that α satisfies the Evidence Principle (Principle 4.3.3). In Section 7.2 we showed that α satisfies the Non-Monotonicity Principle (Principle 4.3.4).

In Section 7.4 we showed that all three elements of $U = \{\neg s_1, \neg s_2, \vee\{s_1, s_2\}\}$ were α-provable; but that the conjunction $\wedge\{\neg s_1, \neg s_2\}$ was not α-provable. Thus α satisfies the Non-Conjunctive Principle (Principle 4.4.2). Theorem 6.1.1(Plausible Conjunction) shows that α satisfies the Plausibly Conjunctive Principle (Principle 4.4.4).

By Theorem 6.2.9(Consistency)(1), both s_1 and s_2 are not α-provable. Thus α satisfies the Not Right Disjunctive Principle (Principle 4.5.2). In Section 7.5 we showed that α satisfies the Not Left Factually Disjunctive Principle (Principle 4.5.5).

By I2 of Definition 5.2.10, φ and α are supraclassical. So by the remark after Principle 4.6.3, they satisfy the Plausible Supraclassicality Principle (Principle 4.6.3).

Theorem 6.1.2(Right Weakening)(1) shows that α has the Strong Right Weakening property. So by the remark after Principle 4.7.5, α satisfies the Plausible Right Weakening Principle (Principle 4.7.5).

Theorem 6.2.9(Consistency)(1) shows that α satisfies the Strong 2-Consistency Principle (Principle 4.8.3). In Section 7.4 we showed that all three elements of $U = \{\neg s_1, \neg s_2, \vee\{s_1, s_2\}\}$ were α-provable. Because U is not satisfiable, α satisfies the Non-3-Consistency Principle (Principle 4.8.4).

By I2 of Definition 5.2.10, φ is a factual proof algorithm. Section 7.3 shows that π and ψ are ambiguity propagating proof algorithms, and β is an ambiguity blocking proof algorithm. Also PPL makes the proof algorithm used explicit. Hence PPL satisfies the Many Proof Algorithms Principle (Principle 4.9.2).

Theorem 5.3.15(Decisiveness) and Theorem 5.3.17 show that φ and α satisfy the Decisiveness Principle (Principle 4.10.2).

7.8 Priority

The truth-value system given in Section 5.4 and Theorem 6.3.1 show that α satisfies the Included Middle Principle (Principle 4.11.2).

Thus PPL satisfies all the principles in Chapter 4 Principles of Plausible Reasoning.

7.8 Priority

In all the examples so far the priority relation $>$ has been empty. The following example shows the utility of the priority relation $>$.

Example 7.8.1 (The Priority Triangle Example). Consider the following four statements.
1) Cephalopods are molluscs.
2) Cephalopods usually do not have external shells.
3) Molluscs usually have external shells.
4) Celine is a cephalopod.

Statement (1) is a fact that can be represented by the clause $\vee\{\neg c, m\}$.
Statement (2) can be represented by the defeasible rule $\{c\} \Rightarrow \neg s$.
Statement (3) can be represented by the defeasible rule $\{m\} \Rightarrow s$.
Statement (4) is a fact that can be represented by the clause c.

As was mentioned at the beginning of Section 5.1, the priority relation, $>$, on rules is used to indicate the more relevant of two rules. In this example, the more specific rule, $\{c\} \Rightarrow \neg s$, is more relevant than the more general rule, $\{m\} \Rightarrow s$, when determining whether Celine has an external shell or not. Therefore it is sensible to require $\{c\} \Rightarrow \neg s > \{m\} \Rightarrow s$.

From (1) and (4) we generate the set of axioms
$Ax = CorRes(\{\vee\{\neg c, m\}, c\}) = Cor(\{\vee\{\neg c, m\}, c, m\}) = \{c, m\}$.
Let $r_{c,m}$ be $\{\} \to \wedge\{c, m\}$,
$r_{c\neg s}$ be $\{c\} \Rightarrow \neg s$, and
r_{ms} be $\{m\} \Rightarrow s$.
The plausible theory $(R, >)$ which models this example is defined as follows: $R = \{r_{c,m}, r_{c\neg s}, r_{ms}\}$ and the priority relation $>$ is defined by $r_{c\neg s} > r_{ms}$.

We want to prove that $\neg s$ is likely. To help with the next evaluation we note the following results.

†) If $\alpha \in Alg$, H is an α-history, and f is a formula then, by P1,
$P(\alpha, H, \{f\}) = P(\alpha, H, f)$.
7.8.e1) $R_d^s[\neg s] = \{r_{c\neg s}\}$.
7.8.e2) $R[s] = \{r_{ms}\}$.
7.8.e3) $Foe(\pi, \neg s, r_{c\neg s}) = \{\}$ by 7.8.e2.

Evaluation 7.8.2. $\pi \models \neg s$
1) $P(\pi, (), \neg s) = For(\pi, (), \neg s, r_{c\neg s})$ by P3, 7.8.e1,
2) $= P(\pi, (\pi r_{c\neg s}), \{c\})$ by P4, 7.8.e3,

3) $= P(\pi,(\pi r_{c\neg s}),c)$ by †,
4) $= +1$ by P2.

So $\pi \models \neg s$, and by Theorem 6.4.7(Proof Algorithm Hierarchy), we have $\psi \models \neg s$, and $\beta \models \neg s$.

7.9 Warning Rules

As was mentioned at the beginning of Section 5.1, warning rules can be used to prevent conclusions that are too risky. A special case of this is to prevent unwanted chaining. Two examples were mentioned; one involving illumination with a red light, and the other involving chaining of defeasible rules. Since none of the examples so far in this chapter have used a warning rule, we shall consider both examples in this section.

7.9.1 The Red Light Example

Example 7.9.1 (The Red Light Example). Consider the following four statements.
1) Blocks that look red usually are red.
2) Blocks that look red when illuminated by a red light might not be red.
3) B is a block that looks red.
4) B is a block that is illuminated by a red light.

Statement (1) can be represented by the defeasible rule $\{lr\} \Rightarrow r$.
Statement (2) can be represented by the warning rule $\{lr, rl\} \rightsquigarrow \neg r$.
Statement (3) is a fact that can be represented by the clause lr.
Statement (4) is a fact that can be represented by the clause rl.
Using a defeasible rule to represent (2) is wrong; because looking red in a red light is not evidence against a block being red; rather it is not sufficient evidence to conclude that a block is probably red.

From (3) and (4) we generate the set of axioms
$Ax = CorRes(\{lr, rl\}) = Cor(\{lr, rl\}) = \{lr, rl\}$.
Let r_{Ax} be $\{\} \rightarrow \wedge \{lr, rl\}$,
r_{lrr} be $\{lr\} \Rightarrow r$, and
$r_{Ax \neg r}$ be $\{lr, rl\} \rightsquigarrow \neg r$.
The rule $r_{Ax \neg r}$ is more specific than r_{lrr}, so it is reasonable to have $r_{Ax \neg r} > r_{lrr}$. However, whether $>$ is empty or is defined by $r_{Ax \neg r} > r_{lrr}$, makes no difference to the evaluations below.

The plausible theory $(R, >)$ which models this example is defined as follows: $R = \{r_{Ax}, r_{lrr}, r_{Ax \neg r}\}$ and the priority relation $>$ is defined by $r_{Ax \neg r} > r_{lrr}$.

We want to prove that both r and $\neg r$ are unlikely. To help with the next evaluation we note the following results.

7.9 Warning Rules

†) If $\alpha \in Alg$, H is an α-history, and f is a formula then, by P1,
$P(\alpha, H, \{f\}) = P(\alpha, H, f)$.
7.9.1.e1) $R_d^s[r] = \{r_{lrr}\}$.
7.9.1.e2) $R[\neg r] = \{r_{Ax\neg r}\}$.
7.9.1.e3) $Foe(\beta, r, r_{lrr}) = \{r_{Ax\neg r}\}$ by 7.8.1.e2.
7.9.1.e4) $R_d^s[r; r_{Ax\neg r}] = \{\}$.

Evaluation 7.9.2. $\beta \not\vdash r$
1) $P(\beta, (), r) = For(\beta, (), r, r_{lrr})$ by P3, 7.9.1.e1,
2) $= \min\{P(\beta, (\beta r_{lrr}), \{lr\}), Dftd(\beta, (), r, r_{lrr}, r_{Ax\neg r})\}$ by P4, 7.9.1.e3,
3) $= \min\{P(\beta, (\beta r_{lrr}), lr), -P(\beta', (\beta' r_{Ax\neg r}), \{lr, rl\})\}$ by †, P5, 7.9.1.e4,
4) $= \min\{+1, -P(\beta', (\beta' r_{Ax\neg r}), \{lr, rl\})\}$ by P2,
5) $= -P(\beta', (\beta' r_{Ax\neg r}), \{lr, rl\})\}$,
6) $= \min\{-P(\beta', (\beta' r_{Ax\neg r}), lr), -P(\beta', (\beta' r_{Ax\neg r}), rl)\}$, by P1,
7) $= \min\{-1, -1\}$, by P2,
8) $= -1$.

So $\beta \not\vdash r$, and by Theorem 6.4.7(Proof Algorithm Hierarchy), $\psi \not\vdash r$, and $\pi \not\vdash r$. Hence r is unlikely.

Evaluation 7.9.3. $\beta \not\vdash \neg r$
1) $P(\beta, (), \neg r) = \max\{\}$ by P3, $R_d^s[\neg r] = \{\}$,
2) $= -1$.

Therefore $\beta \not\vdash \neg r$, and by Theorem 6.4.7(Proof Algorithm Hierarchy), $\psi \not\vdash \neg r$, and $\pi \not\vdash \neg r$. Hence $\neg r$ is unlikely.

7.9.2 Unwanted Chaining

Example 7.9.4 (Unwanted Chaining). Let a be $x \in \{1, 2, 3, 4\}$, b be $x \in \{2, 3, 4\}$, and c be $x \in \{3, 4, 5\}$. Suppose a is true. Then we have the following four statements.
1) a is true.
2) If a then usually b.
3) If b then usually c.
4) If a then c is too risky.

Statement (1) is a fact that can be represented by the clause a.
Statement (2) can be represented by the defeasible rule $\{a\} \Rightarrow b$.
Statement (3) can be represented by the defeasible rule $\{b\} \Rightarrow c$.
Statement (4) can be represented by the warning rule $\{a\} \rightsquigarrow \neg c$.

Without (4), Propositional Plausible Logic (PPL) could prove a is likely, because it is a fact. PPL could then prove that b is likely; because, since a is likely, (2) is supportive evidence for b and there is no evidence against b. Similarly PPL could then prove that c is likely; because, since b is likely, (3) is undisputed evidence supporting c. But c is not likely, nor is $\neg c$.

Using a defeasible rule to represent (4) is wrong; because a does not usually imply $\neg c$; rather a is not sufficient evidence to conclude that c is likely.

From (1) we generate the set of axioms
$Ax = CorRes(\{a\}) = Cor(\{a\}) = \{a\}$.
Let r_a be $\{\} \to a$,
r_{ab} be $\{a\} \Rightarrow b$,
r_{bc} be $\{b\} \Rightarrow c$, and
$r_{a\neg c}$ be $\{a\} \leadsto \neg c$.

The plausible theory $(R, >)$ which models this example is defined as follows: the priority relation $>$ is empty, and $R = \{r_a, r_{ab}, r_{bc}, r_{a\neg c}\}$.

Since there is no evidence for $\neg c$, an evaluation very similar to Evaluation 7.9.3 shows that $\neg c$ is unlikely. We now prove that c is also unlikely. To help with the next evaluation we note the following results.

†) If $\alpha \in Alg$, H is an α-history, and f is a formula then, by P1,
$P(\alpha, H, \{f\}) = P(\alpha, H, f)$.
7.9.2.e1) $R^s_d[c] = \{r_{bc}\}$.
7.9.2.e2) $R[\neg c] = \{r_{a\neg c}\}$.
7.9.2.e3) $Foe(\beta, c, r_{bc}) = \{r_{a\neg c}\}$ by 7.9.2.e2.
7.9.2.e4) $R^s_d[c; r_{a\neg c}] = \{\}$.

Evaluation 7.9.5. $\beta \not\vdash c$
1) $P(\beta, (), c) = For(\beta, (), c, r_{bc})$ by P3, 7.9.2.e1,
2) $= \min\{P(\beta, (\beta r_{bc}), \{b\}), Dftd(\beta, (), c, r_{bc}, r_{a\neg c})\}$ by P4, 7.9.2.e3,
3) $Dftd(\beta, (), c, r_{bc}, r_{a\neg c}) = -P(\beta', (\beta' r_{a\neg c}), \{a\})$ by P5, 7.9.2.e4,
4) $= -P(\beta', (\beta' r_{a\neg c}), a)$ by †,
5) $= -1$ by P2.
6) ∴ $P(\beta, (), c) = -1$ by (5) to (3), and (2) to (1).

So $\beta \not\vdash c$, and by Theorem 6.4.7(Proof Algorithm Hierarchy), $\psi \not\vdash c$, and $\pi \not\vdash c$. Hence c is unlikely.

7.10 Self Attack

All the examples up until now have been intuitively straightforward. But consider the following example.

Example 7.10.1 (A Negative 3-cycle).
1) Presumably a, b, and c all hold.
2) If a then b is unlikely.
3) If b then c is unlikely.
4) If c then a is unlikely.

7.10 Self Attack

The six rules that represent this example, and their names, are as follows.

name	rule	name	rule
r_a	$\{\} \Rightarrow a$	$r_{a \neg b}$	$\{a\} \Rightarrow \neg b$
r_b	$\{\} \Rightarrow b$	$r_{b \neg c}$	$\{b\} \Rightarrow \neg c$
r_c	$\{\} \Rightarrow c$	$r_{c \neg a}$	$\{c\} \Rightarrow \neg a$

The following diagram shows a visualisation of these rules.

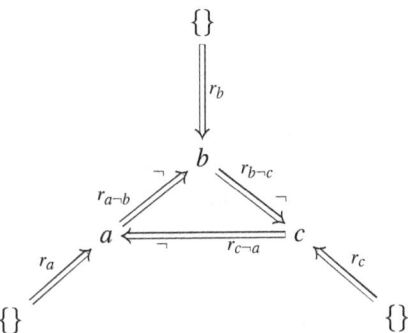

The first thing to notice is the symmetry of this example. If a or $\neg a$ is provable or disprovable then the same should hold for b and for c. The next thing to check is the provability of a.

Assume a is provable. Then there is equal evidence for and against b, and so b is not provable. This is contrary to the symmetry of the example. Continuing around the 3-cycle, we see that there is only evidence for c, since b is not provable, and so c is provable. But now there is equal evidence for and against a, and so a is not provable. This contradicts our assumption that a was provable. It also shows that a is indirect evidence against a, that is, a attacks itself. But before we conclude that a must therefore be not provable, consider the following reasoning.

Assume a is not provable. Then there is only evidence for b, and so b is provable. This is contrary to the symmetry of the example. Continuing around the 3-cycle, we see that there is equal evidence for and against c, since b is provable, and so c is not provable. But now there is only evidence for a, since c is not provable, and so a is provable. This contradicts our assumption that a was not provable.

So we have shown that, a is provable iff a is not provable. Similar reasoning applies to b and to c.

The above reasoning pattern, x iff not x, brings to mind Russell's paradox [56]. Let $R = \{s : s \notin s\}$. The defining property of R is that a set is not an element of itself. If $R \in R$ then R must satisfy the defining property of R, and so $R \notin R$. If $R \notin R$ then R satisfies the defining property of R, and so $R \in R$. Thus $R \in R$ iff $R \notin R$. This contradiction caused a change to the definition of what a set is. Do we need a similar radical change?

The above example is one of a class of examples that cause contradictions. That class may be described as fully supported negative odd-cycles. The example above is a fully supported negative 3-cycle; the supporting rules being r_a, r_b, and r_c. The fully supported negative 1-cycle contains the rule $a \Rightarrow \neg a$; which some may be tempted to exclude from the class of 'legitimate' rules. But excluding the class of fully supported negative odd-cycles seems ad hoc. Before we start excluding rules or examples, we should see if Propositional Plausible Logic (PPL) has trouble with such examples.

Fully supported negative even-cycles do not cause contradictions.

We saw above that intuitively, a is provable iff a is not provable. But does the provability of $\neg a$ cause contradictions?

Assume $\neg a$ is provable. Then there is only evidence for b, and so b is provable. As before there is now equal evidence for and against c, and so c is not provable and $\neg c$ is not provable. But now there is only evidence for a, and so a is provable and $\neg a$ is not provable. This is contrary to both symmetry and the assumption.

However no contradiction follows from just supposing that $\neg a$ is not provable. Similarly for $\neg b$ and $\neg c$. So our intuition leads us to expect that $\neg a$ is not provable, $\neg b$ is not provable, and $\neg c$ is not provable.

Let us now see what PPL does with this example.

The plausible theory $(R, >)$ which models this example is defined as follows: the priority relation $>$ is empty, and $R = \{r_a, r_b, r_c, r_{a \neg b}, r_{b \neg c}, r_{c \neg a}\}$.

We begin with the non-controversial case of $\neg a$. The following preparatory results will help with the forthcoming evaluations.

†) If $\alpha \in Alg$, H is an α-history, and f is a formula then, by P1,

$P(\alpha, H, \{f\}) = P(\alpha, H, f)$.

□) If $\alpha \in Alg$ and H is an α-history, then, by P1, $P(\alpha, H, \{\}) = \min\{\} = +1$.

7.10.e1) $R_d^s[\neg a] = R[\neg a] = \{r_{c \neg a}\}$.
7.10.e2) $R_d^s[a] = R[a] = \{r_a\}$.
7.10.e3) $Foe(\beta, \neg a, r_{c \neg a}) = R[a] = \{r_a\}$ by 7.10.e2.
7.10.e4) If f is any formula and r is any rule then $R_d^s[f; r] = \{\}$.
7.10.e5) $Foe(\beta, a, r_a) = Foe(\beta', a, r_a) = R[\neg a] = \{r_{c \neg a}\}$ by 7.10.e1.
7.10.e6) $R_d^s[c] = R[c] = \{r_c\}$.
7.10.e7) $Foe(\beta', c, r_c) = Foe(\beta, c, r_c) = R[\neg c] = \{r_{b \neg c}\}$.
7.10.e8) $R_d^s[b] = R[b] = \{r_b\}$.
7.10.e9) $Foe(\beta, b, r_b) = Foe(\beta', b, r_b) = R[\neg b] = \{r_{a \neg b}\}$.

Evaluation 7.10.2. $\beta \not\vdash \neg a$

1) $P(\beta, (), \neg a) = For(\beta, (), \neg a, r_{c \neg a})$ by P3, 7.10.e1,
2) $= \min[\{\ldots\} \cup \{Dftd(\beta, (), \neg a, r_{c \neg a}, r_a)\}]$ by P4, 7.10.e3.
3) $Dftd(\beta, (), \neg a, r_{c \neg a}, r_a) = -P(\beta', (\beta' r_a), \{\})$ by P5, 7.10.e4,
4) $= -1$ by □.
5) $\therefore P(\beta, (), \neg a) = -1$ by (4) to (1).

So $\beta \not\vdash \neg a$, and by Theorem 6.4.7(Proof Algorithm Hierarchy), $\psi \not\vdash \neg a$, $\pi \not\vdash \neg a$, and $\varphi \not\vdash \neg a$. Very similar evaluations show that $\beta \not\vdash \neg b$ and $\beta \not\vdash \neg c$. Hence $\psi \not\vdash \neg b$, $\pi \not\vdash \neg b$, $\varphi \not\vdash \neg b$, $\psi \not\vdash \neg c$, $\pi \not\vdash \neg c$, and $\varphi \not\vdash \neg c$.

We shall now consider the seemingly self contradictory case of a.

7.10 Self Attack

Evaluation 7.10.3. $\beta \not\vdash a$

1) $P(\beta,(),a) = For(\beta,(),a,r_a)$ by P3, 7.10.e2,
2) $= \min\{P(\beta,...,\{\}), Dftd(\beta,(),a,r_a,r_{c\neg a})\}$ by P4, 7.10.e5,
3) $= Dftd(\beta,(),\neg a,r_a,r_{c\neg a})$ by □,
4) $= -P(\beta',(\beta'r_{c\neg a}),c)$ by P5, 7.10.e4, †,
5) $= -For(\beta',(\beta'r_{c\neg a}),c,r_c)$ by P3, 7.10.e6,
6) $= -\min\{P(\beta',...,\{\}), Dftd(\beta',(\beta'r_{c\neg a}),c,r_c,r_{b\neg c})\}$ by P4, 7.10.e7,
7) $= -Dftd(\beta',(\beta'r_{c\neg a}),c,r_c,r_{b\neg c})$ by □,
8) $= P(\beta,(\beta'r_{c\neg a},\beta r_{b\neg c}),b)$ by P5, 7.10.e4, †,
9) $= For(\beta,(\beta'r_{c\neg a},\beta r_{b\neg c}),b,r_b)$ by P3, 7.10.e8,
10) $= \min\{P(\beta,...,\{\}), Dftd(\beta,(\beta'r_{c\neg a},\beta r_{b\neg c}),b,r_b,r_{a\neg b})\}$ by P4, 7.10.e9,
11) $= Dftd(\beta,(\beta'r_{c\neg a},\beta r_{b\neg c}),b,r_b,r_{a\neg b})$ by □,
12) $= -P(\beta',(\beta'r_{c\neg a},\beta r_{b\neg c},\beta'r_{a\neg b}),a)$ by P5, 7.10.e4, †,
13) $= -For(\beta',(\beta'r_{c\neg a},\beta r_{b\neg c},\beta'r_{a\neg b}),a,r_a)$ by P3, 7.10.e2,
14) $= -\min\{P(\beta',...,\{\}), Dftd(\beta',(\beta'r_{c\neg a},\beta r_{b\neg c},\beta'r_{a\neg b}),a,r_a,r_{c\neg a})\}$
  by P4, 7.10.e5,
15) $= -Dftd(\beta',(\beta'r_{c\neg a},\beta r_{b\neg c},\beta'r_{a\neg b}),a,r_a,r_{c\neg a})$ by □,
16) $= P(\beta,(\beta'r_{c\neg a},\beta r_{b\neg c},\beta'r_{a\neg b},\beta r_{c\neg a}),c)$ by P5, 7.10.e4, †,
17) $= For(\beta,(\beta'r_{c\neg a},\beta r_{b\neg c},\beta'r_{a\neg b},\beta r_{c\neg a}),c,r_c)$ by P3, 7.10.e6,
18) $= \min\{P(\beta,...,\{\}), Dftd(\beta,(\beta'r_{c\neg a},\beta r_{b\neg c},\beta'r_{a\neg b},\beta r_{c\neg a}),c,r_c,r_{b\neg c})\}$
  by P4, 7.10.e7,
19) $= Dftd(\beta,(\beta'r_{c\neg a},\beta r_{b\neg c},\beta'r_{a\neg b},\beta r_{c\neg a}),c,r_c,r_{b\neg c})$ by □,
20) $= -P(\beta',(\beta'r_{c\neg a},\beta r_{b\neg c},\beta'r_{a\neg b},\beta r_{c\neg a},\beta'r_{b\neg c}),b)$ by P5, 7.10.e4, †,
21) $= -For(\beta',(\beta'r_{c\neg a},\beta r_{b\neg c},\beta'r_{a\neg b},\beta r_{c\neg a},\beta'r_{b\neg c}),b,r_b)$ by P3, 7.10.e8,
22) $= -\min\{P(\beta',...,\{\}),$
    $Dftd(\beta',(\beta'r_{c\neg a},\beta r_{b\neg c},\beta'r_{a\neg b},\beta r_{c\neg a},\beta'r_{b\neg c}),b,r_b,r_{a\neg b})\}$
  by P4, 7.10.e9,
23) $= -Dftd(\beta',(\beta'r_{c\neg a},\beta r_{b\neg c},\beta'r_{a\neg b},\beta r_{c\neg a},\beta'r_{b\neg c}),b,r_b,r_{a\neg b})$ by □,
24) $= P(\beta,(\beta'r_{c\neg a},\beta r_{b\neg c},\beta'r_{a\neg b},\beta r_{c\neg a},\beta'r_{b\neg c},\beta r_{a\neg b}),a)$ by P5, 7.10.e4, †,
25) $= For(\beta,(\beta'r_{c\neg a},\beta r_{b\neg c},\beta'r_{a\neg b},\beta r_{c\neg a},\beta'r_{b\neg c},\beta r_{a\neg b}),a,r_a)$ by P3, 7.10.e2,
26) $= \min\{P(\beta,...,\{\}),$
    $Dftd(\beta,(\beta'r_{c\neg a},\beta r_{b\neg c},\beta'r_{a\neg b},\beta r_{c\neg a},\beta'r_{b\neg c},\beta r_{a\neg b}),a,r_a,r_{c\neg a})\}$
  by P4, 7.10.e5,
27) $= Dftd(\beta,(\beta'r_{c\neg a},\beta r_{b\neg c},\beta'r_{a\neg b},\beta r_{c\neg a},\beta'r_{b\neg c},\beta r_{a\neg b}),a,r_a,r_{c\neg a})$ by □,
28) $= \max\{\}$ by P5, 7.10.e4,
29) $= -1$.
30) $\therefore P(\beta,(),a) = -1$ by (22) to (1).

An examination of this evaluation shows that PPL went round the 3-cycle twice before it was in a never-ending loop. Its loop detecting mechanism then stopped the loop.

So $\beta \not\vdash a$, and by Theorem 6.4.7(Proof Algorithm Hierarchy), $\psi \not\vdash a$, $\pi \not\vdash a$, and $\varphi \not\vdash a$. Very similar evaluations show that $\beta \not\vdash b$ and $\beta \not\vdash c$. Therefore $\psi \not\vdash b$, $\pi \not\vdash b$, $\varphi \not\vdash b$, $\psi \not\vdash c$, $\pi \not\vdash c$, and $\varphi \not\vdash c$.

Therefore if $\alpha \in \{\varphi, \pi, \psi, \beta\}$ and $x \in \{a,b,c\}$ then $\alpha \not\vdash x$ and $\alpha \not\vdash \neg x$. Therefore symmetry is not contradicted. Moreover, from Definition 5.4.1(V4), we have $V(\alpha, x) = \mathbf{u} = V(\alpha, \neg x)$. Thus PPL assigns the undetermined truth value to both x and $\neg x$, which is a sensible outcome for this self-contradictory example. So there is no need for any change or exclusions, as there was for set theory.

Chapter 8
Conclusion

Abstract The sections of Chapter 8 Conclusion are: 8.1 Review, and 8.2 Future Work.

The first section reviews Chapters 2 to 7 in four subsections: 8.1.1 classical propositional logic, 8.1.2 rational reasoning, 8.1.3 plausible reasoning, and 8.1.4 Propositional Plausible Logic (PPL).

The Future Work section suggests some topics to investigate. These topics are divided into four subsections: 8.2.1 Propositional Resolution Logic, 8.2.2 rational reasoning, 8.2.3 plausible reasoning, and 8.2.4 Propositional Plausible Logic (PPL).

8.1 Review

In this section we shall review what has been accomplished in the preceding chapters.

8.1.1 Classical Propositional Logic

A version of classical propositional logic was presented in Chapter 2 Propositional Resolution Logic (PRL). A new set-based, rather than sequence-based, syntax was used that brought more benefits than difficulties. The inference mechanism was resolution, and it was noted that resolution can be applied to meets as well as clauses. Explicit rewriting rules were given that generated the usual conjunctive and disjunctive normal forms for formulas. New versions of these normal forms were defined so that each formula has exactly one stable conjunctive normal form and exactly one stable disjunctive normal form.

8.1.2 Rational Reasoning

It is well known that classical propositional logic is explosive. We regard such behaviour as irrational. Two new satisfiable subsets of an unsatisfiable set of clauses were defined in Section 3.2 to remove this irrationality. But there may be other sensible satisfiable subsets. In an attempt to encompass all such subsets, the properties of these satisfiable subsets were used to define a family of functions called rational clauses functions. From any set of formulas F, a rational clauses function RCl produces a set of clauses $RCl(F)$ by first converting F to an equivalent set of clauses $Cl(F)$ and then selecting a satisfiable subset of $Cl(F)$. We suggest that rather than attempting to prove a formula f from F, as classical propositional logic does, it is more rational to attempt to prove f from $RCl(F)$.

The relationship between tautologies and the concept of 'follows from' is considered in Subsubsection 3.3.2.

When comparing or creating logics it is useful to have a list of properties that they may have. If a logic proves a formula by only referring to a set of formulas then it is sensible to use a consequence function to state the properties. A consequence function takes a set F of formulas and returns all the consequences of F. It happens that none of the consequence functions that are associated with the rational logics above satisfy the various definitions of a consequence function that are in the literature. Even worse, there are functions that satisfy all the definitions of a consequence function in the literature but are certainly not the consequence function of any sensible logic. In Subsection 3.4 we defined what a consequence function should be and showed whether or not it had many properties.

8.1.3 Plausible Reasoning

How do you determine whether a logic does plausible reasoning or not? Chapter 4 gives a partial to answer this question. A definitive answer seems out of reach at the moment.

One way to start to answer the question is to make a list of plausible reasoning examples. If a logic does not get the right answer to all of these examples then clearly the logic does not do plausible reasoning. So these examples act as counter-examples to show that a given logic does not do plausible reasoning. Like all good counter-examples they should be as simple as possible.

Another role that a plausible reasoning example can play is to be a signpost to general principles of plausible reasoning. Five such signpost plausible reasoning examples are presented in Chapter 4, along with 15 principles of plausible reasoning. Three of the 5 signpost examples are based on a finite lottery; namely a 3-lottery, a 4-lottery, and a 7-lottery. Finite lotteries are particularly good examples as they are easy to understand and have exact answers in terms of probabilities.

8.1 Review

The principles are classified as either necessary or desirable. If a logic fails a necessary principle then it does not do plausible reasoning. If a logic fails a desirable principle then it might be regarded as doing plausible reasoning but with restrictions or faults.

One thing that the principles do show is that plausible reasoning is very different from reasoning with just facts. The Non-Conjunctive Principle (Principle 4.4.2) says that the conjunction of two plausible statements may not be plausible. The Non-3-Consistency Principle (Principle 4.8.4) says it is possible for a set of 3 plausible statements to be unsatisfiable. Most logics have only one proof algorithm. But the Many Proof Algorithms Principle (Principle 4.9.2) says there needs to be at least two different proof algorithms to cater for two different intuitions about what is plausible. So there may be more than one right answer to a plausible reasoning example. The Included Middle Principle (Principle 4.11.2) says that a logic for plausible reasoning must have at least 3 truth values. These four principles show that intuitions from factual reasoning must not be imposed on plausible reasoning unless they can be justified.

It is often suggested that logics encompassed by the term 'non-monotonic reasoning' do plausible reasoning. So in Section 4.13 it was shown that, except for Propositional Plausible Logic, every non-monotonic logic of which we are aware, either gets the 3-lottery example (Example 4.1.1) wrong, or fails at least one of the above four principles, or both.

8.1.4 Propositional Plausible Logic (PPL)

Propositional Plausible Logic (PPL) is defined in Chapter 5, investigated in Chapter 6, and shown to satisfy all the principles of Chapter 4 in Chapter 7.

PPL uses strict rules, defeasible rules, warning rules, and a priority relation on the rules to represent a plausible reasoning situation. A formula f is proved by establishing evidence for f and defeating all the evidence against f. Evidence for f is established by proving the set of antecedents of a strict or defeasible rule that has a consequent that implies f. The set of all rules that have a consequent that implies $\neg f$ is all the evidence against f.

A rule is defeated by either a team of rules or by disabling it. A rule s is disabled by showing that the set of antecedents of s cannot be proved. Let us say that the team for f consists of all the strict or defeasible rules that have a consequent that implies f. A rule s is defeated by the team for f iff there is a rule t in the team for f such that t is superior to s and the set of antecedents of t is proved.

If 'implies f' is replaced by 'is f' and 'implies $\neg f$' is replaced by 'is $\neg f$', then the fundamental ideas in the above two paragraphs are due to Nute [69, 70]. Nute used these ideas to define what he called Defeasible Logic. Defeasible Logic only used literals, had only one plausible proof algorithm (which was ambiguity blocking), could get into an endless loop, and had two different proof mechanisms, one to prove a literal, and one to prove that a literal was not provable.

PPL builds on this foundation to produce a logic that uses formulas (not just literals), has 3 plausible proof algorithms (2 ambiguity propagating and one ambiguity blocking), is decisive, and so has no need for a mechanism to prove that a formula is not provable. The details of how this is done are in Chapter 5. Section 5.4 presents a truth theory for PPL with 4 truth values; namely, **a** for ambiguous, **t** for usually true, **f** for usually false, and **u** for undetermined.

Numerous results about PPL are proved in Chapter 5 and Chapter 6. Some of these results increase one's understanding of PPL, while others show that PPL has desirable properties. Theorem 5.3.15 shows that PPL is decisive which is a most desirable property. Another desirable property is Strong Right Weakening, which PPL has by Theorem 6.1.2. Theorem 6.1.2 also shows that Modus Ponens holds for strict rules, which increases our understanding of the behaviour of PPL. The consistency principles are some of the most important, and most surprising, of the principles of plausible reasoning. In particular, the Non-3-Consistency Principle (Principle 4.8.4) says that a set of 3 or more proved formulas may be inconsistent. Theorem 6.2.9 shows that PPL is strongly 2-consistent. Perhaps uniquely, PPL has 7 proof algorithms, so it would be interesting to know how they relate to each other. Theorem 6.4.7 shows that they form a linear hierarchy.

Lemmas and theorems generally state positive results. But negative results can be just as important; these are usually demonstrated by counter-examples. Worked examples also give an impression of how PPL works, and how easy it is to do small examples by hand. Chapter 7 contains many examples worked through in detail. As shown in Section 7.7 these examples demonstrate that PPL can represent and reason with all the signpost examples in Chapter 4. Also in Section 7.7, a combination of proved results and worked examples shows that PPL satisfies all the plausible reasoning principles in Chapter 4. Sections 7.8 and 7.9 show how hitherto unused parts of PPL can be used. Section 7.10 shows that PPL produces an acceptable answer to an example that does not appear to have an intuitive answer. The more one sees how PPL answers small examples, both intuitive and non-intuitive, the more one's confidence increases in the ability of PPL to produce acceptable answers to examples so large that they are difficult to understand.

8.2 Future Work

In this section we shall suggest some topics that would be worthwhile to investigate.

8.2.1 Propositional Resolution Logic (PRL)

In Subsection 2.7.1 Rewrite Functions, six rewrite functions were defined, three were for converting a formula to conjunctive normal form and three were for converting a formula to disjunctive normal form.

One of the properties that a rewrite system may, or may not, have is the Church-Rosser property. Our rewrite system would have the Church-Rosser property if and only if the final result of applying the rewrite functions did not depend on the order in which those functions were applied.

An attempt was made to prove this, but it was taking too long and seemed too difficult. Since this result was not needed, the attempt was abandoned. However it would be interesting to know if our system does have the Church-Rosser property, or whether some modification to our system would yield a system that does have the Church-Rosser property.

8.2.2 Rational Reasoning

In Section 3.2 Four Satisfiable Subsets of a Set of Clauses, four reasonable satisfiable subsets of a, possibly unsatisfiable, set of clauses were defined. Are there any others?

In Subsection 3.3.1 Rational Clauses Functions, a rational clauses function was defined to be any function that satisfied 5 conditions. Are there other useful conditions that could help to define special rational clauses functions?

Section 3.4 Consequence Functions, contains a review of the various definitions of a consequence function that are in the literature. None are satisfactory. So we presented what we think is a satisfactory definition, and proved many of its properties. Are there other useful and satisfactory definitions of a consequence function?

8.2.3 Plausible Reasoning

In Chapter 4 Principles of Plausible Reasoning, many principles of plausible reasoning and several signpost examples were given. Are there any more principles? Are there any more significant examples? Although obvious, these questions are important and point to a substantial hole in our examination of plausible reasoning. That hole is the lack of any definition of plausible reasoning. Our principles and examples are necessary, but may well be not sufficient to characterise plausible reasoning.

A definition or characterisation of plausible reasoning is likely to be difficult. A possible approach would be to create a formal semantics of plausible reasoning. Such a semantics would need to cater for several different, but valid, intuitions. For example the ambiguity blocking or propagating intuitions are different but equally valid and useful.

It is hoped that Chapters 4 to 7 may generate wider interest in plausible reasoning now that Chapter 4 provides some requirements and Propositional Plausible Logic (PPL) provides an example. One approach may be to start with a favourite or likely logic and then see which principles and examples fail. This will provide directions for modifications to the logic.

8.2.4 Propositional Plausible Logic (PPL)

Chapters 5 to 7 define and investigate Propositional Plausible Logic (PPL). But much remains to do.

PPL was designed to be implemented. Indeed Andrew Rock has implemented many early versions of PPL. In particular the version just before PPL was an experimental version with many proof algorithms for testing. This was implemented, and since it contained all the algorithms in PPL, although with different names, in a way PPL has been implemented. But a clean implementation is still required. Moreover there is scope for parallelism in the implementation. Although an analysis of the complexity of PPL might be interesting, it is a very blunt tool and cannot tell whether a particular case is tractable or not, only an implementation of the case can tell. Creating, possibly several, implementations of PPL is probably the most important non-theoretical task to do.

After PPL is implemented all sorts of experiments can be run to see how the system performs, see [20] as an example. The output of the system also needs designing. A simple 'proved' or 'not provable' is insufficient if a reason or justification for the answer is required; but a full trace may be unwieldy. The detailed evaluations in Chapter 7 show that judicious choices were made so that dead ends were avoided. Expecting an implementation to make such choices is probably unrealistic. But it may be feasible to not output the trace of a discovered dead end.

Another approach is to implement PPL as an evaluation assistant, rather than as an autonomous system. This combines the speed and accuracy of a computer with the insight of a person. The input to the assistant would be a formula and the desired proof value. An evaluation is started, when a choice is required the computer pauses to allow the person to make a choice, then it continues until the next choice. Ideally the output would be an evaluation. How an evaluation assistant could exploit any parallelism available is another question that could be considered.

PPL is the propositional version of Plausible Logic, which is yet to be defined. It seemed sensible to get the propositional version right before attempting anything more general. It is an implementational detail to introduce 'removable variables'. That is, variables that are replaced with constants before computation begins. Apparently the introduction of non-removable variables would significantly increase the utility of PPL. Such variables would behave like the variables in Prolog. Defining Plausible Logic by adding non-removable variables to PPL is probably the most important theoretical task to do.

8.2 Future Work

Once PPL is implemented it needs to be used in applications. An obvious application is to use PPL as the inference engine in an expert system. Since PPL can be regarded as a more powerful kind of defeasible logic, the range of already explored applications is fairly extensive. In fact defeasible logics have been used in an expert system, for learning and planning [71], in a robotic dog that plays soccer [14, 15], in a robotic poker player [16], to improve the accuracy of radio frequency identification [29], to model the behaviour of autonomous robots [19], and to facilitate the encoding of software requirements [18] so they can be automatically translated into a programming language [17]. Defeasible logics have been advocated for various applications including modelling regulations and business rules [5], agent negotiations [46], the semantic web [4, 8, 84], modelling agents [49], modelling intentions [48], modelling dynamic resource allocation [50], modelling contracts [45], legal reasoning [51], modelling deadlines [47], and modelling dialogue games [81]. Moreover, defeasible theories, describing policies of business activities, can be mined efficiently from appropriate datasets [57].

References

1. Adams, E.W.: The Logic of Conditionals: An Application of Probability to Deductive Logic. D. Reidel Publishing Co., Dordrecht, Holland (1975)
2. Alchourròn, C.E., Gärdenfors, P., Makinson, D.: On the logic of theory change: Partial meet contraction and revision functions. Journal of Symbolic Logic **50**(2), 510–530 (1985)
3. Antoniou, G.: Nonmonotonic Reasoning. MIT Press (1997)
4. Antoniou, G.: Nonmonotonic rule system on top of ontology layer. In: ISWC, *Lecture Notes in Computer Science*, vol. 2432, pp. 394–398. Springer (2002)
5. Antoniou, G., Billington, D., Maher, M.J.: On the analysis of regulations using defeasible rules. In: Proceedings of the 32nd Hawaii International Conference on Systems Science. IEEE Press (1999)
6. Arlo-Costa, H., Egré, P.: The logic of conditionals. In: E.N. Zalta (ed.) The Stanford Encyclopedia of Philosophy, winter 2016 edn. Metaphysics Research Lab, Stanford University, https://plato.stanford.edu/archives/win2016/entries/logic-conditionals/ (2016)
7. Baral, C.: Knowledge Representation, Reasoning and Declarative Problem Solving. Cambridge University Press (2003)
8. Bassiliades, N., Antoniou, G., Vlahavas, I.: Dr-device: A defeasible logic system for the semantic web. In: 2nd Workshop on Principles and Practice of Semantic Web Reasoning, *Lecture Notes in Computer Science*, vol. 3208, pp. 134–148 (2004)
9. Benferhat, S., Dubois, D., Prade, H.: Representing default rules in possibilistic logic. In: B. Nebel, C. Rich, W. Swartout (eds.) Proceedings of the Third International Conference on the Principles of Knowledge Representation and Reasoning (KR92), pp. 673–684. Morgan Kaufmann Publishers, Inc. (1992)
10. Billington, D.: Defeasible logic is stable. Journal of Logic and Computation **3**(4), 379–400 (1993)
11. Billington, D.: Propositional clausal defeasible logic. In: S. Holldobler, C. Lutz, H. Wansing (eds.) Logics in Artificial Intelligence, *Lecture Notes in Artificial Intelligence*, vol. 5293, pp. 34–47. 11th European Conference on Logics in Artificial Intelligence (JELIA2008), Springer, Dresden, Germany (2008)
12. Billington, D.: A defeasible logic for clauses. In: D. Wang, M. Reynolds (eds.) AI 2011: Advances in Artificial Intelligence 24th Australasian Joint Conference Perth, Australia, December 5-8, 2011 Proceedings, *Lecture Notes in Artificial Intelligence*, vol. 7106, pp. 472–480. Springer (2011)
13. Billington, D.: Principles and examples of plausible reasoning and propositional plausible logic. arXiv:1703.01697v2 [cs.AI] pp. 1–58 (2017)
14. Billington, D., Estivill-Castro, V., Hexel, R., Rock, A.: Non-monotonic reasoning for localisation in robocup. In: Proceedings of the 2005 Australasian Conference on Robotics and Automation (2005). URL http://www.cse.unsw.edu.au/acra2005/proceedings/papers/billington.pdf
15. Billington, D., Estivill-Castro, V., Hexel, R., Rock, A.: Using temporal consistency to improve robot localisation. In: Proceedings of the 10th RoboCup International Symposium, *Lecture Notes in Artificial Intelligence*, vol. 4434, pp. 232–244. Springer (2007)
16. Billington, D., Estivill-Castro, V., Hexel, R., Rock, A.: Architecture for hybrid robotic behavior. In: E. Corchado, X. Wu, E. Oja, A. Herrero, B. Baruque (eds.) Hybrid Artificial Intelligence Systems (HAIS09), *Lecture Notes in Artificial Intelligence*, vol. 5572, pp. 145–156. Springer (2009)
17. Billington, D., Estivill-Castro, V., Hexel, R., Rock, A.: Modelling behaviour requirements for automatic interpretation, simulation and deployment. In: 2nd International Conference on Simulation, Modeling, and Programming for Autonomous Robots (SIMPAR2010), *Lecture Notes in Artificial Intelligence*, vol. 6472, pp. 204–216. Springer (2010)
18. Billington, D., Estivill-Castro, V., Hexel, R., Rock, A.: Non-monotonic reasoning for requirements engineering: State diagrams driven by plausible logic. In: 5th International Conference

on Evaluation of Novel Approaches to Software Engineering (ENASE2010), Communications in Computer and Information Science (CCIS). Springer-Verlag (2010)
19. Billington, D., Estivill-Castro, V., Hexel, R., Rock, A.: Plausible logic facilitates engineering the behaviour of autonomous robots. In: R. Fox, W. Golubski (eds.) The IASTED International Conference on Software Engineering 2010, pp. 41–48. ACTA Press (2010)
20. Billington, D., Rock, A.: Propositional plausible logic: Introduction and implementation. Studia Logica **67**(2), 243–269 (2001)
21. Booth, R., Casini, G., Meyer, T., Varzinczak, I.: On the entailment problem for a logic of typicality. In: Proceedings of the Twenty-Fourth International Joint Conference on Artificial Intelligence (IJCAI 2015), pp. 2805–2811 (2015)
22. Booth, R., Meyer, T., Varzinczak, I.: A propositional typicality logic for extending rational consequence. In: E. Fermé, D. Gabbay, G. Simari (eds.) Trends in Belief Revision and Argumentation Dynamics, Studies in Logic - Logic and Cognitive Systems, vol. 48, pp. 123–154. King's College Publications (2013)
23. Boutilier, C.: Conditional logics of normality: A modal approach. Artificial Intelligence **68**(1), 87–154 (1994)
24. Boutilier, C.: Unifying default reasoning and belief revision in a modal framework. Artificial Intelligence **68**(1), 33–85 (1994)
25. Brewka, G.: Preferred subtheories: An extended logical framework for default reasoning. In: N.S. Sridharan (ed.) Proceedings of the Eleventh International Joint Conference on Artificial Intelligence (IJCAI89), vol. 2, pp. 1043–1048. Morgan Kaufmann Publishers, Inc. (1989)
26. Burgess, J.P.: Quick completeness proofs for some logics of conditionals. Notre Dame Journal of Formal Logic **22**(1) (1981)
27. Caminada, M., Amgoud, L.: On the evaluation of argumentation formalisms. Artificial Intelligence **171**, 286–310 (2007)
28. Crocco, G., Lamarre, P.: On the connection between non-monotonic inference systems and conditional logics. In: B. Nebel, C. Rich, W. Swartout (eds.) Proceedings of the Third International Conference on the Principles of Knowledge Representation and Reasoning (KR92), pp. 565–571. Morgan Kaufmann Publishers, Inc. (1992)
29. Darcy, P., Stantic, B., Derakhshan, R.: Correcting stored rfid data with non-monotonic reasoning. International Journal of Principles and Applications of Information Science and Technology **1**(1), 65–77 (2007)
30. Deagustini, C., Martiinez, M., Falappa, M., Simari, G.: Improving inconsistency resolution by considering global conflicts. In: U.Straccia, A.Cali (eds.) 8th International Conference on Scalable Uncertainty Management, *Lecture Notes in Artificial Intelligence*, vol. 8720, pp. 120–133. Springer (2014)
31. Deagustini, C., Martiinez, M., Falappa, M., Simari, G.: Inconsistency resolution and global conflicts. In: T.Schaub, et al (eds.) 21st European Conference on Artificial Intelligence (ECAI), pp. 991–992. IOS Press (2014)
32. Delgrande, J.P.: On a rule-based interpretation of default conditionals. Annals of Mathematics and Artificial Intelligence **48**, 135–167 (2007)
33. Dubois, D., Lang, J., Prade, H.: Possibilistic logic. In: D.M. Gabbay, C. Hogger, J. Robinson (eds.) Handbook of Logic in Artificial Intelligence and Logic Programming, vol. 3 Nonmonotonic Reasoning and Uncertain Reasoning, pp. 439–513. Oxford Science Publications (1994)
34. Dubois, D., Prade, H.: Conditional objects and non-monotonic reasoning. In: J. Allen, R. Fikes, E. Sandewall (eds.) Proceedings of the Second International Conference on the Principles of Knowledge Representation and Reasoning (KR91), pp. 175–185. Morgan Kaufmann Publishers, Inc. (1991)
35. Dung, P.M.: On the acceptability of arguments and its fundamental role in nonmonotonic reasoning, logic programming, and n-person games. Artificial Intelligence **77**(2), 321–357 (1995)
36. Freund, M., Lehmann, D.: Nonmonotonic reasoning: from finitary relations to infinitary inference operations. Studia Logica **53**(2), 161–201 (1994)

References

37. Gabbay, D.: Theoretical foundations for non-monotonic reasoning in expert systems. In: K. Apt (ed.) Proceedings of the NATO Advanced Study Institute on Logics and Models of Concurrent Systems, pp. 439–457. La Colle-sur-Loup, France (1985)
38. Gabbay, D.M.: Investigations in Modal and Tense Logics with Applications to Problems in Philosophy and Linguistics. Reidel (1976)
39. Garcia, A.J., Simari, G.R.: Defeasible logic programming: an argumentative approach. Theory and Practice of Logic Programming **4**(1,2), 95–138 (2004)
40. Gärdenfors, P.: Knowledge in Flux. Modeling the Dynamics of Epistemic States. MIT Press (1988)
41. Geerts, P., Laenens, E., Vermeir, D.: Defeasible logics. In: D.M. Gabbay, P. Smets (eds.) Handbook of Defeasible Reasoning and Uncertainty Management Systems, vol. 2, pp. 175–210. Kluwer Academic (1998)
42. Geerts, P., Vermeir, D., Nute, D.: Ordered logic: defeasible reasoning for multiple agents. Decision Support Systems **11**, 157–190 (1994)
43. Geffner, H., Pearl, J.: Conditional entailment: bridging two approaches to default reasoning. Artificial Intelligence **53**, 209–244 (1992)
44. Goldszmidt, M., Pearl, J.: System-z+: A formalism for reasoning with variable-strength defaults. In: Proceedings AAAI-91, pp. 399–404 (1991)
45. Governatori, G.: Representing business contracts in ruleml. International Journal of Cooperative Information Systems **14**(2-3), 181–216 (2005)
46. Governatori, G., Dumas, M., ter Hofstede, A.H., Oaks, P.: A formal approach to protocols and strategies for (legal) negotiation. In: Proceedings of the 8th International Conference on Artificial Intelligence and Law, pp. 168–177. ACM Press (2001)
47. Governatori, G., Hulstijn, J., Riveret, R., Rotolo, A.: Characterising deadlines in temporal modal defeasible logic. In: Proceedings of the 20th Australian Joint Conference on Artificial Intelligence, *Lecture Notes in Artificial Intelligence*, vol. 4830, pp. 486–496. Springer (2007)
48. Governatori, G., Padmanabhan, V.: A defeasible logic of policy-based intention. In: T.D. Gedeon, L.C.C. Fung (eds.) AI 2003: Advances in Artificial Intelligence, *Lecture Notes in Artificial Intelligence*, vol. 2903, pp. 414–426. Springer (2003)
49. Governatori, G., Rotolo, A.: Defeasible logic: Agency, intention and obligation. In: A. Lomuscio, D. Nute (eds.) Deontic Logic in Computer Science, *Lecture Notes in Artificial Intelligence*, vol. 3065, pp. 114–128. Springer (2004)
50. Governatori, G., Rotolo, A., Sadiq, S.: A model of dynamic resource allocation in workflow systems. In: K.D. Schewe, H.E. Williams (eds.) Database Technology 2004. Conference Research and Practice of Information Technology, vol. 27, pp. 197–206. Australian Computer Science Association, ACS (2004)
51. Governatori, G., Rotolo, A., Sartor, G.: Temporalised normative positions in defeasible logic. In: Proceedings of the 10th International Conference on Artificial Intelligence and Law, pp. 25–34. ACM Press (2005)
52. Hansson, S.O.: Kernel contraction. The Journal of Symbolic Logic **59**(3), 845–859 (1994)
53. Hawthorne, J., Makinson, D.: The quantitative/qualitative watershed for rules of uncertain inference. Studia Logica **86**, 247–297 (2007)
54. Horty, J.F., Thomason, R.H., Touretzky, D.S.: A skeptical theory of inheritance in nonmonotonic semantic networks. Artificial Intelligence **42**, 311–348 (1990)
55. Hunter, A.: Reasoning with contradictory information using quasi-classical logic. Journal of Logic and Computation **10**(5), 677–703 (2000)
56. Irvine, A.D., Deutsch, H.: Russell's paradox. The Stanford Encyclopedia of Philosophy (2016). URL https://plato.stanford.edu/entries/russell-paradox/
57. Johnston, B., Governatori, G.: An algorithm for the induction of defeasible logic theories from databases. In: K. Schewe, X. Zhou (eds.) Database Technology 2003. Conference Research and Practice of Information Technology, vol. 17, pp. 75–83. Australian Computer Science Association, Australian Computer Science Association (2003)
58. Kraus, S., Lehmann, D., Magidor, M.: Nonmonotonic reasoning, preferential models and cumulative logics. Artificial Intelligence **44**(1-2), 167–207 (1990)

59. Kraus, S., Lehmann, D., Magidor, M.: Nonmonotonic reasoning, preferential models and cumulative logics. arXiv:cs.AI/0202021v1 pp. 1–44 (2002)
60. Lamarre, P.: S4 as the conditional logic of nonmonotonicity. In: J. Allen, R. Fikes, E. Sandewall (eds.) Proceedings of the Second International Conference on the Principles of Knowledge Representation and Reasoning (KR91), pp. 357–367. Morgan Kaufmann Publishers, Inc. (1991)
61. Lehmann, D., Magidor, M.: What does a conditional knowledge base entail? Artificial Intelligence **55**(1), 1–60 (1992)
62. Maier, F., Nute, D.: Ambiguity propagating defeasible logic and the well-founded semantics. In: 10th European Conference on Logics in Artificial Intelligence (JELIA2006), *Lecture Notes in Artificial Intelligence*, vol. 4160, pp. 306–318. Springer (2006)
63. Makinson, D.: General theory of cumulative inference. In: Proceedings of the Second International Workshop on Non-Monotonic Reasoning, *Lecture Notes in Artificial Intelligence*, vol. 346, pp. 1–18. Springer-Verlag (1988)
64. Makinson, D.: General patterns in nonmonotonic reasoning. In: D.M. Gabbay, C. Hogger, J. Robinson (eds.) Handbook of Logic in Artificial Intelligence and Logic Programming, vol. 3 Nonmonotonic Reasoning and Uncertain Reasoning, chap. General Patterns in Nonmonotonic Reasoning, pp. 35–110. Oxford University Press (1994)
65. Makinson, D., Hawthorne, J.: Lossy inference rules and their bounds: a brief review. In: A. Loslow, A. Buchsbaum (eds.) The Road to Universal Logic, vol. 1, pp. 385–408. Springer (2014)
66. Mann, T. (ed.): Australian Law Dictionary, second edn. Oxford University Press (2013)
67. Modgil, S., Prakken, H.: A general account of argumentation with preferences. Artificial Intelligence **195**, 361–397 (2013)
68. Nerode, A., Shore, R.A.: Logic for Applications, 2nd edn. No. ISBN 0-3887-94893-7 in Graduate Texts in Computer Science. Springer (1997)
69. Nute, D.: Defeasible reasoning. In: Proceedings of the 20th Hawaii International Conference on System Science, pp. 470–477. University of Hawaii (1987)
70. Nute, D.: Defeasible Logic, *Handbook of Logic in Artificial Intelligence and Logic Programming*, vol. 3, pp. 353–395. Oxford University Press (1994)
71. Nute, D.: Defeasible logic. In: Proceedings of the 14th International Conference on Applications of Prolog, *Lecture Notes in Artificial Intelligence*, vol. 2543, pp. 151–169. Springer (2003)
72. Nute, D., Cross, C.B.: Conditional logic. In: D.M. Gabbay, F. Guenthner (eds.) Handbook of Philosophical Logic, vol. 4, 2nd edn., pp. 1–98. Kluwer Academic Publishers (2001)
73. Pearl, J.: Probabilistic Reasoning in Intelligent Systems: Networks of Plausible Inference. Morgan Kaufmann Publishers, Inc. (1988)
74. Pearl, J.: System z: A natural ordering of defaults with tractable applications to nonmonotonic reasoning. In: R. Parikh (ed.) Theoretical Aspects of Reasoning about Knowledge (TARK-III), pp. 121–135. Morgan Kaufmann Publishers, Inc. (1990)
75. Poole, D.: A logical framework for default reasoning. Artificial Intelligence **36**, 27–47 (1988)
76. Prakken, H., Sartor, G.: Argument-based extended logic programming with defeasible priorities. Journal of Applied Non-Classical Logics **7**(1-2), 25–75 (1997)
77. Reiter, R.: A logic for default reasoning. Artificial Intelligence **13**, 81–132 (1980)
78. Scott, D.S.: Completeness and axiomatizability in many-valued logic. In: L. Henkin, et al. (eds.) Proceedings of the Tarski Symposium, *Proceedings of Symposia in Pure Mathematics*, vol. XXV, pp. 411–435. American Mathematical Society (1974)
79. Simari, G.R., Loui, R.P.: A mathematical treatment of defeasible reasoning and its implementation. Artificial Intelligence **53**, 125–157 (1992)
80. Tarski, A.: Logic, Semantics, Metamathematics. Papers from 1923 to 1938. Oxford University Press (1956)
81. Thakur, S., Governatori, G., Padmanabhan, V., Lundstrom, J.E.: Dialogue games in defeasible logic. In: Proceedings of the 20th Australian Joint Conference on Artificial Intelligence, *Lecture Notes in Artificial Intelligence*, vol. 4830, pp. 497–506. Springer (2007)

82. Touretzky, D., Horty, J., Thomason, R.: A clash of intuitions: The current state of nonmonotonic multiple inheritance systems. In: J.P. McDermott (ed.) Proceedings of the 10th International Joint Conference on Artificial Intelligence, vol. 1, pp. 476–482. Morgan Kaufmann Publishers, Inc. (1987)
83. Walton, D., Tindale, C., Gordon, T.: Applying recent argumentation methods to some ancient examples of plausible reasoning. Argumentation **28**, 85–119 (2014)
84. Wang, K., Billington, D., Blee, J., Antoniou, G.: Combining description logic and defeasible logic for the semantic web. In: Proceedings of the Rules and Rule Markup Languages for the Semantic Web Workshop (RuleML2004), *Lecture Notes in Computer Science*, vol. 3323, pp. 170–181. Springer (2004)

Index

() empty sequence, 5
$(\alpha : \alpha')$, 177
$(\gamma : \lambda)$, 193
+1 proof value, 158
− evaluation tree minus a set, 167
− minus, 158
− sequence difference, 5
− sequence minus a set, 167
− subtraction of sets, 3
−1 proof value, 158
< strict subclause, 13
< strict submeet, 13
> strict superclause, 13
> strict supermeet, 13
C-Partial Order, 124
$R'[F]$, 160
$R'[f;s]$, 153
$R'[f]$, 152
R^s_d, 152
[..] integer interval notation, 3
.. double dot notation, 3
$\wedge\{\}$ empty meet, verum, 12
$\wedge$ conjunction, 10, 146
$\wedge.$, 28
$\vee\{\}$ empty clause, falsum, 12
$\vee$ disjunction, 10, 146
$\vee.$, 28
$\aleph_0$, 6
$\Diamond$, 30
$\approx$ equivalent proof algorithms, 177
$\cap$ unary intersection, 3
$\cup$ unary union, 3
$\cap$ binary intersection, 3
$\check{(.)}$ check(.), 28
$\vdash_{\overline{c}}$, 108
$\cup$ binary union, 3
$\Rightarrow$ defeasible arrow, 146
$\in$ is an element of, 2
$\equiv$ is equivalent to, 17
$\equiv_C$ C-Equivalence, 120
$\exists x \in X$, 2
$\forall x \in X$, 2
$\geq$ superclause, 13
$\geq$ supermeet, 13
$\gtrsim$ encompasses, 175
$\hat{(.)}$ hat(.), 29
$\leq$ subclause, 13
$\leq$ submeet, 13
$\lesssim$ encompassed, 175
$\neg F$, 21
$\neg$ negation, 10, 21, 146
$\neg^n$, 11
$\sim$, 28
$\sim$ complement of a quasi-formula, 24
ω, 6
$\circledvee$ disjunctive operator, 92
$\vdash$ proof relation, 155
$\rightarrowtail$ rewrite arrow, 56
$\simeq$ isomorphic proof algorithms, 177
$\rightarrow$ strict arrow, 146
$\subset$ strict subset, 2
$\subseteq$ is a subset of, 2
$\sum$ summation, 5
$\times$ Cartesian product, 4
$\rightsquigarrow$ warning arrow, 146
$\{\}$ empty set, 3
f^n, 4
n-lottery, 129
$n$to1 function, 4
$\vdash^D$ D-proves, 102
$\vDash^D$ D-implies, 102

$\vdash^I$ I-proves, 102
$\models^I$ I-implies, 102
$\vdash^N$ N-proves, 102
$\models^N$ N-implies, 102
$\vdash^P$ P-proves, 102
$\models^P$ P-implies, 102
$\vdash^R$ R-proves, 102
$\models^R$ R-implies, 102
$\vdash^S$ S-proves, 102
$\models^S$ S-implies, 102
$\vdash^X$ X-proves, 102
$\models^X$ X-implies, 102
$\vdash$, 74
$\not\models$, 14
$\vdash$ resolution-provable, 74
$\vdash$-consequences function $C_\vdash(.)$, 74
$\models$ implies, 17
$\models$ satisfies, 14
$\models$-consequences function $C_\models(.)$, 19
$|.|$ cardinality of a set, 3
$|.|$ complexity of a quasi-formula, 23
$|.|$ length of a finite sequence, 5
$|.|$ number of children of a node, 7
$|.|$ number of nodes in a tree, 7
+ postpend, 5
+ prepend, 5
++ concatenation, 5
1.2.1 Lemma, 3
1.2.2 Definition, 6
1.2.3 Definition, 7
1.2.4 Definition, 8
1.2.5 Lemma, 8
1.2.6 Lemma, 8
2.2.1 Definition, 10
2.2.2 Definition, 10
2.2.3 Definition, 11
2.2.4 Definition, 11
2.2.5 Definition, 11
2.2.6 Definition, 11
2.2.7 Definition, 12
2.2.8 Definition, 12
2.2.9 Lemma, 12
2.2.10 Definition, 13
2.2.11 Definition, 14
2.2.12 Lemma, 14
2.2.13 Definition, 14
2.2.14 Definition, 14
2.2.15 Definition, 15
2.2.16 Lemma, 15
2.2.17 Definition, 17
2.2.18 Lemma, 17
2.2.19 Lemma, 18

2.2.20 Definition, 19
2.2.21 Lemma, 19
2.3.1 Definition, 21
2.3.2 Lemma, 21
2.3.3 Definition, 22
2.3.4 Definition, 22
2.3.5 Definition, 22
2.3.6 Definition, 23
2.3.7 Lemma, 23
2.3.8 Lemma, 23
2.3.9 Definition, 24
2.3.10 Lemma, 25
2.3.11 Theorem, 26
2.4.1 Definition, 28
2.4.2 Definition, 28
2.4.3 Definition, 28
2.4.4 Definition, 28
2.4.5 Definition, 28
2.4.6 Definition, 29
2.4.7 Definition, 29
2.4.8 Definition, 29
2.4.9 Definition, 29
2.4.10 Lemma, 30
2.4.11 Lemma, 30
2.4.12 Lemma, 33
2.5.1 Definition, 34
2.5.2 Definition, 34
2.5.3 Definition, 35
2.5.4 Definition, 35
2.5.5 Definition, 35
2.5.6 Lemma, 36
2.5.7 Lemma, 44
2.6.1 Definition, 46
2.6.2 Lemma, 47
2.6.3 Lemma, 48
2.6.4 Definition, 49
2.6.5 Lemma, 49
2.6.6 Definition, 50
2.6.7 Lemma, 51
2.6.8 Lemma, 51
2.6.9 Definition, 53
2.6.10 Theorem, 53
2.7.1 Definition, 54
2.7.2 Lemma, 54
2.7.3 Definition, 55
2.7.4 Lemma, 56
2.7.5 Definition, 56
2.7.6 Definition, 56
2.7.7 Example, 57
2.7.8 Lemma, 57
2.7.9 Definition, 58
2.7.10 Lemma, 59
2.7.11 Definition, 64

Index

2.7.12 Lemma, 65
2.8.1 Theorem, 67
2.8.2 Definition, 68
2.8.3 Lemma, 68
2.8.4 Definition, 70
2.8.5 Lemma, 70
2.8.6 Theorem, 73
2.9.1 Definition, 74
2.9.2 Theorem, 74
2.9.3 Definition, 76
2.9.4 Definition, 76
3.2.1 Definition, 79
3.2.2 Theorem, 79
3.2.3 Definition, 82
3.2.4 Lemma, 83
3.2.5 Lemma, 84
3.2.6 Example, 85
3.2.7 Definition, 86
3.2.8 Lemma, 86
3.2.9 Lemma, 86
3.2.10 Definition, 89
3.2.11 Lemma, 90
3.2.12 Definition, 92
3.2.13 Definition, 92
3.2.14 Lemma, 92
3.2.15 Lemma, 93
3.2.16 Lemma, 95
3.2.17 Definition, 98
3.2.18 Lemma, 98
3.3.1 Definition, 101
3.3.2 Lemma, 101
3.3.3 Definition, 102
3.3.4 Theorem, 103
3.3.5 Lemma, 105
3.4.1 Definition, 108
3.4.2 Definition, 109
3.4.3 Definition, 109
3.4.4 Lemma, 110
3.4.5 Example, 112
3.4.6 Example, 113
3.4.7 Definition, 114
3.4.8 Lemma, 114
3.4.9 Definition, 115
3.4.10 Theorem, 115
3.4.11 Lemma, 116
3.4.12 Definition, 117
3.4.13 Lemma, 117
3.4.14 Lemma, 117
3.4.15 Definition, 118
3.4.16 Lemma, 118
3.4.17 Definition, 118
3.4.18 Lemma, 119
3.4.19 Lemma, 119

3.4.20 Example, 120
3.4.21 Definition, 120
3.4.22 Lemma, 121
3.4.23 Definition, 122
3.4.24 Lemma, 122
3.4.25 Lemma, 123
3.4.26 Definition, 124
3.4.27 Lemma, 124
3.4.28 Definition, 124
3.4.29 Lemma, 124
4.1.1 Example, 130
4.2.1 Principle, 130
4.2.2 Principle, 130
4.3.1 Definition, 131
4.3.2 Example, 131
4.3.3 Principle, 132
4.3.4 Principle, 132
4.4.1 Definition, 132
4.4.2 Principle, 132
4.4.3 Definition, 133
4.4.4 Principle, 133
4.5.1 Definition, 133
4.5.2 Principle, 133
4.5.3 Definition, 133
4.5.4 Example, 134
4.5.5 Principle, 134
4.6.1 Definition, 134
4.6.2 Definition, 134
4.6.3 Principle, 134
4.7.1 Definition, 135
4.7.2 Definition, 135
4.7.3 Definition, 135
4.7.4 Definition, 135
4.7.5 Principle, 135
4.8.1 Definition, 136
4.8.2 Definition, 136
4.8.3 Principle, 136
4.8.4 Principle, 136
4.9.1 Example, 137
4.9.2 Principle, 138
4.10.1 Definition, 138
4.10.2 Principle, 138
4.11.1 Example, 139
4.11.2 Principle, 139
5.1.1 Definition, 146
5.1.2 Definition, 146
5.1.3 Definition, 147
5.1.4 Definition, 148
5.1.5 Definition, 148
5.1.6 Lemma, 148
5.1.7 Lemma, 149
5.2.1 Refinement, 151
5.2.2 Definition, 152

5.2.3 Refinement, 152
5.2.4 Definition, 153
5.2.5 Refinement, 153
5.2.6 Refinement, 153
5.2.7 Definition, 154
5.2.8 Definition, 154
5.2.9 Definition, 155
5.2.10 Definition, 155
5.2.11 Definition, 155
5.2.12 Lemma, 156
5.2.13 Lemma, 156
5.2.14 Lemma, 157
5.3.1 Definition, 158
5.3.2 Definition, 159
5.3.3 Definition, 159
5.3.4 Definition, 160
5.3.5 Definition, 160
5.3.6 Definition, 160
5.3.7 Definition, 160
5.3.8 Lemma, 160
5.3.9 Lemma, 161
5.3.10 Lemma, 162
5.3.11 Lemma, 162
5.3.12 Definition, 163
5.3.13 Definition, 163
5.3.14 Definition, 164
5.3.15 Theorem, 164
5.3.16 Definition, 165
5.3.17 Theorem, 165
5.3.18 Definition, 167
5.3.19 Definition, 167
5.3.20 Lemma, 167
5.4.1 Definition, 170
6.1.1 Theorem, 173
6.1.2 Theorem, 174
6.2.1 Definition, 175
6.2.2 Lemma, 175
6.2.3 Definition, 175
6.2.4 Lemma, 175
6.2.5 Definition, 177
6.2.6 Definition, 177
6.2.7 Lemma, 177
6.2.8 Lemma, 180
6.2.9 Theorem, 182
6.3.1 Theorem, 185
6.4.1 Definition, 186
6.4.2 Lemma, 186
6.4.3 Definition, 188
6.4.4 Lemma, 188
6.4.5 Definition, 193
6.4.6 Lemma, 193
6.4.7 Theorem, 195
7.2.1 Evaluation, 201

7.2.2 Evaluation, 201
7.3.1 Evaluation, 203
7.3.2 Evaluation, 203
7.3.3 Evaluation, 203
7.4.1 Evaluation, 204
7.5.1 Evaluation, 207
7.5.2 Evaluation, 207
7.5.3 Evaluation, 207
7.5.4 Lemma, 208
7.5.5 Evaluation, 208
7.5.6 Evaluation, 208
7.5.7 Evaluation, 210
7.5.8 Evaluation, 210
7.5.9 Evaluation, 210
7.5.10 Lemma, 210
7.5.11 Evaluation, 211
7.5.12 Evaluation, 212
7.5.13 Evaluation, 213
7.5.14 Evaluation, 213
7.5.15 Lemma, 213
7.5.16 Evaluation, 214
7.6.1 Evaluation, 216
7.6.2 Lemma, 217
7.6.3 Evaluation, 217
7.6.4 Evaluation, 218
7.6.5 Lemma, 219
7.8.1 Example, 221
7.8.2 Evaluation, 221
7.9.1 Example, 222
7.9.2 Evaluation, 223
7.9.3 Evaluation, 223
7.9.4 Example, 223
7.9.5 Evaluation, 224
7.10.1 Example, 224
7.10.2 Evaluation, 226
7.10.3 Evaluation, 227
1to1 function, 4
1to1 sequence, 5
2to1 function, 4
3-lottery example 4.1.1, 130
4-lottery example 4.11.1, 139

a ambiguous truth value, 170
$\alpha(\gamma:\lambda)$, 193
α-consequences of $\mathcal{P}$ $\mathcal{P}(\alpha)$, 155
α-history, 155
α-provable, 155
acyclic, 5
adequate, 76
Alg, 154
$alg(.)$, 160
algorithm of a node, 160
alphabet for PPL, 146

Index   247

alphabet for PRL, 10
ambiguity blocking, 137
ambiguity propagating, 137
Ambiguity Puzzle 4.9.1, 137
ambiguous, 137
ancestor, 7
AND, 14
$AndOr(.)$, 58
antisymmetric, 5
arc, 7
$arrow(r)$, 146
Atm set of propositional atoms, 10, 146
atom, 10, 146
Ax, 148
$Ax(R)$, 148
axioms, 148

β, 154
β', 154
bijection, 4
binary operation, 4
binary operator, 4
binary relation, 4

$C_{\vdash P}(.)$, 102
$C_{\vDash P}(.)$, 102
$C_{\vdash L}(.)$, 102
$C_{\vDash L}(.)$, 102
$C_{\vdash N}(.)$, 102
$C_{\vDash N}(.)$, 102
$C_{\vdash P}(.)$, 102
$C_{\vDash P}(.)$, 102
$C_{\vdash R}(.)$, 102
$C_{\vDash R}(.)$, 102
$C_{\vdash S}(.)$, 102
$C_{\vDash S}(.)$, 102
$C_{\vdash X}(.)$, 102
$C_{\vDash X}(.)$, 102
$C_{\vdash}(.)$ $\vdash$-consequences function, 74
$C_{\vDash}(.)$ $\vDash$-consequences function, 19
$\check{C}$ check C, 28
$\check{c}$ check c, 28
$c(R)$, 147
$c(r)$ the consequent of r, 146
C-antisymmetric, 124
C-Antisymmetry, 124
C-Equivalence $\equiv_C$, 120
C-Partial Order, 124
cardinal number, 6
cardinality of a set $|.|$, 3

Cartesian product $\times$, 4
Cautious Monotonicity, 109, 119
cautiously monotonic, 109
check $\check{}$, 28
child, 7, 29
$Cl(.)$, 68
Classical Closure, 109, 122
Classical consequence function Cn, 109
clause, 12
$ClCor(.)$, 70
closed formula, 128
Cls set of clauses, 12
$Cls(.)$, 12
$ClsC_{\vDash}(.)$, 19
Cn the classical consequence function, 109
cnf-conversion, 56
cnf-formula, 54
cnf-terminal formula, 64
$CnfFml$ set of all cnf-formulas, 54
$cnfNext(.)$, 56
$cnfRank(.)$, 58
$CnfTerminals$, 64
co-algorithm, 154
compact, 109
Compactness, 109, 118
Compactness Theorem 2.3.11, 26
complement of a quasi-formula $\sim$, 24
complete, 53
complexity of a quasi-formula, 23
component, 11
concatenation ++, 5
conjunction $\wedge$, 10
conjunctive normal form, 54
conjunctive proof algorithm, 132
connectives for PPL, 146
connectives for PRL, 10
consequent, 146
contingent, 15
$Contrad$, 15
contradiction, 15
conversion, 56
$Cor(.)$ core, 46
$CorCl(.)$, 70
$CorClsC_{\vDash}(.)$, 49
core $Cor(.)$, 46
$CorMtsImp(.)$, 49
$CorRes(.)$, 50
countable, 7
countably infinite, 7
$Cp(.)$ set of components, 11
$Ctge(.)$, 46
cumulative, 109

Cumulativity, 109, 119
Cut, 109, 119
cyclic, 5

D-implies $\models^D$, 102
D-proves $\vdash^D$, 102
dc, 55
$DCl(.)$, 89
dd, 55
Ddc, 55
Ddd, 55
De Morgan's law, 5, 21
decisive, 138
Decisiveness Principle 4.10.2, 138
Decisiveness Theorem, 164
dedication, v
defeasible arrow $\Rightarrow$, 146
defeasible rule, 146
denumerable, 7
derivable, 29
derivation, 29
descendant, 7, 30
$DFrom(.)$, 102
$Dftd()$ defeated, 165
disabled, 152
disjunction, 13
disjunction $\vee$, 10
disjunctive normal form, 54
disjunctive operator $\oslash$, 92
disproof, 164
distributes over, 5
distributivity, 5
Divisibility, 92, 109, 123
divisible, 92, 109
$Dncc$, 55
$Dncd$, 55
$Dndc$, 55
$Dndd$, 55
dnf-conversion, 56
dnf-formula, 54
dnf-terminal formula, 64
$DnfFml$ set of all dnf-formulas, 54
$dnfNext(.)$, 56
$dnfRank(.)$, 58
$DnfTerminals$, 64
double dot notation .. , 3
$Dub(.)$, 82
$Dub^*(.)$, 82
$DubCl(.)$, 89

element of $\in$, 2
empty clause, falsum $\vee\{\}$, 12
empty meet, verum $\wedge\{\}$, 12

empty sequence (), 5
empty set $\{\}$, 3
encompassed $\lesssim$, 175
encompasses $\gtrsim$, 175
equivalence relation, 5
equivalent proof algorithms $\approx$, 177
$Errcl(.)$ error clauses in, 82
$Errfulcl(.)$ errorful clauses in, 82
$Errlit(.)$ error literals of, 82
error clauses in $Errcl(.)$, 82
error literals of $Errlit(.)$, 82
errorful clauses in $Errfulcl(.)$, 82
Essential Classicality, 114
essentially classical, 114
evaluation tree, 159
Evidence Principle 4.3.3, 132
exclusive disjunction, 13
explosive logic, 77

F the false truth value, 14
f probably false, 170
fact, 149
$Fact(S)$, 128
factual proof algorithm, 130
falsifiable, 14
falsifies, 14
falsum, empty clause $\vee\{\}$, 12
φ, 154
φ', 154
finite ordinal, 6
finite sequence, 5
Fml set of all formulas, 10
$Foe(.,.,.)$, 154
$For()$ evidence for, 165
formula, 10
function, 4

generalised-rule, 146
generated, 7

$H(\alpha := \pi')$, 175
$H(\alpha : \alpha')$, 177
$H(\gamma : \lambda)$, 193
$H(\pi := \psi)$, 186
$H(\psi := \beta)$, 188
has a singleton, 34
has no singletons, 34
hat $\hat{\ }$, 29
height of a rooted tree, 8
$Hist(.)$, 160
history, 155
history of a node, 160

Index 249

I-implies $\models^I$, 102
I-proves $\vdash^I$, 102
$ICl(.)$, 89
Idempotence, 109, 117
idempotent, 109
idempotent binary operation, 4
idempotent function, 4
iff, 2
$IFrom(.)$, 102
$\bigcap MaxSatCl(.)$, 89
$Imp(.)$ semantic implicant function, 19
Included Middle Principle 4.11.2, 139
Inclusion, 109, 117
inclusive, 109
inclusive disjunction, 13
inclusive function, 4
inferior, 147, 148
infinite ordinal, 6
infinite sequence, 5
integer interval, 3
integer interval notation $[\,..\,]$, 3
integers, set of all, $\mathbb{Z}$, 3
intersection binary $\cap$, 3
intersection unary $\bigcap$, 3
isomorphic proof algorithms $\simeq$, 177

$\lambda(\alpha : \alpha')$, 177
label of a node, 11
leaf, 7
lean reduct, 163
Left C-Equivalence, 120
Left Absorption, 109, 122
Left disjunction, 92
left disjunctive, 92
Left Equivalence, 109, 120
Left Factual Disjunction Example 4.5.4, 134
left factually disjunctive proof algorithm, 133
length of a finite sequence $|.|$, 5
length of a path, 7
$level(.)$, 8
Lit set of literals, 12
$Lit(.)$, 12
literal, 12

$\hat{M}$ hat M, 29
$\hat{m}$ hat m, 29
Many Proof Algorithms Principle 4.9.2, 138
max, 4
maximal satisfiable subsets, 79
$MaxSat(.)$, 79
meet, 12
$Min(.)$, 46
min, 4

$MinCtge(.)$, 46
minimal unsatisfiable subsets, 79
$MinUnsat(.)$, 79
monotonic, 109
monotonic function, 5
monotonic proof algorithm, 131
Monotonicity, 109, 118
$Mt(.)$, 68
Mts set of meets, 12
$Mts(.)$, 12
$MtsImp(.)$, 19

$\mathbb{N}$ set of all natural numbers, 3
$N(\diamond,F)$, 35
$N(\diamond,f)$, 35
$N(s,.)$ number of singletons, 35
$N(Set,F)$, 35
$N(Set,f)$, 35
n-consistent, 136
N-implies $\models^N$, 102
N-proves $\vdash^N$, 102
name of a node, 11
natural numbers, set of all, $\mathbb{N}$, 3
ncc, 55
ncd, 55
$NCl(.)$, 89
ndc, 55
ndd, 55
negation $\neg$, 10, 21
negation normal form, 22
negative even-cycles, 226
negative odd-cycles, 225
Negative Triangle example 7.10.1, 224
$NFrom(.)$, 102
$Nil(.)$, 79
$NilCl(.)$, 89
$nnf(.)$, 22
nnf-formula, 22
nnf-quasi-formula, 22
$NnfFml$ set of all nnf-formulas, 22
$NnfQFml$ set of all nnf-quasi-formulas, 22
no-singletons $ns(.)$, 35
node, 7
Non-3-Consistency Principle 4.8.4, 136
Non-Conjunctive Principle 4.4.2, 132
non-conjunctive proof algorithm, 132
non-monotonic proof algorithm, 131
Non-Monotonicity example 4.3.2, 131
Non-Monotonicity Principle 4.3.4, 132
NOT, 14
Not Left Factually Disjunctive Principle 4.5.5, 134
Not Right Disjunctive Principle 4.5.2, 133

$ns(.)$ no-singletons, 35
$nsMin(.)$, 46
$nsMinCtge(.)$, 46
number of children of a node $|.|$, 7
number of nodes in a tree $|.|$, 7
number of singletons $N(s,.)$, 35

onto function, 4
$op(.)$, operation, 158
operation, $op(.)$, 158
OR, 14
Or-rule, 91
$OrAnd(.)$, 58
ordinal number, 6

$\mathscr{P}(.)$ power set of, 3
$P(.,.,.)$ proof function, 165
$\mathcal{P}(\alpha)$ the α-consequences of $\mathcal{P}$, 155
P-implies $\models^P$, 102
P-proves $\vdash^P$, 102
parent, 7
partial order, 5
path, 7, 29
$PCl(.)$, 89
$PFrom(.)$, 102
π, 154
π', 154
$Plaus(\mathcal{S})$, 128
Plausible Conjunction Theorem, 173
plausible description, 148
plausible proof algorithm, 130
plausible reasoning, 128
Plausible Right Weakening Principle 4.7.5, 135
plausible right weakening property, 135
Plausible Supraclassicality Principle 4.6.3, 134
plausible supraclassicality property, 134
plausible theory, 160
plausible truth values **a t f u**, 170
plausible-reasoning situation, 128
plausible-structure, 128
Plausibly Conjunctive Principle 4.4.4, 133
plausibly conjunctive proof algorithm, 133
plausibly supraclassical proof algorithm, 134
positive integers, set of all, $\mathbb{Z}^+$, 3
positive train of rules, 160
postpend +, 5
power set of $\mathscr{P}(.)$, 3
PPL propositional plausible logic, 160
pre-cumulative, 86
pre-cumulativity, 86
pre-divisibility, 98
pre-divisible, 98
Precision Principle 4.2.2, 130

prefix, 5
prepend +, 5
priority relation, 148
Priority Triangle Example 7.8.1, 221
PRL propositional resolution logic, 76
proof, 164
proof algorithms, 155
proof function $P(.,.,.)$, 165
proof relation $\vdash$, 155
proof value $+1$ -1, 158
proof value of, $pv(.)$, 158
proper prefix, 5
proper suffix, 5
propositional plausible logic (PPL), 160
propositional resolution logic (PRL), 76
propositionally adequate logic, 76
ψ, 154
ψ', 154
$Pure(.)$ pure clauses in, 82
pure clauses in $Pure(.)$, 82
$PureCl(.)$, 89
$pv(.)$, proof value of, 158

$QFml$ set of quasi-formulas, 11
quasi-formula, 11

r_{ax}, 148
r is applicable, 55
R-implies $\models^R$, 102
R-proves $\vdash^R$, 102
rational, 114
Rational Classicality, 114
rational clauses function, 101
rational consequence function, 115
Rationality, 114
rationally classical, 114
$RCl(.)$, 101
R_d, 147
$Reach(.)$, 29
reach, 29
Red Light Example 7.9.1, 222
reduct, 163
reflexive, 5, 109
Reflexivity, 109
replace-singletons $rs(.)$, 34
Representation Principle 4.2.1, 130
$Res(.)$, 28, 29
$res(.;.,.)$, 28
$ResLit(.)$, 28
resolution-child, 28
resolution-derivable, 28, 29
resolution-derivation, 28, 29
resolution-descendant, 29

Index   251

resolution-provable $\vdash$, 74
resolvable, 28
resolvent, 28, 29
rewrite arrow $\rightarrowtail$, 56
$RFrom(.)$, 102
Right Absorption, 109, 120
right disjunctive proof algorithm, 133
Right Weakening, 122
right weakening property, 135
Right Weakening Theorem, 174
root, 7
rooted tree, 7
R_s, 147
$rs(.)$ replace-singletons, 34
$Rul(.)$, 147
$Rul(.)$, 160
$Rul(.,.)$, 147
rule, 146
Russell's paradox, 225
R_w, 147

S-implies $\models^S$, 102
S-proves $\vdash^S$, 102
$Sat(.)$, 79
satisfiable, 14
Satisfiable Compactness, 118
Satisfiable Left Equivalence, 120
Satisfiable Monotonicity, 118
Satisfiable Right Absorption, 120
Satisfiable Tautology-free Inclusion, 117
satisfies $\models$, 14
$scnf(.)$, 68
scnf-formula, 68
$sdnf(.)$, 68
sdnf-formula, 68
Self Attack, 224
semantic implicant function $Imp(.)$, 19
sequence, 5
sequence difference $-$, 5
$Set(.)$, 5
set of antecedents, 146
$SFrom(.)$, 102
singleton, 3
sound, 53
stable, 50
stable conjunctive normal form, 68
stable conjunctive normal form of f, 73
stable disjunctive normal form, 68
stable disjunctive normal form of f, 73
stem, 28
strict arrow $\rightarrow$, 146
strict rule, 146
strict subclause $<$, 13

strict subformula, 11
strict submeet $<$, 13
strict subset $\subset$, 2
strict superclause $>$, 13
strict supermeet $>$, 13
Strong 2-Consistency Principle 4.8.3, 136
strong right weakening property, 135
strongly n-consistent, 136
Subclassicality, 124
subclause $\leq$, 13
subformula, 11
$Subformula(.)$, 11
$Subformula(.,.)$, 11
$Subj(.)$, subject, 158
subject, $Subj(.)$, 158
submeet $\leq$, 13
subsequence, 5
subset $\subseteq$, 2
subtraction of sets $-$, 3
subtree, 7
successor ordinal, 6
suffix, 5
$Sup(.)$, 13
superclause $\geq$, 13
superior, 147, 148
supermeet $\geq$, 13
supraclassical proof algorithm, 134
Supraclassicality, 124
symmetric, 5

T the true truth value, 14
t probably true, 170
$T(\ )$, 163
$t(.)$, tag, 158
$T(\alpha:\alpha')$, 177
$T(\gamma:\lambda)$, 193
$T[\ ]$, 163
tag, $t(.)$, 158
$Taut$, 15, 128
tautology, 15
tautology-free, 15
Tautology-free Classical Closure, 122
tautology-free classically closed, 122
Tautology-free Left Absorption, 122
tautology-free left absorptive, 122
team defeat, 152
$Thm(F)$, 130
$Thm(\mathcal{L},\alpha,\mathcal{S})$, 130
$T_l(...)$, 167
$T_l[...]$ lean reduct of $T[...]$, 167
T_p, 7
train of rules, 160
transitive, 5, 124

transitive set, 6
Transitivity, 124
Tree(.) tree of a formula, 11
truth (veracity) function $V(.,.)$, 170

u undetermined truth value, 170
uncountable, 7
unfalsifiable, 14
union binary $\cup$, 3
union unary $\bigcup$, 3
Unit Classical Closure, 122
unit clause, 12
unit meet, 12
Unsat(.), 79
unsatisfiable, 14
Unwanted Chaining Example 7.9.4, 223

$V(.,.)$ truth (veracity) function, 170
Val, 14
Val(.), 76
valuation, 14
verum, empty meet $\wedge\{\}$, 12

warning arrow $\rightsquigarrow$, 146
warning rule, 146
weak right weakening property, 135

X-implies $\models^X$, 102
X-proves $\vdash^X$, 102
XCl(.), 89, 101
XFrom(.), 102

$\mathbb{Z}$ set of all integers, 3
$\mathbb{Z}^+$ set of all positive integers, 3

www.ingramcontent.com/pod-product-compliance
Lightning Source LLC
Chambersburg PA
CBHW050135170426
43197CB00011B/1843